METHODS IN MOLECULAR BIOLOGY

Series Editor
John M. Walker
School of Life and Medical Sciences
University of Hertfordshire
Hatfield, Hertfordshire, AL10 9AB, UK

For further volumes:
http://www.springer.com/series/7651

Proteomics in Systems Biology

Methods and Protocols

Edited by

Jörg Reinders

Institute of Functional Genomics, University of Regensburg, Regensburg, Germany

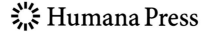 Humana Press

Editor
Jörg Reinders
Institute of Functional Genomics
University of Regensburg
Regensburg, Germany

ISSN 1064-3745   ISSN 1940-6029   (electronic)
Methods in Molecular Biology
ISBN 978-1-4939-3339-6   ISBN 978-1-4939-3341-9   (eBook)
DOI 10.1007/978-1-4939-3341-9

Library of Congress Control Number: 2015957400

Springer New York Heidelberg Dordrecht London

Printed on acid-free paper

Humana Press is a brand of Springer
Springer Science+Business Media LLC New York is part of Springer Science+Business Media (www.springer.com)

Preface

Proteomics techniques have constantly been developed further through the last decade and have been applied successfully for all kinds of samples and biological or medical questions. Nowadays, they are established methods in many research labs, and proteomic studies can be accomplished with good reliability and coverage on a routine basis. Therefore, proteomics can be used as a powerful tool in functional genomics and systems biology studies. Current challenges are thus the implementation of proteomic analyses in these comprehensive studies. This applies for both sample generation and preparation to ensure consistency over several levels of analyses like genomics, transcriptomics, and metabolomics and integration of the multilevel data to generate biological knowledge. This book gives an overview of contemporary quantitative proteomics methods and data interpretation approaches and also gives examples of how to implement proteomics into systems biology.

Regensburg, Germany   *Jörg Reinders*

Contents

Preface. . . . .   *v*
Contributors. . . . .   *ix*

1 Multiplexed Quantitative Proteomics for High-Throughput
   Comprehensive Proteome Comparisons of Human Cell Lines.   1
   Amanda Edwards and Wilhelm Haas

2 Sample Preparation Approaches for iTRAQ Labeling and Quantitative
   Proteomic Analyses in Systems Biology   15
   Christos Spanos and J. Bernadette Moore

3 Two Birds with One Stone: Parallel Quantification of Proteome
   and Phosphoproteome Using iTRAQ   25
   Fiorella A. Solari, Laxmikanth Kollipara, Albert Sickmann,
   and René P. Zahedi

4 Selected Reaction Monitoring to Measure Proteins of Interest
   in Complex Samples: A Practical Guide   43
   Yuehan Feng and Paola Picotti

5 Monitoring PPARG-Induced Changes in Glycolysis by Selected Reaction
   Monitoring Mass Spectrometry.   57
   Andreas Hentschel and Robert Ahrends

6 A Targeted MRM Approach for Tempo-Spatial Proteomics Analyses   75
   Annie Moradian, Tanya R. Porras-Yakushi, Michael J. Sweredoski,
   and Sonja Hess

7 Targeted Phosphoproteome Analysis Using Selected/Multiple Reaction
   Monitoring (SRM/MRM)   87
   Jun Adachi, Ryohei Narumi, and Takeshi Tomonaga

8 Testing Suitability of Cell Cultures for SILAC-Experiments
   Using SWATH-Mass Spectrometry   101
   Yvonne Reinders, Daniel Völler, Anja-K. Bosserhoff, Peter J. Oefner,
   and Jörg Reinders

9 Combining Amine-Reactive Cross-Linkers and Photo-Reactive Amino Acids
   for 3D-Structure Analysis of Proteins and Protein Complexes   109
   Philip Lössl and Andrea Sinz

10 Tissue MALDI Mass Spectrometry Imaging (MALDI MSI) of Peptides   129
   Birte Beine, Hanna C. Diehl, Helmut E. Meyer, and Corinna Henkel

11 Ethyl Esterification for MALDI-MS Analysis of Protein Glycosylation   151
   Karli R. Reiding, Emanuela Lonardi, Agnes L. Hipgrave Ederveen,
   and Manfred Wuhrer

12 Characterization of Protein *N*-Glycosylation by Analysis
   of ZIC-HILIC-Enriched Intact Proteolytic Glycopeptides.   163
   Gottfried Pohlentz, Kristina Marx, and Michael Mormann

13 Simple and Effective Affinity Purification Procedures for Mass
    Spectrometry-Based Identification of Protein-Protein Interactions
    in Cell Signaling Pathways.   181
    Julian H.M. Kwan and Andrew Emili

14 A Systems Approach to Understand Antigen Presentation
    and the Immune Response   189
    *Nadine L. Dudek, Nathan P. Croft, Ralf B. Schittenhelm,
    Sri H. Ramarathinam, and Anthony W. Purcell*

15 Profiling of Small Molecules by Chemical Proteomics   211
    Kilian V.M. Huber and Giulio Superti-Furga

16 Generating Sample-Specific Databases for Mass Spectrometry-Based
    Proteomic Analysis by Using RNA Sequencing   219
    Toni Luge and Sascha Sauer

17 A Proteomic Workflow Using High-Throughput De Novo Sequencing
    Towards Complementation of Genome Information for Improved
    Comparative Crop Science   233
    *Reinhard Turetschek, David Lyon, Getinet Desalegn, Hans-Peter Kaul,
    and Stefanie Wienkoop*

18 From Phosphoproteome to Modeling of Plant Signaling Pathways   245
    Maksim Zakhartsev, Heidi Pertl-Obermeyer, and Waltraud X. Schulze

19 Interpretation of Quantitative Shotgun Proteomic Data   261
    *Elise Aasebø, Frode S. Berven, Frode Selheim, Harald Barsnes,
    and Marc Vaudel*

20 A Simple Workflow for Large Scale Shotgun Glycoproteomics.   275
    *Astrid Guldbrandsen, Harald Barsnes, Ann Cathrine Kroksveen,
    Frode S. Berven, and Marc Vaudel*

21 Systemic Analysis of Regulated Functional Networks.   287
    *Luis Francisco Hernández Sánchez, Elise Aasebø, Frode Selheim,
    Frode S. Berven, Helge Ræde, Harald Barsnes, and Marc Vaudel*

Index *311*

Contributors

ELISE AASEBØ • *Proteomics Unit, Department of Biomedicine, University of Bergen, Bergen, Norway*

JUN ADACHI • *Laboratory of Proteome Research, National Institute of Biomedical Innovation, Health and Nutrition, Osaka, Japan*

ROBERT AHRENDS • *Leibniz-Institut für Analytische Wissenschaften—ISAS—e.V., Dortmund, Germany*

HARALD BARSNES • *Proteomics Unit, Department of Biomedicine, University of Bergen, Bergen, Norway; KG Jebsen Center for Diabetes Research, Department of Clinical Science, University of Bergen, Bergen, Norway*

BIRTE BEINE • *Leibniz-Institut für Analytische Wissenschaften—ISAS—e.V., Dortmund, Germany*

FRODE S. BERVEN • *Proteomics Unit, Department of Biomedicine, University of Bergen, Bergen, Norway; KG Jebsen Centre for Multiple Sclerosis Research, Department of Clinical Medicine, University of Bergen, Bergen, Norway; Norwegian Multiple Sclerosis Competence Centre, Department of Neurology, Haukeland University Hospital, Bergen, Norway*

ANJA-K. BOSSERHOFF • *Institute of Pathology, University of Regensburg, Regensburg, Germany; Institute of Biochemistry, Emil-Fischer-Zentrum, Friedrich-Alexander-University Erlangen-Nürnberg, Erlangen, Germany*

NATHAN P. CROFT • *Department of Biochemistry and Molecular Biology, School of Biomedical Sciences, Monash University, Clayton, VIC, Australia*

GETINET DESALEGN • *Department of Crop Sciences, University of Natural Resources and Life Sciences, Vienna, Austria*

HANNA C. DIEHL • *Clinical Proteomics, Medizinisches Proteome-Center, Ruhr-University, Bochum, Germany*

NADINE L. DUDEK • *Department of Biochemistry and Molecular Biology, School of Biomedical Sciences, Monash University, Clayton, VIC, Australia*

AMANDA EDWARDS • *Center for Cancer Research, Massachusetts General Hospital, Charlestown, MA, USA; Department of Medicine, Harvard Medical School, Boston, MA, USA*

ANDREW EMILI • *Department of Molecular Genetics, Donnelly Centre for Cellular and Biomolecular Research, University of Toronto, Toronto, ON, Canada*

YUEHAN FENG • *Department of Biology, Institute of Biochemistry, ETH Zurich, Zurich, Switzerland*

ASTRID GULDBRANDSEN • *Proteomics Unit, Department of Biomedicine, University of Bergen, Bergen, Norway; KG Jebsen Centre for Multiple Sclerosis Research, Department of Clinical Medicine, University of Bergen, Bergen, Norway*

WILHELM HAAS • *Center for Cancer Research, Massachusetts General Hospital, Charlestown, MA, USA; Department of Medicine, Harvard Medical School, Boston, MA, USA*

CORINNA HENKEL • *Leibniz-Institut für Analytische Wissenschaften—ISAS—e.V., Dortmund, Germany*

ANDREAS HENTSCHEL • *Leibniz-Institut für Analytische Wissenschaften—ISAS—e.V., Dortmund, Germany*

LUIS FRANCISCO HERNÁNDEZ SÁNCHEZ • *Graduate Program in Optimization, Universidad Autónoma Metropolitana Azcapotzalco, Mexico City, Mexico*

SONJA HESS • *Proteome Exploration Laboratory, California Institute of Technology, Pasadena, CA, USA*

AGNES L. HIPGRAVE EDERVEEN • *Center for Proteomics and Metabolomics, Leiden University Medical Center, Leiden, The Netherlands*

KILIAN V.M. HUBER • *Structural Genomics Consortium, University of Oxford, Oxford, UK; Target Discovery Institute, University of Oxford, Oxford, UK*

HANS-PETER KAUL • *Department of Crop Sciences, University of Natural Resources and Life Sciences, Vienna, Austria*

LAXMIKANTH KOLLIPARA • *Leibniz-Institut für Analytische Wissenschaften—ISAS—e.V., Dortmund, Germany*

ANN CATHRINE KROKSVEEN • *Proteomics Unit, Department of Biomedicine, University of Bergen, Bergen, Norway; KG Jebsen Centre for Multiple Sclerosis Research, Department of Clinical Medicine, University of Bergen, Bergen, Norway*

JULIAN H.M. KWAN • *Department of Molecular Genetics, Donnelly Centre for Cellular and Biomolecular Research, University of Toronto, Toronto, ON, Canada*

EMANUELA LONARDI • *Center for Proteomics and Metabolomics, Leiden University Medical Center, Leiden, The Netherlands*

PHILIP LÖSSL • *Department of Pharmaceutical Chemistry and Bioanalytics, Institute of Pharmacy, Martin-Luther University Halle-Wittenberg, Halle/Saale, Germany; Biomolecular Mass Spectrometry and Proteomics, Netherlands Proteomics Center, Bijvoet Center for Biomolecular Research and Utrecht Institute for Pharmaceutical Sciences, Utrecht University, Utrecht, The Netherlands*

TONI LUGE • *Otto Warburg Laboratory, Max Planck Institute for Molecular Genetics, Berlin, Germany*

DAVID LYON • *Department of Ecogenomics and Systems Biology, University of Vienna, Vienna, Austria*

KRISTINA MARX • *Bruker Daltonik GmbH, Bremen, Germany*

HELMUT E. MEYER • *Leibniz-Institut für Analytische Wissenschaften—ISAS—e.V., Dortmund, Germany*

J. BERNADETTE MOORE • *Department of Nutritional Sciences, Faculty of Health and Medical Sciences, University of Surrey, Guildford Surrey, UK*

ANNIE MORADIAN • *Proteome Exploration Laboratory, California Institute of Technology, Pasadena, CA, USA*

MICHAEL MORMANN • *Institute for Hygiene, University of Münster, Münster, Germany*

RYOHEI NARUMI • *Laboratory for Synthetic Biology, RIKEN Quantitative Biology Center, Kobe, Japan*

PETER J. OEFNER • *Institute of Functional Genomics, University of Regensburg, Regensburg, Germany*

HEIDI PERTL-OBERMEYER • *Plant Systems Biology, Plant Physiology, University of Hohenheim, Stuttgart, Germany*

PAOLA PICOTTI • *Department of Biology, Institute of Biochemistry, ETH Zurich, Zurich, Switzerland*

GOTTFRIED POHLENTZ • *Institute for Hygiene, University of Münster, Münster, Germany*

TANYA R. PORRAS-YAKUSHI • *Proteome Exploration Laboratory, California Institute of Technology, Pasadena, CA, USA*

ANTHONY W. PURCELL • *Department of Biochemistry and Molecular Biology, School of Biomedical Sciences, Monash University, Clayton, VIC, Australia; The Department of Biochemistry and Molecular Biology, The Bio21 Molecular Science and Biotechnology Institute, University of Melbourne, Melbourne, VIC, Australia*

HELGE RÆDER • *KG Jebsen Center for Diabetes Research, Department of Clinical Science, University of Bergen, Bergen, Norway; Department of Pediatrics, Haukeland University Hospital, Bergen, Norway*

SRI H. RAMARATHINAM • *Department of Biochemistry and Molecular Biology, School of Biomedical Sciences, Monash University, Clayton, VIC, Australia*

KARLI R. REIDING • *Center for Proteomics and Metabolomics, Leiden University Medical Center, Leiden, The Netherlands*

JÖRG REINDERS • *Institute of Functional Genomics, University of Regensburg, Regensburg, Germany*

YVONNE REINDERS • *Department of Biochemistry I, University of Regensburg, Regensburg, Germany*

SASCHA SAUER • *Otto Warburg Laboratory, Max Planck Institute for Molecular Genetics, Berlin, Germany*

RALF B. SCHITTENHELM • *Department of Biochemistry and Molecular Biology, School of Biomedical Sciences, Monash University, Clayton, VIC, Australia*

WALTRAUD X. SCHULZE • *Plant Systems Biology, Plant Physiology, University of Hohenheim, Stuttgart, Germany*

FRODE SELHEIM • *Proteomics Unit, Department of Biomedicine, University of Bergen, Bergen, Norway*

ALBERT SICKMANN • *Leibniz-Institut für Analytische Wissenschaften—ISAS—e.V., Dortmund, Germany; Department of Chemistry, College of Physical Sciences, University of Aberdeen, Aberdeen, Scotland, UK*

ANDREA SINZ • *Department of Pharmaceutical Chemistry and Bioanalytics, Institute of Pharmacy, Martin-Luther University Halle-Wittenberg, Halle/Saale, Germany*

FIORELLA A. SOLARI • *Leibniz-Institut für Analytische Wissenschaften—ISAS—e.V., Dortmund, Germany*

CHRISTOS SPANOS • *Department of Nutritional Sciences, Faculty of Health and Medical Sciences, University of Surrey, Guildford Surrey, UK*

GIULIO SUPERTI-FURGA • *CeMM Research Center for Molecular Medicine of the Austrian Academy of Sciences, Vienna, Austria*

MICHAEL J. SWEREDOSKI • *Proteome Exploration Laboratory, California Institute of Technology, Pasadena, CA, USA*

TAKESHI TOMONAGA • *Laboratory of Proteome Research, National Institute of Biomedical Innovation, Health and Nutrition, Osaka, Japan*

REINHARD TURETSCHEK • *Department of Ecogenomics and Systems Biology, University of Vienna, Vienna, Austria*

MARC VAUDEL • *Proteomics Unit, Department of Biomedicine, University of Bergen, Bergen, Norway*

DANIEL VÖLLER • *Institute of Pathology, University of Regensburg, Regensburg, Germany*

STEFANIE WIENKOOP • *Department of Ecogenomics and Systems Biology, University of Vienne, Vienna, Austria*

MANFRED WUHRER • *Center for Proteomics and Metabolomics, Leiden University Medical Center, Leiden, The Netherlands; Division of BioAnalytical Chemistry, VU University Amsterdam, Amsterdam, The Netherlands*

RENÉ P. ZAHEDI • *Leibniz-Institut für Analytische Wissenschaften—ISAS—e.V., Dortmund, Germany*

MAKSIM ZAKHARTSEV • *Plant Systems Biology, Plant Physiology, University of Hohenheim, Stuttgart, Germany*

Chapter 1

Multiplexed Quantitative Proteomics for High-Throughput Comprehensive Proteome Comparisons of Human Cell Lines

Amanda Edwards and Wilhelm Haas

Abstract

The proteome is the functional entity of the cell, and perturbations of a cellular system almost always cause changes in the proteome. These changes are a molecular fingerprint, allowing characterization and a greater understanding of the effect of the perturbation on the cell as a whole. Monitoring these changes has therefore given great insight into cellular responses to stress and disease states, and analytical platforms to comprehensively analyze the proteome are thus extremely important tools in biological research. Mass spectrometry has evolved as the most relevant technology to characterize proteomes in a comprehensive way. However, due to a lack of throughput capacity of mass spectrometry-based proteomics, researchers frequently use measurement of mRNA levels to approximate proteome changes. Growing evidence of substantial differences between mRNA and protein levels as well as recent improvements in mass spectrometry-based proteomics are heralding an increased use of mass spectrometry for comprehensive proteome mapping. Here we describe the use of multiplexed quantitative proteomics using isobaric labeling with tandem mass tags (TMT) for the simultaneous quantitative analysis of five cancer cell proteomes in biological duplicates in one mass spectrometry experiment.

Key words Quantitative proteomics, Multiplexing, Isobaric labels, TMT

1 Introduction

Proteins are the primary functional units of the cell, and as such, information about their abundance, interaction partners, and modifications is critical for understanding both healthy and abnormal cellular function. Traditionally, such work has been accomplished on a protein-by-protein basis through genetic or biochemical techniques. More recently, large-scale approaches attempting to monitor an entire proteome—all proteins expressed in a cell or tissue—in one step have become accessible [1, 2]. Such a holistic approach allows identification of proteome imbalances and changes in functional networks, enabling us to study and probe the state of a cell in an unbiased and rapid fashion.

Jörg Reinders (ed.), *Proteomics in Systems Biology: Methods and Protocols*, Methods in Molecular Biology, vol. 1394, DOI 10.1007/978-1-4939-3341-9_1, © Springer Science+Business Media New York 2016

Mass spectrometry has emerged as the leading platform to rapidly characterize whole proteomes, primarily driven by improvements in both MS sensitivity and throughput in the last 15 years. However, the technology has traditionally lagged behind in throughput capacity when compared to genomics platforms, such as DNA-microarrays or next-generation sequencing technology, to study cellular expression profiles. Therefore, mRNA expression profiles are still the main source for estimations of protein-level changes for most researchers [3, 4]. Yet evidence is accumulating that significant differences exist between mRNA- and protein-level changes in different cell or tissue states [1, 5–9]. There is thus an enormous need for improved mass spectrometry-based proteomics technology to enable direct protein-level measurements, with a throughput comparable to that provided by genomics technologies.

Mass spectrometry-based proteomics has been used as a quantitative tool since the late 1990s with the introduction of accurate relative quantification using stable isotopes. One of the earliest approaches was the employment of isotope-coded affinity tags (ICAT). ICAT enables a chemical incorporation of differential stable isotopes into two different samples and, in parallel, a reduction of proteome complexity by enrichment of cysteine-containing peptides [10]. An approach to incorporate stable isotopes metabolically through heavy isotope-labeled amino acids—stable isotope labeling in cell culture (SILAC)—was first described in 2002 [11] and is still widely used. The commercial availability of high-performance mass spectrometers [12, 13] optimized for use in combination with liquid chromatography further contributed to the propagation of quantitative proteomics, and new strategies for chemical incorporation of stable isotopes into peptides, such as reductive dimethylation [14, 15], arose. Each of the above methods relies on the use of full MS data from intact peptide ions to perform quantitative analysis. Each peptide in the two samples of interest is detected in its light and heavy form, leading to an increase in the signal complexity in the full MS spectra. This increase in signal complexity necessarily decreases the overall sensitivity of the approach and complicates the quantitative analysis of individual peptides. Consequently, although more is theoretically possible [16], the number of samples routinely compared simultaneously using these methods is limited to three [17].

A very elegant strategy to remove this roadblock in multiplexing MS proteomics was first described in 2003 through the use of isobaric tags to incorporate stable isotopes into proteomics samples [18]. These tags consist of three regions: a mass reporter ion, a mass balancer region, and a reactive terminal amino group. To quantify different protein levels in different biological samples, peptide mixtures are labeled with different forms of the tag by allowing the tag to react with amino groups at the N-terminus or lysine residues of a peptide. Importantly, each tag has the same mass, as the chemical structures only differ in the distribution of heavy stable isotopes between reporter and balancer regions. Thus differentially labeled peptides migrate together through the chromatographic separation and are indistinguishable in MS1 scans. However, during MS2 fragmentation, the mass reporter ions (with a unique mass for each tag) separate from the parent tag, and their relative intensities represent the relative abundance of the original peptides in the measured samples. There are two commercial sources for isobaric tags: isobaric tags for relative and absolute quantitation (ITRAQ) reagents (Sciex) that allow the analysis of up to eight samples [19, 20], and tandem mass tag (TMT) reagents (Pierce) that enable a simultaneous quantification of up to ten samples [21, 22]. An early caveat of the isobaric labeling strategy was a limitation in the achievable accuracy and reproducibility of quantitative results due to co-isolation and fragmentation of contaminant ions with the ions targeted for identification and quantification. Solutions to overcome this limitation were presented in the form of applying ion-ion chemistry for removing contaminant ions [23] or by separating identification and quantification of a peptide ion, performing the identification based on MS2 data but shifting the quantification to an MS3 experiment as an additional gas-phase enrichment and fragmentation step [24]. The MS3 method was further optimized to increase sensitivity and throughput, and this MultiNotch MS3 method [25] is now implemented as a synchronous precursor selection (SPS)-supported MS3 method on the Orbitrap Fusion mass spectrometer (Thermo Fisher Scientific). We believe that multiplexed quantitative proteomics is a tool that will prove to be indispensable in studying complex biological systems and disease states requiring the analysis of many samples.

This chapter describes the workflow for using 10-plex tandem mass tag (TMT) reagents for isobaric labeling-based multiplexed quantitative proteomics to comprehensively map proteomes of human cell lines. We routinely apply this protocol to quantify approximately 8000 proteins simultaneously in ten samples, occupying 36 h of mass spectrometry time, or less than 4 h to quantitatively characterize the proteome of a human cell line.

2   Materials

2.1   Cell Culture

1. Cell lines: This protocol is applicable for the proteomic analysis of any adherent human cell line. Detached cell lines can also be used, with modifications to the cell culture protocols.

2. Cell media: Use culture media appropriate for the chosen cell lines.

3. 1× sterile phosphate-buffered saline (PBS).

4. 0.25 % Trypsin.

2.2 Cell Lysis	1. Lysis buffer: 75 mM NaCl, 50 mM HEPES (pH 8.5), 10 mM sodium pyrophosphate, 10 mM NaF, 10 mM β-glycerophosphate, 10 mM sodium orthovanadate, 10 mM phenylmethanesulfonylfluoride (PMSF), Roche Complete Protease Inhibitor EDTA-free tablet, 3 % SDS.

2.3 Sample Preparation for Mass Spectrometry

1. HPLC-grade methanol.
2. HPLC-grade chloroform.
3. HPLC-grade water.
4. HPLC-grade acetonitrile (ACN).
5. HPLC-grade acetone.
6. 1 M Dithiothreitol (DTT) in 50 mM HEPES (pH 8.5).
7. 1 M Iodoacetamide (IAA) in 50 mM HEPES (pH 8.5).
8. Digestion buffer: 1 M urea, 50 mM HEPES (pH 8.5).
9. Lysyl endopeptidase (LysC) (Wako Chemicals, 10 AU, resuspended in 2 mL HPLC-grade water, stored at –80 °C).
10. Trypsin (Promega, sequencing grade, 0.4 μg/μL, stored at –80 °C).
11. 1 % and 10 % trifluoroacetic acid (TFA), 99.5 % purity.
12. 0.5 % Acetic acid.
13. 40 % ACN, 0.5 % acetic acid.
14. 80 % ACN, 0.5 % acetic acid.
15. 5 % ACN, 5 % formic acid.
16. 50 % ACN, 5 % formic acid.
17. Bicinchoninic acid (BCA) protein assay kit (Pierce).
18. Bovine serum albumin.
19. Tandem Mass Tag™ 10-plex reagent set (Pierce).
20. 200 mM HEPES (pH 8.5), 30 % anhydrous ACN.
21. 200 mM HEPES (pH 8.5), 5 % hydroxylamine.

2.4 High-pH Reversed-Phase High-Pressure Liquid Chromatography

1. HpRP buffer A: 5 % ACN, 10 mM ammonium bicarbonate.
2. HpRP buffer B: 90 % ACN, 10 mM ammonium bicarbonate.

2.5 Mass Spectrometry

1. MS buffer A: 3 % ACN, 0.125 % formic acid.
2. MS buffer B: 0.125 % formic acid in ACN.

2.6 Equipment

1. Minicentrifuge.
2. 1 cc syringes.
3. 21-gauge needles.

4. Vacuum manifold.

5. SepPak 1 cc (50 mg) C18 Cartridges (Waters).

6. Savant SC100 SpeedVac Concentrator.

7. High-pressure liquid chromatography system (ex: Agilent 1260 Infinity Quaternary LC System).

8. Agilent ZORBAX Extend-C18 column (4.6 mm × 250 mm, 5 μm particle size).

9. Deep-well 96-well plates.

10. Orbitrap Fusion (Thermo Fisher Scientific).

11. Easy-nLC 1000 (Thermo Fisher Scientific).

12. Resins: Magic C4 resin (5 μm, 100 Å, Michrom Bioresources), Maccel C18AQ resin (3 μm, 200 Å, Nest Group), and GP-C18 (1.8 μm, 120 Å, Sepax Technologies).

3 Methods

3.1 Cell Culture

1. Grow each of the five cell lines to 90 % confluence in duplicate in 10 cm² dishes (a total of ten samples).

2. Prior to collecting the cells, wash gently twice with 2.5 mL pre-warmed sterile 1× PBS (*see* **Note 1**).

3. Add 2 mL pre-warmed trypsin to each 10 cm² dish, covering the cell layer completely. Incubate for 5 min at 37 °C. Add 3 mL pre-warmed media to each 10 cm² dish, and collect each cell mixture in 15 mL Falcon tubes.

4. Pellet cells by centrifuging at $500 \times g$ for 5 min. Discard the supernatant. Wash the cell pellet once with sterile 1× PBS (*see* **Note 2**).

3.2 Cell Lysis

Note: All subsequent steps should be performed at room temperature, as the SDS in the lysis buffer will precipitate at cold temperatures.

1. Resuspend each cell pellet in 0.5 mL lysis buffer, pipetting up and down to disrupt the cell pellet (*see* **Note 3**).

2. Lyse the cells by passing the resuspended cells ten times through a 21-gauge needle (*see* **Note 4**). Transfer the suspension to 1.5 mL Eppendorf tubes.

3. Clear away cellular debris by centrifuging at $16,000 \times g$ for 5 min. Collect the supernatant (*see* **Note 5**).

3.3 Reduction, Alkylation, and Precipitation of Proteins

1. Add 2.5 μL of 1 M DTT to each sample (final concentration of DTT = 5 mM), and vortex thoroughly. Centrifuge briefly at $3000 \times g$ to bring all contents to the bottom of the tube. Incubate at 56 °C for 30 min (*see* **Note 6**).

2. Cool the tubes on ice for 3 min.

3. Add 7.5 µL of 1 M IAA to each sample (final concentration of IAA = 15 mM), and vortex thoroughly. Centrifuge briefly at $3000 \times g$. Incubate in the dark for 20 min (*see* **Note 7**).

4. Add 2.5 µL of 1 M DTT to each sample to quench the reaction, and vortex thoroughly. Centrifuge briefly at $3000 \times g$. Incubate in the dark for 15 min.

5. Transfer each sample to a 15 mL Falcon tube.

6. Begin protein precipitation by adding 2.0 mL methanol to each tube ($4 \times$ the initial lysate volume) (*see* **Note 8**). Vortex, and centrifuge at $1300 \times g$ for 3 min.

7. Add 0.5 mL chloroform to each tube ($1 \times$ the initial lysate volume), and vortex. Make sure to disrupt any pellet fully (*see* **Note 9**). Centrifuge at $1300 \times g$ for 3 min.

8. Add 1.5 mL water to each tube ($3 \times$ the initial lysate volume), and vortex. Again, make sure that pellet is fully disrupted. Centrifuge at $1300 \times g$ for 3 min.

9. At this stage, precipitated proteins should form a white disk between the aqueous and organic phases. Carefully remove all aqueous and organic supernatant.

10. Wash the protein pellet with 2.0 mL methanol. Centrifuge at $1300 \times g$ for 3 min.

11. Remove the supernatant, and place the protein pellet on ice. Add 1.5 mL ice-cold acetone to each pellet. Disrupt the pellet, and centrifuge at $1300 \times g$ at 4 °C for 3 min. Remove the supernatant, and repeat once (*see* **Note 10**).

12. Dry the precipitated protein in open tubes at 56 °C for 15 min or at 37 °C for 60 min, until pellets are completely dry. Cool the pellet on ice for 5 min (*see* **Note 11**).

3.4 Digestion of Proteins

1. Resuspend the protein pellet in 0.5 mL digestion buffer (*see* **Note 12**).

2. Add 2.5 µL LysC stock (5 µg) to each pellet. Vortex. Centrifuge briefly at $3000 \times g$. Incubate overnight at room temperature, agitating gently on a tabletop vortexer.

3. Add 12.5 µL trypsin stock (5 µg) to each tube. Vortex. Centrifuge briefly at $3000 \times g$. Incubate at 37 °C for 6 h.

4. Acidify the reaction with 25 µL 10 % TFA (final concentration TFA = 0.5 %). Vortex. Centrifuge at $16,000 \times g$ for 5 min, and collect the supernatant (*see* **Note 13**).

3.5 Cleanup of Sample Using SepPak Columns (See Notes 14 and 15)

1. Place 50 mg C18 SepPak columns on the vacuum manifold, and wash with 5×1 mL ACN.

2. Wash with 5×1 mL 80 % ACN, 0.5 % acetic acid.

3. Wash with 5×1 mL 0.1 % TFA.

4. Apply sample to column, and pull over column at a slow speed (*see* **Note 16**).

5. Wash with 5×1 mL 0.1 % TFA.

6. Wash with 1 mL 0.5 % acetic acid, and allow the column to go completely dry.

7. Remove SepPak columns from the vacuum manifold, and place in clean 2 mL Eppendorf tubes. Add 0.75 mL of 40 % ACN, 0.5 % acetic acid to each column. Using a 1 mL syringe, push solution through the column slowly. Add 0.75 mL of 80 % ACN, 0.5 % acetic acid to each column, and push the solution through the column, allowing the column to go completely dry (*see* **Note 17**).

8. Dry the eluted peptides in a SpeedVac.

3.6 Quantify and Aliquot Digested Peptides

1. Resuspend each sample in 0.5 mL 50 % ACN, 5 % formic acid. Vortex. Centrifuge briefly at $3000 \times g$. Sonicate for 5 min.

2. (*See* **Note 18**) Prepare standard bovine serum albumin (BSA) stocks ranging from 25 to 2000 µg/mL, with 50 % ACN, 5 % formic acid as the buffer.

3. Add 10 µL of BSA standard or sample into 96-well plate wells (*see* **Note 19**). Add 200 µL of working reagent to each well, and mix thoroughly.

4. Cover the plate with plastic wrap, and incubate at 37 °C for 30 min. Remove the plate, and allow cooling to room temperature for 5 min.

5. Measure the absorbance at 562 nM, and use the standard curve to determine the protein concentration of each sample.

6. Prepare 50 µg aliquots of each sample, and dry the peptides in a SpeedVac.

3.7 TMT Labeling of Peptides

1. Resuspend the TMT reagent according to the manufacturer's instructions in anhydrous acetonitrile (*see* **Note 20**).

2. Resuspend the peptides in 50 µL 200 mM HEPES (pH 8.5), 30 % anhydrous ACN (*see* **Note 21**). Vortex. Centrifuge briefly at $3000 \times g$. Sonicate for 5 min.

3. Add 5 µL of TMT reagent to each peptide solution, with 1 TMT label used for each of the ten samples (126, 127n, 127c, 128n, 128c, 129n, 129c, 130n, 130c, and 131).

4. Incubate the reaction mixtures at room temperature for 1 h.

5. Quench the reaction by adding 6 µL of 200 mM HEPES (pH 8.5), 5 % hydroxylamine. Incubate at room temperature for 15 min.

6. Acidify the mixture by adding 50 µL of 1 % TFA. Combine all ten samples into one sample, as they are now all distinctly labeled.

7. De-salt the mixture over a 50 mg C18 SepPak column (*see* Subheading 3.5 above for details).

8. Dry the peptides in a SpeedVac.

3.8 Fractionation of Peptides

1. Resuspend the peptides in 0.5 mL 5 % ACN, 5 % formic acid. Vortex. Centrifuge briefly at $3000 \times g$. Sonicate for 5 min.

2. Fractionate the sample by high pH reverse-phase high-pressure liquid chromatography (HpRP) using a two-buffer gradient system at a flow rate of 500 μL/min. Load the sample in 0 % HpRP buffer B for 2 min, and then separate the peptides using a linear gradient from 20 to 35 % HpRP buffer B over 60 min. Wash the column with 100 % HpRP buffer B for 5 min, and re-equilibrate the column with 100 % HpRP buffer A for 10 min. Monitor peptide elution by UV absorption at a wavelength of 220 nm and collect a total of 96 fractions in a deep-well 96-well plate from 11.5 to 69.5 min (*see* **Note 22**).

3. Combine the 96 fractions into 12 fractions, based on the schematic in Fig. 1.

4. Dry the fractions in a SpeedVac.

3.9 Mass Spectrometry Analysis

The details of the LC-MS2/MS3 methods will depend on the instrumentation available. Here, we describe a method using an Easy-nLC 1000 (Thermo Fisher Scientific) with chilled autosampler and an Orbitrap Fusion mass spectrometer (Thermo Fisher Scientific)

1. *Sample preparation*: Resuspend each fraction in 8 μL 5 % ACN, 5 % formic acid and sonicate to ensure full suspension of all peptides. Inject 3 μL of each sample for chromatographic separation and mass spectrometry analysis.

2. *Nanospray liquid chromatography method*: Separate peptides over a 100 μm inner diameter microcapillary column, packed in-house with 0.5 cm of Magic C4 resin, 0.5 cm of Maccell C18 resin, and 29 cm of GP-C18 resin. Use a 6–25 % gradient of MS buffer B over 165 min at 300 nL/min to elute the peptides. End the gradient with a 10-min wash with 100 % MS buffer B to clear all remaining peptides off the column, and re-equilibrate the column with 9 μL of 100 % MS buffer A to prepare the column for subsequent runs.

3. *Mass spectrometry method*: Begin acquisition with a full MS1 spectrum acquired in the Orbitrap, and use synchronous precursor selection to isolate the ten highest intensity peptides for MS2 analysis. Following CID fragmentation of these peptides, perform MS2 scans in the linear ion trap. Once again, use synchronous precursor selection to isolate the ten highest intensity peptides for MultiNotch MS3 analysis [25]. Following HCD fragmentation of the peptides, perform MS3 scans in the Orbitrap for maximum sensitivity (*see* **Note 23**).

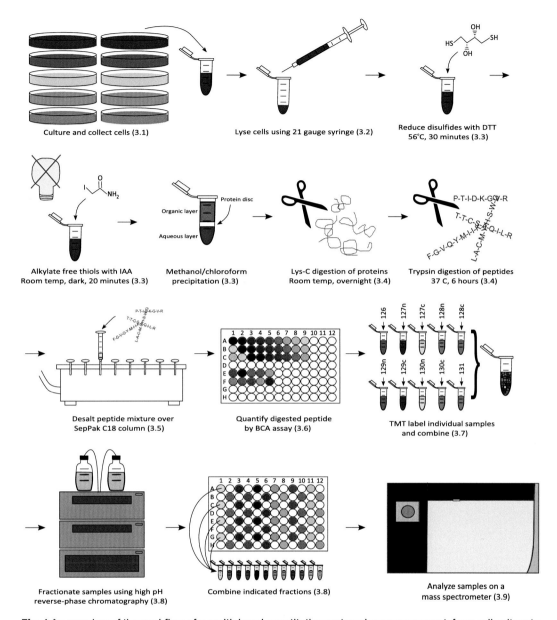

Fig. 1 An overview of the workflow of a multiplexed quantitative proteomics measurement, from cell culture to mass spectrometer

3.10 Data Analysis

As above, the details of the data analysis will depend on the specific search algorithms and software used. While we use Sequest [26] to match peptide spectra to sequences, a variety of other options are available (e.g., Mascot, X!Tandem). However, some parameters should be universally applied.

1. Specific search parameters include digestion enzyme, static peptide modifications, variable peptide modifications, and precursor ion tolerance. In this case, select trypsin as the enzyme,

requiring all matching peptides to have termini consistent with tryptic cleavage, allowing at most two missed cleavages. Static modifications include the TMT label on the N-terminus and lysine residues (229.162932 Da) as well as carbamidomethylation (57.021464 Da) on cysteine residues. Oxidation of methionine (15.994915 Da) should be set as a variable modification. Set the precursor m/z ion tolerance to 50 ppm.

2. Several online servers provide complete or near-complete protein databases for a variety of species, including UniProt, Ensembl, and RefSeq, against which MS2 spectra can be searched. We use UniProt databases, and we apply a target-decoy database search strategy to accurately estimate the false discovery rate of peptide and protein identifications [27, 28]. This requires compiling a concatenated database with a target component including the organism-specific protein sequence database as well as that of known contaminants such as porcine trypsin or other proteases used in the sample preparation. The second and so-called decoy component includes the same sequences but in reversed—or flipped—order, where every protein sequence starts with the original C-terminus of the original sequence. A practical protocol for the use of this database for estimating the FDR of a proteomics dataset is described elsewhere [29]. Filter the final results of peptides as well as protein identifications to an FDR of at most 1 %. Several algorithms are used for filtering proteomics data. We use a linear discriminant analysis to assess the FDR of MS2 spectra assignments to peptide sequences (peptide-spectral matches, PSMs). For a protein identification FDR filter, we also rely on the target-decoy database search strategy by using a posterior error histogram with protein FDR estimations that are based on combining probabilities of correct assignments for each PSM for all peptides matching a protein sequence [30].

3. In order to accurately quantify a peptide, signal-to-noise values and isolation specificity must exceed a background threshold. These values will depend on the instrument being used. To quantify a protein level, sum all reporter ion intensities from all peptides assigned to that protein.

4. Normalization of the data allows correction for slight preparation errors or MS anomalies. We recommend a two-step normalization procedure. Begin by normalizing all protein intensities to the ratio of the average intensity of that protein to all median protein intensities. This will bring all protein values closer to one another, allowing for more unbiased downstream statistical testing. Secondly, normalize all protein intensities to the ratio of the median protein intensities for a given TMT channel to the median of all protein intensities. This will account for any slight mixing errors from each sample.

4   Notes

1. Different cell types will adhere with different strengths, so one must be careful not to dislodge the cells prematurely, depending on cell type.

2. The cell pellets can be frozen here at −80 °C for future use.

3. A 5:1 ratio of lysis buffer:cell pellet or greater is helpful here in order to make the lysate less viscous.

4. It works best to do this slowly so as not to create excess bubbles.

5. Depending on the viscosity of the lysate, there may or may not be a visible cell pellet that clearly separates from the supernatant. If there is not, move forward with the entire mixture.

6. The goal of the DTT is to reduce all disulfide bonds.

7. The goal of the IAA is to alkylate free thiols.

8. This precipitation technique requires enough protein to visualize the protein pellet. If the protein output is too low, a TCA precipitation is preferred.

9. If the pellet is difficult to disrupt, rake the tube against an Eppendorf tube rack until it is dislodged.

10. While loss in the previous precipitation steps will be unbiased, loss at this stage will be biased towards hydrophobic peptides and should be carefully avoided.

11. The protein pellets can be frozen here at −80 °C for future use.

12. Depending on the size, it may be difficult to fully resuspend the pellet. Use a small pestle to grind the pellet and disrupt as best as possible.

13. Any pellet is undigested protein. The pellet can be stored at −20 °C and further digested in the future in the case of low peptide yield.

14. SepPak columns are used for larger peptide amounts. For low amounts (<10 μg), use StageTips (packed with C18 solid-phase extraction disks [Empore]) instead to minimize peptide loss.

15. Do not let column go completely dry until indicated.

16. It may be easier to open other ports on the vacuum manifold to allow a slower draw on the column.

17. Make sure to add the 40 % ACN solution first, followed by the 80 % ACN solution, to prevent too many peptides from eluting at once and clogging the column.

18. Prepare the protein quantification assay of your choice. We use the BCA assay from Pierce.

19. More concentrated samples may be diluted 1:3–1:10 to fall into the standard curve range.

20. It is very important here that the ACN used is anhydrous, as hydrated ACN will reduce the labeling efficiency.

21. The manufacturer recommends performing TMT labeling using a triethylammonium bicarbonate (TEAB) buffer. However, it has been shown previously [24] that using this buffer produces unidentified and unwanted site reaction products (in particular, singly charged ions with m/z of 303.26, 317.26, and 331.29) of high intensity in the LC-MS2/MS3 chromatograms. The formation of these products is avoided by using the described buffer conditions.

22. The given retention time range is based on the described system. This range may differ slightly between different HPLC systems and the fraction collection retention time frame should be adjusted accordingly.

23. The Orbitrap Fusion allows acquisition of high-resolution, MultiNotch MS3 data [25]. As demonstrated previously [24, 25], this approach reduces the observed interference effect in quantitation at the MS2 level. However, not all instrumentation allows for this approach. Other approaches to decrease interference in the quantitation include TMT_C quantitation [31] and ion-ion chemistry for removing contaminant ions [23]. If these approaches are untenable, one must use the resulting quantitative data with appropriate caution.

References

1. Kim M, Pinto S, Getnet D et al (2014) A draft map of the human proteome. Nature 509:575–581

2. Wilhelm M, Schlegl J, Hahne H et al (2014) Mass-spectrometry-based draft of the human proteome. Nature 509:582–587

3. Wang Z, Gerstein M, Snyder M (2009) RNA-Seq: a revolutionary tool for transcriptomics. Nat Rev Genet 10:57–63

4. Li G, Burkhardt D, Gross C et al (2014) Quantifying absolute protein synthesis rates reveals principles underlying allocation of cellular resources. Cell 157:624–635

5. Torres E, Dephoure N, Panneerselvam A et al (2010) Identification of aneuploidy-tolerating mutations. Cell 143:71–83

6. Stingele S, Stoehr G, Peplowska K et al (2012) Global analysis of genome, transcriptome and proteome reveals the response to aneuploidy in human cells. Mol Syst Biol 8:608

7. Dephoure N, Hwang S, O'Sullivan C et al (2014) Quantitative proteomic analysis reveals posttranslational responses to aneuploidy in yeast. Elife 3, e03023

8. Wu Y, Williams E, Dubuis S et al (2014) Multilayered genetic and omics dissection of mitochondrial activity in a mouse reference population. Cell 158:1415–1430

9. Zhang B, Wang J, Wang X et al (2014) Proteogenomic characterization of human colon and rectal cancer. Nature 513:382–387

10. Gygi S, Rist B, Gerber S et al (1999) Quantitative analysis of complex protein mixtures using isotope-coded affinity tags. Nat Biotechnol 17:994–999

11. Ong S, Blagoev B, Kratchmarova I et al (2002) Stable isotope labeling by amino acids in cell culture, SILAC, as a simple and accurate approach to expression proteomics. Mol Cell Proteomics 1:376–386

12. Syka J, Marto J, Bai D et al (2004) Novel linear quadrupole ion trap/FT mass spectrometer: performance characterization and use in the comparative analysis of histone H3 post-translational modification. J Proteome Res 3:621–626

13. Olsen J, de Godoy L, Li G et al (2005) Parts per million mass accuracy on an Orbitrap mass

spectrometer via lock mass injection into a C-trap. Mol Cell Proteomics 12:2010–2021

14. Hsu J, Huang S, Chow N et al (2003) Stable-isotope dimethyl labeling for quantitative proteomics. Anal Chem 75:6843–6852

15. Wilson-Grady J, Haas W, Gygi S (2013) Quantitative comparison of the fasted and re-fed mouse liver phosphoproteomes using lower pH reductive dimethylation. Methods 61: 277–286

16. Wu Y, Wang F, Liu Z et al (2014) Five-plex isotope dimethyl labeling for quantitative proteomics. Chem Commun (Camb) 50: 1708–1710

17. Blagoev B, Ong S, Kratchmarova I et al (2004) Temporal analysis of phosphotyrosine-dependent signaling networks by quantitative proteomics. Nat Biotechnol 22:1139–1145

18. Thompson A, Schäfer J, Kuhn K et al (2003) Tandem mass tags: a novel quantification strategy for comparative analysis of complex protein mixtures by MS/MS. Anal Chem 75: 1895–1904

19. Ross P, Huang Y, Marchese J et al (2004) Multiplexed protein quantitation in Saccharomyces cerevisiae using amine-reactive isobaric tagging reagents. Mol Cell Proteomics 3:1154–1169

20. Choe L, D'Ascenzo M, Relkin N et al (2007) 8-Plex quantitation of changes in cerebrospinal fluid protein expression in subjects undergoing intravenous immunoglobulin treatment for Alzheimer's disease. Proteomics 7:3651–3660

21. McAlister G, Huttlin E, Haas W et al (2012) Increasing the multiplexing capacity of TMTs using reporter ion isotopologues with isobaric masses. Anal Chem 84:7469–7478

22. Weekes M, Tomasec P, Huttlin E et al (2014) Quantitative temporal viromics: an approach to investigate host-pathogen interaction. Cell 157:1460–1472

23. Wenger C, Lee M, Hebert A et al (2011) Gas-phase purification enables accurate, multiplexed proteome quantification with isobaric tagging. Nat Methods 8:933–935

24. Ting L, Rad R, Gygi S et al (2011) MS3 eliminates ratio distortion in isobaric multiplexed quantitative proteomics. Nat Methods 8: 937–940

25. McAlister G, Nusinow D, Jedrychowski M et al (2014) MultiNotch MS3 enables accurate, sensitive, and multiplexed detection of differential expression across cancer cell line proteomes. Anal Chem 86:7150–7158

26. Eng J, McCormack A, Yates J (1994) An approach to correlate tandem mass spectral data of peptides with amino acid sequences in a protein database. J Am Soc Mass Spectrom 5:976–989

27. Peng J, Elias J, Thoreen C et al (2003) Evaluation of multidimensional chromatography coupled with tandem mass spectrometry (LC/LC-MS/MS) for large-scale protein analysis: the yeast proteome. J Proteome Res 1:43–50

28. Elias J, Gygi S (2007) Target-decoy search strategy for increased confidence in large-scale protein identifications by mass spectrometry. Nat Methods 4:207–214

29. Elias J, Gygi S (2009) Target-decoy search strategy for mass spectrometry-based proteomics. Methods Mol Biol 604:55–71

30. Huttlin E, Jedrychowski M, Elias J et al (2010) A tissue-specific atlas of mouse protein phosphorylation and expression. Cell 143:1174–1189

31. Wühr M, Haas W, McAlister G et al (2012) Accurate multiplexed proteomics at the MS2 level using the complement reporter ion cluster. Anal Chem 84:9214–9221

Chapter 2

Sample Preparation Approaches for iTRAQ Labeling and Quantitative Proteomic Analyses in Systems Biology

Christos Spanos and J. Bernadette Moore

Abstract

Among a variety of global quantification strategies utilized in mass spectrometry (MS)-based proteomics, isobaric tags for relative and absolute quantitation (iTRAQ) are an attractive option for examining the relative amounts of proteins in different samples. The inherent complexity of mammalian proteomes and the diversity of protein physicochemical properties mean that complete proteome coverage is still unlikely from a single analytical method. Numerous options exist for reducing protein sample complexity and resolving digested peptides prior to MS analysis. Indeed, the reliability and efficiency of protein identification and quantitation from an iTRAQ workflow strongly depend on sample preparation upstream of MS. Here we describe our methods for: (1) total protein extraction from immortalized cells; (2) subcellular fractionation of murine tissue; (3) protein sample desalting, digestion, and iTRAQ labeling; (4) peptide separation by strong cation-exchange high-performance liquid chromatography; and (5) peptide separation by isoelectric focusing.

Key words Proteomics, Mass spectrometry, iTRAQ, Subcellular fractionation, High-performance liquid chromatography, Isoelectric focusing

1 Introduction

Quantitative analysis of protein expression, function, and subcellular localization is fundamental to network biology. Mass spectrometry (MS)-based quantitative proteomic approaches have evolved rapidly in the last 15 years and are generating datasets essential for systems biology and the modeling of biological networks [1]. Discovery applications in MS-based proteomics have largely employed untargeted strategies where proteins in one or more samples (diseased or treated) are quantified relative to the amount of proteins in a separate sample (normal or control). Quantification of the peptides/proteins can either be done by comparative analysis of spectral features in a "label-free" workflow or alternatively be accomplished through isotopic labeling or the incorporation of a differential mass tag. Isobaric tags for relative and absolute

Jörg Reinders (ed.), *Proteomics in Systems Biology: Methods and Protocols*, Methods in Molecular Biology, vol. 1394, DOI 10.1007/978-1-4939-3341-9_2, © Springer Science+Business Media New York 2016

quantitation (iTRAQ) are widely used amine-specific, stable-isotope reagents which can be used to label the N terminus and ε side chain of lysines of peptides generated by tryptic digestion of extracted proteins. The reagents were designed for "multiplexing," and four or eight separate iTRAQ labels are available permitting the simultaneous analysis of multiple samples. The labels consist of a low-mass reporter group, a balance group, and an amine-reactive group; they have isobaric masses in MS mode (145 and 305 Da for 4-plex or 8-plex reagents), but upon fragmentation release low-mass reporter ions (m/z values of 114.1, 115.1, 116.1, 117.1 for 4-plex, plus 113.1, 118.1, 119.1, 121.1 for 8-plex) allowing quantification at the MS/MS level. Unlike metabolic incorporation of stable isotopes, a key advantage to iTRAQ labels is that samples from any source, including patient material, can be chemically labeled. This fact, in combination with the ability to simultaneously analyze multiple samples, has likely contributed to the widespread use of these reagents [2, 3]. It should be noted that quantitation by iTRAQ is not perfect; contamination during precursor ion selection, specific to MS/MS quantitation, results in compression of the iTRAQ ratio and underestimation of relative protein abundance estimates; and variance is higher for low-intensity signals [4, 5]. Data processing and instrument-specific approaches aimed at addressing issues related to the precision and accuracy of iTRAQ quantitation continue to evolve and have been reviewed elsewhere [2, 3].

Technological advances in high-resolution MS instrumentation means that almost complete coverage of unicellular organisms such as yeast is now possible [6] and comprehensive analysis of mammalian proteomes is envisaged as feasible in the near future [7]. However, the required technology is not widely available and currently most researchers will find that proteome coverage is dependent on reducing sample complexity and their choice of multiple sample preparation steps including protein separation, digestion, and peptide fractionation steps. Proteomic workflows can be complex and the role of error propagation through multiple handling steps should not be underestimated [8]. Critical to the success of an iTRAQ experiment is the use of equal sample amounts, reproducible protein digestion, and efficient peptide labeling. Protein samples may be pre-fractionated either by subcellular fractionation or based on size by polyacrylamide gel electrophoresis. Proteins are most typically digested using trypsin, although other proteases can be used [9], and digested peptides are separated by either high-performance liquid chromatography (HPLC) or isoelectric focusing.

We have used iTRAQ reagents to examine differential protein expression in fatty acid-treated hepatocarcinoma cells and liver tissue mice fed a high-fat diet. Here we describe our methods for: (1) total protein extraction from immortalized cells; (2) subcellular fractionation of murine liver tissue; (3) protein sample desalting, digestion, and iTRAQ labeling; (4) peptide separation by strong cation-exchange (SCX) HPLC; and (5) peptide separation by isoelectric focusing.

2   Materials

All reagents should be of analytical grade and solvents either HPLC or LC-MS grade. All solutions should be prepared with distilled, deionized water (ddH$_2$O), typically 18.2 MΩ·cm at 25 °C, except for LC-MS/MS buffers which require LC-MS-grade H$_2$O.

2.1   Protein Extraction from Cells

1. Phosphate-buffered saline (PBS; 1×): 137 mM NaCl, 2.7 mM KCl, 10 mM Na$_2$HPO$_4$.

2. Radioimmunoprecipitation assay buffer (RIPA): 150 mM NaCl, 1.0 % IGEPAL® CA-630, 0.5 % sodium deoxycholate, 0.1 % SDS, and 50 mM Tris, pH 8.0.

3. Protease inhibitor cocktail-EDTA free (PI): Mixture of several protease inhibitors inhibiting serine, cysteine, but not metalloproteases.

4. Low-speed swinging bucket centrifuge.

5. Refrigerated table top microcentrifuge.

6. QIAshredder columns (QIAGEN, UK).

7. BCA assay.

2.2   Protein Extraction from Liver Tissue

1. HEPES, EDTA, mannitol (HEM) buffer: 20 mM HEPES, 1 mM EDTA, 300 mM mannitol.

2. Protease inhibitor cocktail EDTA free (PI): Mixture of several protease inhibitors inhibiting serine, cysteine, but not metalloproteases.

3. 10 ml glass Dounce homogenizer with pestle attached to electric drill.

4. Refrigerated swinging bucket centrifuge.

5. Ultracentrifuge.

6. Appropriate polyallomer tubes for ultracentrifugation.

7. Amicon Ultra-15 Centrifugation Filter Device (NMWL of 3 kDa; Millipore, USA).

8. BCA assay (Pierce Scientific, UK).

2.3   Protein Desalting, Digestion, and iTRAQ Labeling

1. 2 ml ZEBA columns (Pierce Scientific, UK; *see* **Note 6**).

2. RIPA buffer and PI as before.

3. Vacuum centrifuge (Eppendorf, UK).

4. Low-bind microcentrifuge tubes (*see* **Note 7**).

5. Bovine serum albumin (BSA) as control to monitor digestion and future LC separation efficiency.

6. Triethylammonium bicarbonate (TEAB) buffer, 1 M pH 8.5 (Sigma Aldrich, UK).

7. iTRAQ Reagents Multiplex Kit (ABSciex, UK).

8. Shaking dry heat block (Eppendorf, UK).

9. Mass spectrometry-grade trypsin gold (Promega, UK).

10. Sonication water bath.

2.4 Peptide Separation by Strong Cation-Exchange HPLC

1. HPLC instrument (Hewlett Packard 1100 series) with autosampler (Agilent Technologies 1200).

2. Polysulfoethyl A chromatography column (100×94 2.1 mm—300 Å; Hichrom Ltd, UK).

3. Guard cartridge (Hichrom Ltd, UK; *see* **Note 9**).

4. Buffer A: 10 mM KH_2PO_4 (pH 2.75), 25 % acetonitrile (*see* **Note 10**).

5. Buffer B: 10 mM KH_2PO_4, 1 M KCl (pH 2.75), 25 % acetonitrile (*see* **Note 10**).

6. Vacuum centrifuge (Eppendorf, UK).

2.5 Peptide Separation by Isoelectric Focusing

1. 3100 OFFGEL Fractionator (Agilent Technologies, UK).

2. OFFGEL High Resolution Kit, pH 3–10, 12 samples, 24 fractions (Agilent Technologies, UK).

3. Vacuum centrifuge (Eppendorf, UK).

2.6 Sample Cleanup Prior to Mass Spectrometry

1. 0.1 % (v/v) trifluoroacetic acid (TFA; Sigma Aldrich, UK) in HPLC-grade H_2O.

2. ZipTips C18 with 0.6 μl bed of chromatography media (Millipore, USA).

3. "Wet solution": 25 % acetonitrile in HPLC-grade H_2O.

4. "Equilibration and wash solution": 0.1 % TFA in HPLC-grade H_2O.

5. "Elution solution": 50 % acetonitrile/0.1 % TFA in HPLC-grade H_2O.

3 Methods

3.1 Total Protein Extraction from Hepatocellular Carcinoma Cell Line

1. Add protease inhibitor cocktail to RIPA buffer to a final concentration of 1.5×; keep on ice. Bring a microcentrifuge for the $10,000 \times g$ spin to 4 °C.

2. Following your preferred cell treatment, detach cells from flask surface using trypsin and centrifuge cell suspension for 5 min at $200 \times g$ RT.

3. Wash cell pellets with 1× PBS and centrifuge again for 5 min at 1200×g.

4. Resuspend cell pellet in 350 μl RIPA buffer containing PI and leave on ice for 10 min.

5. Vortex lysed cells, then place on QIAshredder column, and centrifuge for 2 min at 10,000×g.

6. Measure protein concentration of resulting eluate using BCA assay.

3.2 Membrane and Cytosol Protein Extraction from Mouse Liver Tissue

1. Bring both an ultracentrifuge and swinging bucket centrifuge to 4 °C.

2. Add protease inhibitor cocktail to of ice-cold HEM buffer (*see* **Note 1**) to a final concentration of 1.5×; keep on ice.

3. From frozen liver lobes (*see* **Note 2**) excise approximately 50 mg of liver with a razor blade on dry ice. Place into Dounce homogenizer with 6 ml of ice-cold HEM buffer and homogenize using electric drill (*see* **Note 3**).

4. Transfer sample to sterile 15 ml conical tube and keep on ice while repeating for all biological replicates. Rinse pestle and container in between each sample with ddH$_2$O.

5. Centrifuge samples in swinging bucket centrifuge at 2000×g for 10 min at 4 °C.

6. Transfer supernatant (containing membrane and cytosol proteins) to polyallomer ultracentrifuge tubes on ice; add ~4 ml of HEM buffer with PI and balance tubes within 1/10th of a gram of each other. Discard pellet (*see* **Note 4**).

7. Centrifuge at 100,000×g for 30 min at 4 °C.

8. Transfer the supernatant, representing the cytosolic fraction, into Amicon Ultra-15 Centrifugation Filter Device and centrifuge for 1 h at 4000×g at 4 °C to collect concentrated cytosolic protein (>3 kDa).

9. Re-homogenize (*see* **Note 5**) the membrane-enriched pellet in 0.5–1.0 ml of fresh HEM buffer with PI.

10. Determine protein concentration using BCA assay before aliquoting samples and storing at −80 °C.

3.3 Protein Desalting/Buffer Exchange, Digestion, and iTRAQ Labeling

1. Remove storage solution from 2 ml ZEBA columns (*see* **Note 6**) by centrifuging at 2000×g for 2 min. For tissue samples use RIPA for buffer exchange; for cells use ddH$_2$O for desalting. Add 1 ml RIPA buffer with PI or ddH$_2$O to column resin and centrifuge for 2 min at 2000×g; repeat three times. Then add samples to column resin and centrifuge for 2 min at 2000×g. Process BSA control alongside protein samples up until iTRAQ labeling (**step 9**).

2. Collect the resulting desalted protein eluate and quantify protein concentration by BCA assay.

3. Aliquot 100 μg of protein into low-bind microcentrifuge tubes (*see* **Note 7**) from each treatment and dry in vacuum centrifuge at 45 °C; dried sample can be stored at –80 °C prior to digestion.

4. If stored at –80 °C bring dried protein samples to RT, and then add 40 μl of 1 M, pH 8.5, TEAB buffer (*see* **Note 8**) and 2 μl of 2 % sodium dodecyl sulfate (SDS; denaturing agent from iTRAQ kit). Vortex thoroughly to solubilize, and then pulse spin to bring the sample to the bottom of the tube.

5. Add 4 μl of *tris*(2-carboxyethyl)phosphine (TCEP; reducing agent from iTRAQ kit); vortex thoroughly, then pulse spin, and incubate for 1 h at 60 °C.

6. Add 1 μl of methyl methanethiosulfonate (MMTS; cysteine blocking reagent from iTRAQ kit); vortex thoroughly, then pulse spin, and incubate for 10 min at RT.

7. Add 10 μl of 1 μg/μl trypsin solution (reconstituted in 50 mM acetic acid) vortex thoroughly, then pulse spin, and incubate, shaking, at 37 °C for 12–16 h.

8. Post-digestion, dry down samples by vacuum centrifugation at 45 °C and prepare 100 μl/sample 7:3 (v/v) ethanol (100 %): TEAB (0.5 M) solution.

9. Resuspend peptide digests in 80 μl of ethanol:TEAB by vortexing and then sonicate for 5 min at RT in sonicating water bath. Repeat to ensure complete solubilization.

10. Bring required iTRAQ reagent vials (114, 115, 116, 117) to RT; pulse spin, add each peptide digest to the appropriate iTRAQ reagent vial, and vortex thoroughly.

11. Rinse each sample tube with the additional 20 μl of ethanol:TEAB solution and transfer to the correct iTRAQ reagent vials. Vortex, pulse spin, and then incubate for 1 h at RT.

12. Combine iTRAQ-labeled samples into single tube; rinse empty tubes with one aliquot of 100 μl of 50 % acetonitrile and add to combined tube.

13. Dry down pooled sample via vacuum centrifugation at 45 °C.

3.4 Peptide Separation by Strong Cation-Exchange HPLC

1. The dried, digested, and iTRAQ-labeled sample (one tube) was resuspended in 100 μl of buffer A.

2. Set up guard cartridge (*see* **Note 9**) before HPLC column.

3. Condition the HPLC column by introducing buffer A (*see* **Note 10**) for 30 min at a flow rate of 1.5 ml/min.

4. Use digested BSA sample to check both the efficiency of the tryptic digestion and the separation efficiency of the column using the stepped gradient (Table 1).

Table 1
The stepped gradient that was followed during the SCX approach. The pressure was set to 400 bar

Time (min)	% Buffer B	Flow rate (ml/min)
0.00	0.00	1.00
0.01	0.00	0.80
10.00	0.00	0.80
10.01	0.10	0.80
10.02	6.00	0.80
12.02	6.00	0.80
12.03	9.00	0.80
14.03	9.00	0.80
14.04	11.00	0.80
16.04	11.00	0.80
16.05	13.00	0.80
18.05	13.00	0.80
18.06	15.00	0.80
20.06	15.00	0.80
20.07	17.00	0.80
22.07	17.00	0.80
22.08	19.00	0.80
24.08	19.00	0.80
24.09	23.00	0.80
26.09	23.00	0.80
26.10	27.00	0.80
28.10	27.00	0.80
28.11	33.00	0.80
30.12	33.00	0.80
30.13	40.00	0.80
32.13	40.00	0.80
33.14	50.00	0.80
38.14	50.00	0.80
39.15	100.00	0.80
42.16	100.00	0.80

(continued)

Table 1
(continued)

Time (min)	% Buffer B	Flow rate (ml/min)
42.17	6.00	0.80
43.17	6.00	0.80
43.18	0.00	0.50
50.01	0.00	0.10

5. Inject the entire volume of sample with a 100 μl injection volume and collect fractions in 2-min time slices; use LC results to assess how many fractions contain sample (*see* **Note 11**).

6. Dry down collected fractions by vacuum centrifugation at 45 °C.

3.5 Peptide Separation by Isoelectric Focusing

1. Make up "peptide OFFGEL stock solution (1.25×)" and "peptide-immobilized pH gradient (IPG) strip rehydration solution" with the appropriate ampholytes depending on which IPG strips you are using according to kit instructions (*see* **Note 12**). Assemble the IPG strips, frames, and electrodes according to the manufacturer's protocol.

2. Resuspend by vortexing the dried peptide samples in 3.6 ml 1× OFFGEL stock solution (e.g., 2.88 ml 1.25× OFFGEL stock solution + 0.72 ml dH$_2$O). Load 150 μl of this sample in each of the 24 wells.

3. Samples should be focused using a maximum current of 50 μA and the focusing is stopped after the total voltage reaches 50 kVh (60 h). During the focusing, pads on the electrodes should be exchanged to prevent evaporation (*see* **Note 13**).

4. After focusing, 50–150 μl of peptide fractions should be recovered for each well and transferred to individual microfuge tubes. To recover as much as possible of the focused peptides, 150 μl of HPLC-grade methanol should be added to each well, incubated for 15 min without voltage, and then added to the appropriate tube.

5. Dry down collected fractions by vacuum centrifugation at 45 °C.

3.6 Sample Cleanup Prior to Mass Spectrometry

1. Resuspend peptide samples (fractionated by either SCX or IEF) in 120 μl 0.1 % TFA.

2. Equilibrate the ZipTip by first aspirating 10 μl of the "wet solution" and discarding to waste, repeat once; then aspirate 10 μl of the "equilibration and wash solution" and discarding to waste, repeating three times.

3. Bind peptides to ZipTip by cycling (aspirate-dispense-aspirate-dispense) ten times with your sample.

4. Wash sample by aspirating 10 μl of the "equilibration and wash solution" and discarding to waste; repeat three times.

5. Elute sample by aspirating 10 μl of the "elution solution" and collecting in microfuge tubes; repeat six times until 70 μl is collected in all.

6. Dry down collected cleaned up fractions by vacuum centrifugation at 45 °C. Resuspend in 10 μl formic acid just prior to nLC-ESI-MS/MS analysis.

4 Notes

1. An alternative buffer for tissue lysis may be applied as deemed appropriate.

2. We recommend at the time of study termination excising the same liver lobe from all animals, immediate snap freezing in liquid nitrogen, and then keeping on dry ice until transfer to −80 °C.

3. In our case we used the lowest speed setting on a very powerful bench-top electric drill and used ten, slow, up-and-down strokes for each sample. This may vary depending on your setup; just keep consistent for all samples.

4. Pellet contains nuclear proteins and cellular debris; you may wish to use differential detergent fractionation here to isolate nuclear proteins as well.

5. We use pestle manually here in addition to pipetting.

6. Methanol-chloroform and other precipitation methods may be used, but we have experienced better recovery and reproducibility using Zeba columns for buffer exchange and sample concentration.

7. Drying down by vacuum centrifuge may result in sample adhering on the walls of the tube. To minimize any protein loss and to make resuspension easier, low-bind tubes should be used in all drying, by vacuum centrifuge, steps.

8. The TEAB buffer that is included in the iTRAQ kit is 0.5 M. We have had better recovery of the sample from the dry phase using 1 M TEAB for resuspension of the dried sample before the iTRAQ labeling process begins. For the next step and whenever required, the TEAB buffer provided by the iTRAQ kit should be used.

9. Although not necessary, it is advised to use a guard cartridge attached in front of the HPLC column. In case of blockage due to sample impurities the column remains unaffected and the guard cartridge is cheaper to replace.

10. For strong cation exchange, the pH of the buffers should be tightly controlled below or equal to 3. Whenever buffers need to be replaced, the pH levels should be exactly the same as previously.

11. In our case, separating peptide digests of total protein from HuH7 cells, we had sample in first 17 fractions.

12. The OFFGEL stock solution contains glycerol; this can make later sample resuspension difficult and interfere with the subsequent performance of the mass spectrometer by blocking the nLC lines. It has been suggested that the glycerol can be omitted or reduced but we have not tested this.

13. The total run for the OFFGEL fractionation is around 60 h. In order to prevent water evaporation from the pads that will lead to insufficient peptide separation, upper electrode pads should be changed every 15 h with fresh pads wetted in ddH$_2$O.

Acknowledgements

This work was supported by the Biotechnology and Biological Sciences Research Council, with a studentship grants (BB/D526853/1) to C.S. and a project grant (BB/I008195/1) to J.B.M.

References

1. Moore JB, Weeks ME (2011) Proteomics and systems biology: current and future applications in the nutritional sciences. Adv Nutr 2:355–364

2. Evans C, Noirel J, Ow SY et al (2012) An insight into iTRAQ: where do we stand now? Anal Bioanal Chem 404:1011–1027

3. Christoforou AL, Lilley KS (2012) Isobaric tagging approaches in quantitative proteomics: the ups and downs. Anal Bioanal Chem 404:1029–1037

4. Ow SY, Salim M, Noirel J et al (2009) iTRAQ underestimation in simple and complex mixtures: "the good, the bad and the ugly". J Proteome Res 8:5347–5355

5. Karp NA, Huber W, Sadowski PG et al (2010) Addressing accuracy and precision issues in iTRAQ quantitation. Mol Cell Proteomics 9:1885–1897

6. Nagaraj N, Kulak NA, Cox J et al (2012) System-wide perturbation analysis with nearly complete coverage of the yeast proteome by single-shot ultra HPLC runs on a bench top Orbitrap. Mol Cell Proteomics 11:M111. 013722

7. Mann M, Kulak NA, Nagaraj N et al (2013) The coming age of complete, accurate, and ubiquitous proteomes. Mol Cell 49:583–590

8. Kruger T, Lehmann T, Rhode H (2013) Effect of quality characteristics of single sample preparation steps in the precision and coverage of proteomic studies—a review. Anal Chim Acta 776:1–10

9. Vandermarliere E, Mueller M, Martens L (2013) Getting intimate with trypsin, the leading protease in proteomics. Mass Spectrom Rev 32:453–465

Chapter 3

Two Birds with One Stone: Parallel Quantification of Proteome and Phosphoproteome Using iTRAQ

Fiorella A. Solari, Laxmikanth Kollipara, Albert Sickmann, and René P. Zahedi

Abstract

Altered and abnormal levels of proteins and their phosphorylation states are associated with many disorders. Detection and quantification of such perturbations may provide a better understanding of pathological conditions and help finding candidates for treatment or biomarkers. Over the years, isobaric mass tags for relative quantification of proteins and protein phosphorylation by mass spectrometry have become increasingly popular. One of the most commonly used isobaric chemical tags is iTRAQ (isobaric tag for relative and absolute quantitation). In a typical iTRAQ-8plex experiment, a multiplexed sample amounts for up to 800 μg of peptides. Using state-of-the-art LC-MS approaches, only a fraction (~5 %) of such a sample is required to generate comprehensive quantitative data on the global proteome level, so that the bulk of the sample can be simultaneously used for quantitative phosphoproteomics. Here, we provide a simple and straightforward protocol to perform quantitative analyses of both proteome and phosphoproteome from the same sample using iTRAQ.

Key words iTRAQ, Phosphopeptide enrichment, Quantitative proteomics, Quantitative phosphoproteomics

1 Introduction

Quantitative analysis of proteins and their phosphorylation sites can provide valuable insights into altered molecular mechanisms and signaling pathways and thus can be used to study abnormal behavior, such as in cancer [1]. For almost a decade, isobaric mass tags have been used extensively for relative quantification in proteomics [2–8] and are particularly relevant for biological samples that are not accessible to metabolic labeling, such as human tissues and body fluids. Chemical labeling with iTRAQ reagents [9] enables relative quantification of up to eight different multiplexed conditions/samples in a single LC-MS run, and can provide sample amounts (up to 0.8 mg) that are considerably higher than what is required for conventional proteomics experiments using

Jörg Reinders (ed.), *Proteomics in Systems Biology: Methods and Protocols*, Methods in Molecular Biology, vol. 1394,
DOI 10.1007/978-1-4939-3341-9_3, © Springer Science+Business Media New York 2016

state-of-the-art equipment and workflows (even with fractionation no more than 50 μg). Therefore, the remaining >90 % of the multiplexed sample can be effectively used to analyze low-abundance posttranslational modifications (PTM), such as phosphorylation.

Here, we describe a protocol based on iTRAQ-8plex labeling that allows for both global and phosphoproteome analyses. For the global proteome analysis, 5 % of the multiplexed sample is fractionated using high-pH reversed-phase chromatography [10], whereas the rest of the sample is used for phosphopeptide enrichment with titanium dioxide (TiO_2) beads [11] followed by hydrophilic interaction liquid chromatography (HILIC) [12] for fractionation of phosphopeptides prior to LC-MS. Thus, with as little as 40 μg for proteome and 760 μg for phosphoproteome analysis, it is possible to quantify thousands of proteins and phosphopeptides from the same sample.

2   Materials

All solutions and buffers should be prepared using ultrapure deionized water. Alternatively, LC-MS-grade water (Biosolve) can be used.

2.1   Cell Lysis of Biological Samples

1. Biological sample obtained from primary cells or tissue, e.g., 100 μg of fibroblasts.

2. Lysis buffer: 50 mM Tris–HCl (pH 7.8, adjust with HCl), 150 mM sodium chloride (NaCl) containing 1 % (w/v) sodium dodecyl sulfate (SDS). Add one tablet of phosphatase inhibitor cocktail phosSTOP (Roche Diagnostics) to 10 mL of buffer.

3. Benzonase (Novagen) and 1 M magnesium chloride ($MgCl_2$) solution.

4. Refrigerated benchtop centrifuge (Eppendorf) and a vortex mixer.

5. Determination of protein concentration: Bicinchoninic acid (BCA) protein assay.

2.2   Carbamido-methylation, Protein Precipitation, and Enzymatic Protein Digestion

1. Reducing agent: 2 M dithiothreitol (DTT) solution. Stock can be stored at –40 °C.

2. Alkylating agent: 1 M iodoacetamide (IAA) solution. Freshly prepared. See Note 1.

3. Organic solvents: Ethanol and acetone. Store both solvents at –40 °C prior to usage.

4. Centrifuge: Refrigerated benchtop centrifuge (Eppendorf).

5. Enzyme: Sequencing Grade Modified Trypsin (Promega). Dissolve the lyophilized trypsin in 50 mM ammonium bicarbonate (NH_4HCO_3), pH 7.8 to get a final concentration of 1 mg/mL.

6. Digestion buffer: 0.2 M Guanidine hydrochloride (GuHCl), 5 % acetonitrile (ACN), 2 mM calcium chloride (CaCl$_2$), trypsin solution [1 mg/mL], and 50 mM NH$_4$HCO$_3$ (pH 7.8).

7. Stop digestion: 10 % trifluoroacetic acid (TFA).

2.3 Digestion Control by Monolithic Reversed-Phase Chromatography

1. HPLC: UltiMate 3000 rapid separation liquid chromatography (RSLC) (Thermo Fisher Scientific) or similar HPLC system.

2. HPLC column: PepSwift monolithic trap column 200 μm × 5 mm and PepSwift monolithic capillary column 200 μm × 5 cm (both Thermo Fisher Scientific).

3. HPLC buffer A: 0.1 % TFA.

4. HPLC buffer B: 0.08 % TFA, 84 % ACN.

2.4 Solid-Phase Extraction Cartridges for Sample Cleanup

1. Solid-phase extraction cartridge (SPEC) sorbent material: C18 AR 4 mg columns (Agilent).

2. Vacuum manifold system and vacuum centrifuge (SpeedVac).

3. Wetting/activating buffer: 100 % ACN.

4. Equilibration and wash buffer: 0.1 % TFA.

5. Elution buffer: 60 % ACN in 0.1 % TFA.

2.5 iTRAQ Labeling

1. iTRAQ reagents: 8 plex kit (113–119, 121) from AB SCIEX.

2. Dissolution buffer: 0.5 M triethylammonium bicarbonate (TEAB), pH 8.5.

3. Reagent dilution solvent: 100 % isopropanol, LC-MS grade.

4. Thermomixer.

2.6 Phosphopeptide Enrichment with Titanium Dioxide (TiO$_2$) Beads

1. Metal oxide affinity chromatography (MOAC) material: Titanium dioxide (TiO$_2$) beads, 5 μm diameter (Titansphere, GL Sciences).

2. Loading buffer 1: 80 % ACN, 5 % TFA, and 1 M glycolic acid [13]. *See* **Note 2**.

3. Washing buffer 1: 80 % ACN, 1 % TFA.

4. Washing buffer 2: 10 % ACN, 0.1 % TFA.

5. Elution buffer: 1 % ammonium hydroxide (NH$_4$OH), pH 11.3.

6. Loading buffer 2: 70 % ACN, 2 % TFA.

7. Washing buffer 3: 50 % ACN, 0.1 % TFA.

8. Phosphopeptide purification: C8 Empore Disc (3 M).

9. Acidifying buffer: 100 % formic acid (FA) and 10 % TFA.

10. C18 SPE: Oligo R3 beads (Applied Biosystems) and C18 Empore Disc (3 M).

11. Air-filled syringe.

2.7 Sample Fractionation Using Chromatography

2.7.1 High-pH Reversed-Phase Chromatography

1. HPLC: UltiMate 3000 liquid chromatography (Thermo Fisher Scientific) or similar HPLC system.
2. HPLC column: RP C18, 0.5 mm × 15 cm, 5 μm (Biobasic, Thermo Scientific).
3. HPLC buffer A: 10 mM ammonium formate (NH_4HCO_2), pH 8.0.
4. HPLC buffer B: 84 % (v/v) in 10 mM NH_4HCO_2, pH 8.0.

2.7.2 Hydrophilic Interaction Chromatography

1. HPLC: UltiMate 3000 rapid separation liquid chromatography (RSLC) (Thermo Fisher Scientific) or similar HPLC system.
2. HPLC column: Polar-phase TSK gel, 150 μm × 15 cm, 5 μm, self-packed.
3. HPLC solvent A: 0.1 % TFA, 98 % ACN.
4. HPLC solvent B: 0.1 % TFA.

2.8 LC-MS/MS

1. HPLC: UltiMate 3000 nano RSLC system (Thermo Fisher Scientific) with nano-UV cell for quality control or a similar HPLC system.
2. HPLC column: Acclaim PepMap100 C18 trap column 100 μm × 2 cm, and Acclaim PepMap100 C18 main column 75 μm × 50 cm (both Thermo Fisher Scientific).
3. HPLC loading buffer: 0.1 % TFA.
4. HPLC solvent A: 0.1 % FA.
5. HPLC solvent B: 0.1 % FA, 84 % ACN.
6. Mass spectrometer: Q-Exactive mass spectrometer (Thermo Fisher Scientific) or other MS systems that can provide high mass accuracy and high resolution for both MS and MS/MS scans, as well as beam-type CID fragmentation (or HCD in Thermo instruments).

2.9 Data Analysis

1. Data analysis software: Proteome Discoverer (version 1.4, Thermo Scientific) using the following nodes.
   (a) Search algorithms: Mascot [14] (version 2.4.1, Matrix Science), Sequest [15], and MS Amanda [16].
   (b) Quantification: Reporter ion quantifier.
   (c) Phosphorylation site assignment: PhosphoRS [17].
   (d) False discovery rate (FDR) estimator: Peptide validator.

3 Methods

Make sure that you treat all samples completely the same way during all mentioned steps until labeling, in order to keep the technical variation as low as possible. Any differences during sample treatment may affect the quantitative results.

3.1 Cell Lysis and Determination of Protein Concentration

1. The volume of the lysis buffer depends on the number of cells. Typically, 100 μL of buffer is required to solubilize for, e.g., 10^6 HeLa cells. Promote homogenization by mechanical/shearing forces, induced by pipetting and ultrasonication.

2. To hydrolyze the DNA, add 3 μL of benzonase per 200 μL of cell lysate plus 2 mM $MgCl_2$ and incubate at 37 °C for 30 min.

3. Clarify the lysates by centrifugation. Use a precooled (4 °C) centrifuge and spin down the sample tubes at 18,000 *rcf* for 30 min.

4. Collect the supernatant in a new LoBind Eppendorf tube and determine the protein concentration using BCA assay according to the manufacturer's instructions. *See* **Note 3**.

3.2 Carbamido-methylation, Sample Cleanup, and Proteolytic Digestion

1. Reduce cysteines by incubating the cell lysates with 10 mM of DTT at 56 °C for 30 min. After incubation, cool down the tubes to room temperature (RT).

2. Without further delay, add the alkylating agent (IAA) to a final concentration of 30 mM and incubate at RT for 30 min in the dark.

3. Organic solvent protein precipitation: *See* **Note 4**.

   (a) Take an aliquot of the cell lysate corresponding to 100 μg of protein and dilute tenfold with cold (−40 °C) ethanol, i.e., one part of sample plus nine parts of ethanol, and vortex briefly.

   (b) Store the sample at −40 °C for 1 h. In the meantime, cool the centrifuge to 4 °C.

   (c) Place the samples in the precooled (4 °C) centrifuge and spin down the tubes at 12,000 *rcf* for 30 min. Discard the supernatant.

   (d) Place 100 μL of cold acetone onto the protein pellet, briefly vortex, and centrifuge as mentioned above for 10 min.

   (e) Remove the supernatant and allow the protein pellet to dry. *See* **Note 5**.

4. Digestion:

   (a) Resolubilize the protein pellet with 6 M GuHCl prepared in 50 mM NH_4HCO_3, pH 7.8. For 100 μg of protein, use 20 μL of 6 M GuHCl buffer.

   (b) Add 100 % ACN and 1 M $CaCl_2$ solution to get final concentrations of 5 % and 2 mM, respectively.

   (c) Vortex the tubes to solubilize the pellet completely.

   (d) Dilute GuHCl concentration from 6 M to 0.2 M with 50 mM NH_4HCO_3, pH 7.8.

   (e) Take an aliquot of 1 μg of protein and label as "before digest."

(f) Add trypsin solution [1 mg/mL] to get 1:20 (w/w) ratio of enzyme to protein and incubate the sample tubes at 37 °C for 14 h [18]. *See* **Note 6**.

(g) Stop the enzyme activity by adding 10 % TFA to a final concentration of 1 %.

3.3 Digestion Control with Monolithic Reversed-Phase Chromatography

1. Take an aliquot of 1 μg of peptides. Label as "after digest."

2. For all samples measure "before digest" and "after digest" on an RP monolithic HPLC system to evaluate the digestion efficiency and reproducibility. *See* **Note 7** [19].

3. If the samples look well digested and reproducible, then continue with Subheading 3.5; otherwise consider repeating the sample preparation.

3.4 SPEC Sample Cleanup (Desalting)

1. To improve the reproducibility of this step, use a vacuum manifold system for peptide desalting. Use C18 AR 4 mg material (Agilent) tips.

    (a) Activation: Three times with 100 μL of 100 % ACN.

    (b) Equilibration: Three times with 100 μL of 0.1 % TFA.

    (c) Sample loading: Place the sample on the material and reload the flow through three times.

    (d) Washing: Three times with 100 μL of 0.1 % TFA.

    (e) Peptides elution: Two times with 100 μL of 60 % ACN in 0.1 % TFA

2. To control the reproducibility of the desalting procedure, take a small but equal volume (e.g., 2 μL) of eluate from each sample and dry it completely in the SpeedVac. Snap freeze the remaining volume using liquid nitrogen and store the frozen eluates at −80 °C until further use. *See* **Note 8**.

3. Measure the samples (2 μL) using nano-LC with UV detection (214 nm) or LC-MS. If they look reproducible, take 100 μg aliquots from each sample, dry them in the SpeedVac, and proceed with iTRAQ labeling.

3.5 iTRAQ Labeling

1. Resolubilize each peptide pellet (~100 μg) in 30 μL of dissolution buffer (0.5 M TEAB, pH 8.5).

2. Perform the labeling reaction as per the manufacturer's instructions (iTRAQ-8plex, AB SCIEX).

3. After incubation, combine all the eight differentially labeled samples in a new 1.5 mL LoBind Eppendorf tube.

4. Divide the multiplexed sample into two parts. Use one part corresponding to ~5 % (~40 μg) for complete (global) proteome analysis and the second part (~95 % or 760 μg) for the enrichment of phosphopeptides.

5. Dry both parts completely in the SpeedVac and store the pellets at −80 °C until further use.

3.6 Fractionation at High-pH RP for Complete Proteome Analysis

Pre-fractionation can be conducted at pH 10 [20] albeit high pH conditions affect the column stability due to the hydrolysis of siloxane groups in the fused silica [21, 22]. Therefore, we recommend performing the RP fractionation at pH 8.0; nevertheless, pH 10 yields slightly higher orthogonality.

1. Solubilize the dried multiplexed sample (~40 µg) in buffer A.

2. Perform peptide fractionation on a Biobasic C18, 0.5 mm × 15 cm column using an UltiMate 3000 liquid chromatography (LC) system (Thermo Fisher Scientific) with the following gradient:

   0–3 % B in 15 min, 3–50 % B in 65 min, 50–60 % B in 10 min, 95 % B hold for 5 min, 95 %–3 % B in 5 min, and finally re-equilibrate the column with 3 % B for 20 min.

3. Collect 16 fractions at 1-min interval from 15 to 70 min in a concatenation mode (Fig. 1). To minimize sample losses, collect fractions directly in HPLC sample inlets.

4. Finally, dry the collected fractions in the SpeedVac.

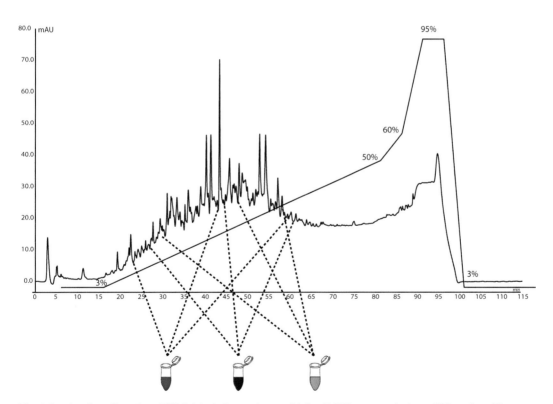

Fig. 1 Pre-fractionation of an iTRAQ-labeled sample on a high-pH C18 reversed-phase (RP) system. The peptides are separated using a binary gradient (buffer A: 10 mM NH_4HCO_2, buffer B: 10 mM, NH_4HCO_2, 84 % ACN, both pH 8.0) ranging from 3 to 50 % buffer B in 65 min. In total, 16 fractions are collected at 1-min intervals using a concatenation approach. Separation of the peptides on high-pH RP columns and subsequent concatenation of the eluted fractions reduce sample complexity, improve selectivity, and increase proteome coverage [23]. *See* **Note 10**

3.7 Phosphopeptide Enrichment Using TiO₂ Beads

The following protocol for the enrichment of phosphopeptides is based on the workflow published by Larsen and co-workers [24].

1. Resolubilize the multiplexed sample (~760 µg) in 1 mL of loading buffer 1.

2. Adjust the amount of TiO₂ beads according to the starting material. Use a bead-to-peptide ratio of 6:1, e.g., weigh in 4.56 mg of TiO₂ beads for ~760 µg of sample. Wash the beads with 200 µL of loading buffer 1. Pellet the beads and discard the supernatant.

3. To the washed beads, add 200 µL of loading buffer 1 and vortex to obtain a homogenous suspension.

4. Add appropriate amount of beads to the peptides and incubate the peptide-bead suspension for 10 min on a vortex mixer at low speed. Afterwards, centrifuge at 3000 *rcf* for 1 min to pellet the beads. Collect the supernatant in a LoBind Eppendorf tube.

5. Repeat the enrichment process in a stepwise manner (Fig. 2) by treating the supernatant with freshly prepared suspension of

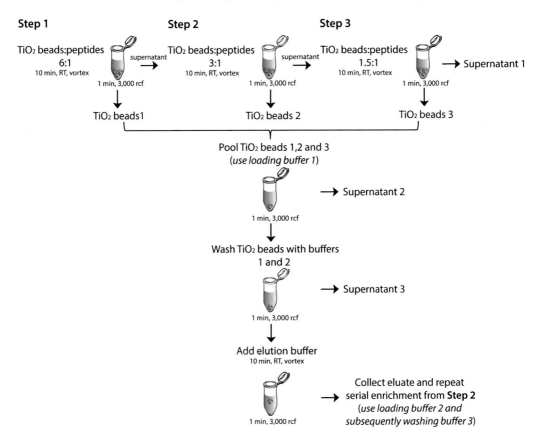

Fig. 2 Serial enrichment of phosphopeptides from an iTRAQ-labeled sample using TiO₂ beads. It is recommendable to perform a second round of enrichment (from Step 2) to enhance the phosphoproteome coverage. The supernatants (1–3), which contain mostly non-phosphorylated peptides, can be combined and used for, e.g., global proteome analysis. *See* **Note 9**

TiO$_2$ beads. However, use bead-to-peptide ratios of 3:1 and 1.5:1 for the subsequent enrichment steps, i.e., 2.28 mg and 1.14 mg of TiO$_2$ beads, respectively. Collect the supernatant in a LoBind Eppendorf tube.

6. Pool the beads from all the three enrichment steps in one LoBind Eppendorf tube using 100 μL of loading buffer 1. Centrifuge as above to pellet the beads.

7. Collect the supernatant and combine with the one obtained from **step 5**.

8. Wash the beads with 100 μL of TiO$_2$ washing buffer 1, vortex, and centrifuge as in **step 4** to pellet the beads. This step is important to remove non-phosphorylated peptides, which bind in a HILIC mode unspecifically to TiO$_2$.

9. Afterwards, perform another washing step using 100 μL of washing buffer 2 and pellet the beads as mentioned above. Collect the supernatant and combine with the one obtained from **step 7**. *See* **Note 9**.

10. Finally, dry the beads completely in the SpeedVac.

11. Resuspend the beads in 100 μL of elution buffer, briefly vortex, and incubate the sample on a vortex mixer at low speed for 10 min to elute the phosphopeptides from the beads.

12. Transfer the eluate containing the phosphopeptides into a new LoBind Eppendorf tube and acidify the sample with 2 μL of 10 % TFA and 8 μL of 100 % FA.

13. Dry the sample completely in the SpeedVac.

14. To increase the phosphoproteome coverage, a second serial enrichment step with TiO$_2$ beads should be performed. However, use loading buffer 2 instead of 1 and repeat **steps 2–5**.

15. Next, wash the beads with washing buffer 3, centrifuge to pellet the beads, and remove the supernatant. Dry the beads in the SpeedVac.

16. Afterwards, elute the phosphopeptides from the beads using 100 μL of TiO$_2$ elution buffer. Vortex and incubate the samples on a vortex mixer at low speed for 10 min.

17. In the meantime prepare a C8 stage tip; cut 5–6 mm from the bottom of a 2–200 μL pipette tip. Use it as a stamp to excise a small piece from the C8 Disc Empore. Place this piece on the top of a new 2–200 μL pipette tip. Use a gel loader tip to push down the piece of C8 material to the bottom of the tip and finally make sure that the C8 material is properly fixed [25].

18. After 10 min, pellet the beads by centrifugation at 3000 *rcf* for 1 min and transfer the eluate onto the self-made C8 stage tip (*see* above) to remove possible residues of beads.

19. For the following steps, use an air-filled syringe to allow the passage of the liquid through the material.

20. Recover the flow-through containing the phosphopeptides into a LoBind Eppendorf tube.

21. To further elute the phosphopeptides that might be still attached to the beads, add 30 μL of TiO$_2$ elution buffer to the beads, vortex, and centrifuge to pellet the beads at 3000 *rcf* for 1 min. Collect the eluate and transfer onto the previously used C8 stage tip and recover the flow-through in the same LoBind Eppendorf tube (**step 20**).

22. For maximum recovery of the phosphopeptides, place 2–3 μL of 30 % ACN over the C8 tip and collect the flow through in the LoBind Eppendorf tube (**step 21**).

23. Finally, prior to the SPE acidify the eluate with 2 μL of 10 % TFA and 8 μL of 100 % FA. Make sure that the pH is around 2.0.

3.8 SPE Cleanup of the Phosphopeptide-Enriched Sample

1. Prepare a C18 stage tip following the instructions from **step 17**, but use C18 material instead of C8 material.

2. Weigh in 5 mg of Oligo R3 material and add 200 μL of 70 % ACN in 0.1 % TFA.

3. Fill the tip with 10 μL of R3 material (the height of the R3 material should be between 2 and 3 mm).

4. For the following steps, use an air-filled syringe to allow the passage of the liquid through the material.

5. Activate the material with 50 μL of 100 % ACN. Repeat this step twice.

6. Equilibrate the material with 50 μL of 0.1 % TFA. Repeat this step twice.

7. Load the acidified phosphopeptides (from **step 23** of Subheading 3.8) onto the stage tip. Reload the flow-through once more.

8. Wash the material with 50 μL of 0.1 % TFA. Repeat this step twice.

9. Finally, elute the phosphopeptides from the material with 50 μL of 98 % ACN in 0.1 % TFA, in an HPLC vial, and directly proceed with HILIC fractionation.

3.9 HILIC Fractionation

1. Perform separation of the TiO$_2$-enriched phosphopeptide sample on a polar-phase column (TSK gel) 150 μm × 15 cm, 5 μm using an UltiMate 3000 liquid chromatography (LC) system (Thermo Fisher Scientific) with the following gradient:
   1 % B for 20 min, 1–15 % B in 1.5 min, 15–40 % B in 37.5 min, 40–80 % B in 5 min, 80 % B hold for 5 min, and finally 80–1 % B for 6 min.

2. Collect ten fractions at 3.5-min intervals, as depicted in Fig. 3.

3. Dry the collected fractions completely in the SpeedVac.

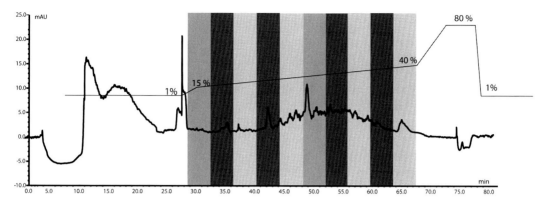

Fig. 3 An off-line-HILIC fractionation of phosphopeptides. The peptides (previously enriched using TiO_2 beads) are separated using a binary gradient (solvent A: 0.1 % TFA, 98 % ACN, solvent B: 0.1 % TFA) ranging from 15 to 40 % of solvent B in 37.5 min. HILIC fractionation of a pre-enriched sample improves the specificity and increases the phosphoproteome coverage [26]

3.10 LC-MS/MS Analysis

1. LC conditions.

    (a) Complete proteome: Dissolve each peptide fraction (from Subheading 3.7) in an appropriate volume of HPLC loading buffer (0.1 % TFA). Load each fraction with 0.1 % TFA and a flow rate of 20 μL/min for 10 min onto the trap column, followed by separation of peptides on the main column using a binary gradient ranging from 3 to 42 % B in 140 min at a flow rate of 250 nL/min at 60 °C.

    (b) Phosphoproteome: Use similar HPLC conditions as above, i.e., columns, buffers, and flow rates. Owing to the reduced complexity as compared to the global proteome analysis, analyze each fraction using a 55-min gradient, ranging from 3 to 42 % B.

2. MS conditions

    (a) Operate the Q-Exactive mass spectrometer in data-dependent acquisition mode.

    (b) Complete proteome: Acquire MS survey scans at a resolution of 70,000 with a target value of 1×10^6 ions and maximum injection time of 120 ms. Acquire MS/MS scans of 15 most abundant ions (Top 15) using 2×10^5 ions as target value and a maximum fill time of 120 ms. Use a normalized collision energy (NCE) of 30 and a dynamic exclusion of 20 s. Set the first fixed mass 100 m/z to allow a good signal intensity of the first reporter ion (i.e., 113) and select only precursor ions with charge state between +2 and +5 for MS/MS fragmentation (Fig. 4).

    (c) Phosphoproteome analysis: Same as above, but using an NCE of 35 and a dynamic exclusion of 12 s. *See* **Note 10**.

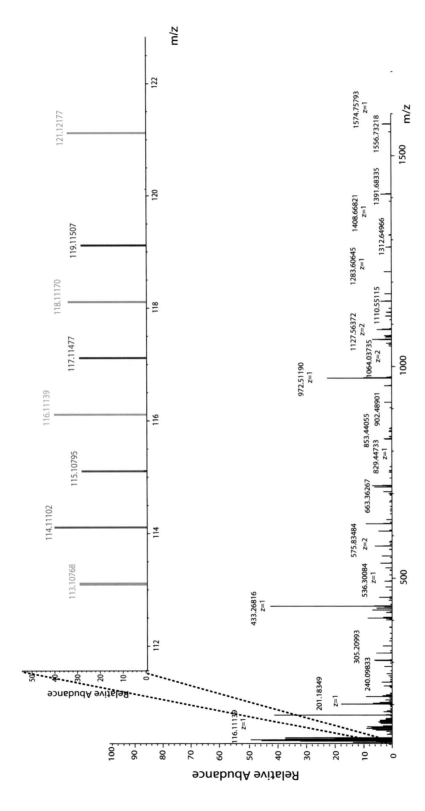

Fig. 4 MS/MS spectrum of the phosphopeptide iTRAQ-QPGLRQPsPSHDGSLSPLQDR (s = phosphorylated serine) acquired at 17,500 resolution in the Orbitrap mass analyzer. The low-mass region contains the iTRAQ 8plex reporter ions that represent the relative abundances of the same peptide in eight different biological samples

3.11 Data Analysis

1. Search the mass spectrometry (MS) raw data against a human Uniprot database (*see* **Note 11**) with Proteome Discoverer (PD) software. Use a mere "target" database as PD generates random decoy hits on the fly.

2. In PD, use the spectrum selector node for precursor ion selection to process MS raw data with default settings.

3. Use the reporter ion quantifier node and select iTRAQ-8plex as the quantification method. *See* **Note 12**.

4. To maximize the number of peptide spectrum matches (PSMs), include different search algorithms (Mascot, SEQUEST, and MS Amanda) using the same set of parameters, i.e., precursor and fragment ion tolerances of 10 ppm and 0.02 Da for MS and MS/MS, respectively; trypsin as enzyme with a maximum of two missed cleavages; carbamidomethylation of Cys, iTRAQ-8plex on N-terminus, and Lys as fixed modifications; oxidation of Met, and phosphorylation (only for the phosphoproteome data) of Ser, Thr, and Tyr as variable modifications.

5. Use the peptide validator node to filter the data with a false discovery rate (FDR) of 1 %.

6. For the phosphoproteome data analysis, add the PhosphoRS [17] (version 3.1) node to score localization probabilities for the identified phosphorylation sites.

3.12 Data Evaluation

1. Complete proteome:

   (a) Export the list of unique proteins from PD to a Microsoft Excel spread sheet. *See* **Note 13**.

   (b) To improve the reliability of the relative protein quantification, select only those proteins that are identified with at least two unique peptides.

   (c) PD only provides ratios; that is, only seven values are given in relation to a selected reference (here 113) sample. To allow a statistic comparison in case of comparing four against four samples, create an artificial 113/113 ratio of 1.0 for each protein.

   (d) For each protein, calculate the median by taking the ratios from different iTRAQ channels (113/113, ..., 121/113).

   (e) Then, divide the iTRAQ ratio of each protein by the previously calculated median to obtain relative abundance per individual channel.

   (f) Use a Student's t-test to determine p-values for the respective sample groups/conditions.

   (g) Calculate average ratios for the replicates and determine regulated proteins, e.g., by selecting those that have p-values < 0.05 and fold changes > 2.

2. Phosphoproteome:

(a) Export the phosphopeptide PSM list from PD to Microsoft Excel. *See* **Note 13**.

(b) Select only unique phospho-PSMs that were used for quantification.

(c) Calculate the normalization factor as described previously (from Subheading 3.12) and normalize each iTRAQ channel (*see* Subheading 3.12, **steps 4(a)–(d)**).

(d) With the help of ready-to-use Excel macro provided by Mechtler lab (http://ms.imp.ac.at/?goto=phosphors), determine the confident phosphorylation sites for each peptide, as well as the position of the phosphorylation within the protein sequence. (The analysis should be done in the same Excel worksheet.)

(e) In Excel concatenate: (1) peptide sequence, (2) protein accession, and (3) phospho-RS phosphorylation site (considering only those with probabilities ≥95 %) to define distinguishable phosphopeptide PSMs. *See* **Note 14**.

(f) Sort the data according to the concatenated row to group phospho-PSMs that belong together. For those, determine the median normalized abundance values per iTRAQ channel, as well as the relative standard deviation.

(g) If the experiment consists of replicates, determine student's *t*-test and significantly change phosphopeptides as done for the global proteome.

4 Notes

1. IAA solution is unstable and light sensitive. Therefore, prepare the stock solution freshly and use it within 1 h after preparation. In addition, buffers that either contain sulfhydryls or are not slightly alkaline (pH 7.5–8.0) should be avoided. Moreover, excess of IAA or non-buffered IAA reagent can lead to the alkylation of amines (lysine, N-termini), thioethers (methionine), imidazoles (histidine), and carboxylates (aspartate, glutamate).

2. Glycolic acid improves the selectivity to enrich phosphopeptides using TiO_2, by reducing unspecific binding from non-phosphorylatedpeptides [13].

3. Besides BCA, other calorimetric assays such as Bradford or modified-Lowry could be used for the determination of protein concentration. Irrespective of the method, the assay should provide comparably accurate protein concentrations as it is decisive for calculating the amount of trypsin (for digestion) and TiO_2 beads (for phosphopeptide enrichment).

4. The use of cold organic solvents (ethanol and acetone) for protein precipitation is a fast and easy procedure that yields reproducible results. Nevertheless, membrane filter-based sample preparation protocols such as filter-aided sample preparation (FASP) [27, 28] are widely used nowadays. However, FASP involves urea, which could lead to undesired side reactions, especially carbamylation of primary amines on N-terminus and Lys side chains [29].

5. Avoid contamination with keratin at this step and do not dry the pellet in an oven or a thermomixer. Instead, air-dry the protein pellet under the laminar flow hood.

6. It has been shown that the presence of phosphoamino acids (Ser and Thr) near to the cleavage site of trypsin could impair cleavage. Moreover, 50 mM TEAB buffer also shows slightly reduced cleavage efficiency, compared to NH_4HCO_3 buffer. Therefore, for enhancing digestion efficiency, thus improve the coverage of the phosphoproteome, and achieve better reproducibility, it is recommended to use 1:20 (w/w) trypsin-to-protein ratio and perform the proteolytic digestion in 50 mM NH_4HCO_3 buffer [18].

7. It is most important to quality control the enzymatic digests as this step is critical for the subsequent quantitative analysis [19]. The monolithic HPLC system provides a rapid and direct comparison of the samples as it can be used to measure proteins and peptides. Additionally, monolithic columns are more robust and sensitive in contrast to other techniques that are used for digestion control such as SDS-PAGE followed by Coomassie or silver staining.

8. Resolubilize the dried peptides in HPLC loading buffer (0.1 % TFA) and analyze each sample on a nanoLC system with UV detection (214 nm) or if possible MS-coupling, using a 90-min gradient ranging from 3 to 42 % of HPLC buffer B (84 % ACN in 0.1 % FA). Compare UV and/or total ion chromatogram (TICs) intensities of all samples and if necessary correct the amounts to compensate for systematic errors such that each sample has identical starting material before labeling with iTRAQ reagents.

9. The supernatant of the enrichment steps contains mostly non-phosphorylated peptides. It is possible to keep it and combine it with the supernatant from the washing steps; therefore, after SPE this sample can again be used for global proteome analysis. In general, it is possible to use the entire multiplexed sample for phosphopeptide enrichment, and just use the loading and washing supernatants for the global proteome analysis (after cleaning by SPE C18). However, we recommend using the completely "unbiased" sample prior to TiO_2 incubation

and take into account losing 5 % of the sample amount for phosphopeptide enrichment instead as the loss in sensitivity might be negligible.

10. Due to the depletion of unmodified peptides and the HILIC fractionation, the resulting peptide fractions are not highly complex. Therefore, set the dynamic exclusion to 12 s on the Q-Exactive mass spectrometer (Thermo Fisher Scientific) and reduce gradient length.

11. Note down the date and the source of the download, as well as the number of target (forward) sequences.

12. Select the vendor-provided isotope correction factors for iTRAQ-8plex reagents in the reporter ion node of Proteome Discoverer software.

13. Before exporting the unique protein list from the Proteome Discoverer, apply the data reduction filters such as high confidence corresponding to an FDR <1 % on the PSM level and peptide search engine rank of 1.

14. This step is necessary to allow separating phospho-PSMs in clearly distinguishable groups, such that PSMs that are identified with the same sequence but different positions of the phosphorylation site are differentiated.

References

1. Macek B, Mann M, Olsen JV (2009) Global and site-specific quantitative phosphoproteomics: principles and applications. Annu Rev Pharmacol Toxicol 49(1):199–221

2. Batalha IL, Lowe CR, Roque ACA (2012) Platforms for enrichment of phosphorylated proteins and peptides in proteomics. Trends Biotechnol 30(2):100–110

3. Boja ES, Phillips D, French SA et al (2009) Quantitative mitochondrial phosphoproteomics using iTRAQ on an LTQ-Orbitrap with high energy collision dissociation. J Proteome Res 8(10):4665–4675

4. Jones AE, Nühse T (2011) Phosphoproteomics using iTRAQ. In: Dissmeyer N, Schnittger A (eds) Plant kinases. Humana Press, New York, pp 287–302

5. Mertins P, Udeahi ND, Clauser KR et al (2012) iTRAQ labeling is superior to mTRAQ for quantitative global proteomics and phosphoproteomics. Mol Cell Proteomics 11(6):M111.014423

6. Liang X, Fonnum G, Hajivandi M et al (2007) Quantitative comparison of IMAC and TiO2 surfaces used in the study of regulated, dynamic

7. Beck F, Geiger J, Gambaryan S et al (2013) Time-resolved characterization of cAMP/PKA-dependent signaling reveals that platelet inhibition is a concerted process involving multiple signaling pathways. Blood 123(5):e1–e10

8. Wu Y-B, Dai J, Yang XL et al (2009) Concurrent quantification of proteome and phosphoproteome to reveal system-wide association of protein phosphorylation and gene expression. Mol Cell Proteomics 8(12):2809–2826

9. Ross PL, Huang YN, Marchese JN et al (2004) Multiplexed protein quantitation in Saccharomyces cerevisiae using amine-reactive isobaric tagging reagents. Mol Cell Proteomics 3(12):1154–1169

10. Gilar M, Olivova P, Daly AE et al (2005) Two-dimensional separation of peptides using RP-RP-HPLC system with different pH in first and second separation dimensions. J Sep Sci 28(14):1694–1703

11. Thingholm TE, Jorgensen TJ, Jensen ON et al (2006) Highly selective enrichment of phos-

protein phosphorylation. J Am Soc Mass Spectrom 18(11):1932–1944

phorylated peptides using titanium dioxide. Nat Protoc 1(4):1929–1935

12. Alpert AJ (1990) Hydrophilic-interaction chromatography for the separation of peptides, nucleic acids and other polar compounds. J Chromatogr 499:177–196

13. Thingholm TE, Larsen MR (2009) The use of titanium dioxide micro-columns to selectively isolate phosphopeptides from proteolytic digests. Methods Mol Biol 527:57–66

14. Perkins DN, Pappin DJ, Creasy DM et al (1994) Probability-based protein identification by searching sequence databases using mass spectrometry data. Electrophoresis 20(18): 3551–3567

15. Geer LY, Markey SP, Kowalak JA et al (2004) Open mass spectrometry search algorithm. J Proteome Res 3(5):958–964

16. Eng JK, McCormack AL, Yates JR (1994) An approach to correlate tandem mass spectral data of peptides with amino acid sequences in a protein database. J Am Soc Mass Spectrom 5(11):976–989

17. Taus T, Kocher T, Pichler P et al (2011) Universal and confident phosphorylation site localization using phosphoRS. J Proteome Res 10(12):5354–5362

18. Dickhut C, Feldmann I, Lambert J et al (2014) Impact of digestion conditions on phosphoproteomics. J Proteome Res 13(6):2761–2770

19. Burkhart JM, Schumbrutzki C, Wortelkamp S et al (2012) Systematic and quantitative comparison of digest efficiency and specificity reveals the impact of trypsin quality on MS-based proteomics. J Proteomics 75(4): 1454–1462

20. Toll H, Oberacher H, Swart R et al (2005) Separation, detection, and identification of peptides by ion-pair reversed-phase high-performance liquid chromatography-electrospray ionization mass spectrometry at high and low pH. J Chromatogr A 1079(1–2):274–286

21. Batth TS, Francavilla C, Olsen JV (2014) Off-line high-pH reversed-phase fractionation for in-depth phosphoproteomics. J Proteome Res 13(12):6176–6186

22. Castillo A, Roig-Navarro AF, Pozo OJ (2007) Secondary interactions, an unexpected problem emerged between hydroxyl containing analytes and fused silica capillaries in anion-exchange micro-liquid chromatography. J Chromatogr A 1172(2):179–185

23. Yang F, Shen Y, Camp DG et al (2012) High pH reversed-phase chromatography with fraction concatenation for two-dimensional proteomic analysis. Expert Rev Proteomics 9(2):129–134

24. Larsen MR, Thingholm TE, Jensen ON et al (2005) Highly selective enrichment of phosphorylated peptides from peptide mixtures using titanium dioxide microcolumns. Mol Cell Proteomics 4(7):873–886

25. Dickhut C, Radau S, Zahedi RP (2014) Fast, efficient, and quality-controlled phosphopeptide enrichment from minute sample amounts using titanium dioxide. In: Martins-de-Souza D (ed) Shotgun proteomics. Springer, New York, pp 417–430

26. Engholm-Keller K, Birck P, Storling J et al (2012) TiSH—a robust and sensitive global phosphoproteomics strategy employing a combination of TiO2, SIMAC, and HILIC. J Proteomics 75(18):5749–5761

27. Wisniewski JR, Zougman A, Nagaraj N et al (2009) Universal sample preparation method for proteome analysis. Nat Methods 6(5): 359–362

28. Manza LL, Stamer SL, Ham AJ et al (2005) Sample preparation and digestion for proteomic analyses using spin filters. Proteomics 5(7):1742–1745

29. Kollipara L, Zahedi RP (2013) Protein carbamylation: in vivo modification or in vitro artefact? Proteomics 13(6):941–944

Chapter 4

Selected Reaction Monitoring to Measure Proteins of Interest in Complex Samples: A Practical Guide

Yuehan Feng and Paola Picotti

Abstract

Biology and especially systems biology projects increasingly require the capability to detect and quantify specific sets of proteins across multiple samples, for example the components of a biological pathway through a set of perturbation-response experiments. Targeted proteomics based on selected reaction monitoring (SRM) has emerged as an ideal tool to this purpose, and complements the discovery capabilities of shotgun proteomics methods. SRM experiments rely on the development of specific, quantitative mass spectrometric assays for each protein of interest and their application to the quantification of the protein set in various biological samples. SRM measurements are multiplexed, namely, multiple proteins can be quantified simultaneously, and are characterized by a high reproducibility and a broad dynamic range. We provide here a practical guide to the development of SRM assays targeting a set of proteins of interest and to their application to complex biological samples.

Key words Selected reaction monitoring, Targeted proteomics, Protein quantitation, Assay design, Assay validation

---

1 Introduction

In the last two decades mass spectrometry (MS)-based proteomics has evolved into a powerful technique for large-scale protein analysis, and is now routinely applied to identify and quantify proteins in a variety of biological samples [1]. Classical, unbiased proteomics approaches, commonly referred to as shotgun proteomics, are used in discovery-based projects to generate an inventory of the protein content of a sample and to identify proteins with varying abundance across different perturbing conditions. In these workflows, proteins are extracted from biological samples, denatured, and proteolytically cleaved with a specific enzyme (typically trypsin). The resulting peptide mixtures are subjected to liquid chromatography (LC) separation, at the end of which peptide ions are transferred to the gas phase and enter the mass spectrometer. In shotgun approaches, peptide ions are then subjected to data-dependent MS

Jörg Reinders (ed.), *Proteomics in Systems Biology: Methods and Protocols*, Methods in Molecular Biology, vol. 1394,
DOI 10.1007/978-1-4939-3341-9_4, © Springer Science+Business Media New York 2016

acquisition, where fragmentation spectra are generated from selected peptide precursors based on their measured intensity. Peptide identifications from MS spectra are validated by statistical models and translated into protein identifications. While extremely powerful in the large-scale characterization of the protein content of biological samples, shotgun approaches are not ideal when specific proteins need to be measured across various samples, as their intrinsic semi-stochastic nature might affect the detection and consistent quantification of the proteins of interest across the sample set. To address this limitation, a parallel proteomics workflow involving selected reaction monitoring (SRM) mass spectrometry was proposed and has quickly developed into the gold standard for the targeted detection and quantification of specific proteins in complex biological matrices [2].

SRM-based proteomics experiments share an identical sample preparation workflow with shotgun approaches, until the mass spectrometric analysis. An SRM measurement starts with the definition of a set of proteins of interest, for example all the components of a protein complex or a biological pathway, a set of biomarker candidate proteins, or, in general, proteins that are relevant to address a given biological question. For each of these proteins then specific, sensitive, and quantitative mass spectrometric assays based on SRM are developed and subsequently applied to the measurement of the target proteins in biological samples. SRM assays consist of MS coordinates that enable the selective measurement on a triple-quadrupole mass spectrometer (QqQ) of each target protein, through measurement of some of its representative peptides. These coordinates consist of pairs of mass-to-charge (m/z) values (so-called SRM transitions) that are selected with the first (Q1) and second (Q3) analyzer of a QqQ, to isolate a peptide ion and corresponding fragment ions generated upon fragmentation of the peptide in a collision cell (q2). The detector placed at the end of the quadrupole series counts ions matching the defined m/z values and returns a signal intensity over the chromatographic time. The area under the resulting (SRM) peak is proportional to the amount of the protein initially contained in the sample. SRM measurements are characterized by a high reproducibility and a broad dynamic range (up to 4.5 orders of magnitude). SRM data acquisition is also multiplexed; that is, several SRM transitions can be sequentially measured within an MS duty cycle, thus allowing for the concurrent quantification of multiple peptides and proteins. When targeting many transitions, however, the time spent measuring each transition (dwell time) will be reduced, resulting in a lower signal-to-noise of the associated peaks, and thus lower sensitivity for the target peptides. To address this issue, the scheduled acquisition of SRM transitions was devised, where transitions for a given peptide are monitored only for a short time interval, centered around the known elution time of the peptide. This decreases the number of concurrent SRM transitions measured at a given chromatographic time, allowing optimizing cycle time and dwell time while measuring larger numbers of peptides per MS analysis. Using this approach, up to 150 proteins or several hundred peptides [3] can be simultaneously monitored in a 30-min LC-SRM run. Once developed, SRM assays are stable and applicable to measuring the target proteins in a variety of samples at high throughput. The application of SRM in biological and biomedical projects has been further promoted by a growing number of open-access databases [4] providing experimentally validated SRM coordinates and by automated or semiautomated data analysis tools [5, 6].

In the following, we provide a basic protocol for developing SRM assays and conducting SRM measurements, assuming that a list of proteins of interest has been defined.

2 Materials

2.1 Sample Preparation

1. Denaturation buffer: 8 M Urea, 0.1 M ammonium bicarbonate (AmBic), pH 8.0.

2. Estimation of total protein amount: Bicinchoninic acid (BCA) Protein Assay Kit (Thermo Scientific, Rockford, IL, USA).

3. Reduction buffer: 1 M Tris(2-carboxyethyl)phosphine hydrochloride (TCEP–HCl).

4. Alkylation buffer: 1 M Iodoacetamide (IAA).

5. Proteases: Endoproteinase Lys-C from *Lysobacter enzymogenes* and porcine trypsin (both sequencing grade).

6. Peptide desalting: Acetonitrile; 0.1 % formic acid; 60 % acetonitrile; Sep-Pak tC18 Vac Cartridges (Waters, Milford, MA, USA).

2.2 LC-SRM/MS

1. Buffer A: HPLC-grade water with 0.1 % (v/v) formic acid.

2. Buffer B: HPLC-grade acetonitrile with 0.1 % (v/v) formic acid.

3. A QqQ spectrometer equipped with a nano-electrospray ion source and interfaced with a liquid chromatography (LC) system operating in the nanoliter/min flow rate range.

4. A chromatographic column for nano-LC separation, packed with C18 resin (20 cm length × 75 μm diameter).

5. *Optional*

   (a) iRT Kit (Biognosys, Switzerland).

3   Methods

3.1   Sample Preparation

1. Depending on the sample type under investigation (e.g., bacteria, yeast, mammalian cells, human tissue, or plasma) different protein extraction procedures are used. After protein extraction, estimate the concentration of the protein solution using the BCA assay. Typical protein extraction buffers may contain reagents that interfere with the BCA assay (e.g., DTT) or with the subsequent MS analysis (e.g., detergents such as sodium dodecyl sulfate or NP-40) and should thus be omitted or diluted prior to the BCA assay.

2. The extraction buffer should contain a chaotropic agent at high concentration, such as 8 M urea, to achieve complete protein denaturation. If the protein extract derives from a protein precipitation step (for example using cold acetone), add the denaturation buffer to the protein pellet to a final protein concentration of 2–3 mg/ml.

3. Adjust the pH of the protein extract to ~8.

4. Add the reduction buffer to a final concentration of 5 mM TCEP–HCl, followed by incubation for 30 min at 37 °C to reduce disulfide bridges.

5. Add the alkylation buffer to a final IAA concentration of 40 mM, followed by incubation at 25 °C in the dark for 45 min to alkylate free cysteine residues.

6. Dilute the protein solution with 0.1 M AmBiC to a final concentration of 6 M urea to prevent denaturation of the protease used in the first digestion step.

7. To achieve completeness of the digestion reaction, two proteases are sequentially applied, Lys-C (cleaving at the C-terminus of lysine) followed by trypsin (cleaving at the C-terminus of lysine and arginine). Add first Lys-C to the protein solution to an enzyme-to-substrate (E:S) ratio of 1:100, and incubate the reaction mixture at 37 °C for 4 h. Lys-C tolerates harsher conditions (e.g., higher denaturant concentration) compared to trypsin. Therefore, dilute the proteolytic mixture again prior to addition of trypsin to a final concentration of 2 M urea. Add then trypsin to an E:S ratio of 1:100 and incubate the proteolytic mixture at 37 °C overnight.

8. Stop the tryptic digestion by addition of formic acid to pH <3.

9. Desalt the peptide mixture using Sep-Pak tC18 Vac cartridges, according to the manufacturer's instruction, and using acetonitrile for the initial wash of the cartridges, 0.1 % formic acid for cartridge equilibration and peptide desalting, and 60–80 % acetonitrile for peptide elution. Evaporate the peptide mixture eluted in 60–80 % acetonitrile in a vacuum centrifuge to dryness.

10. Resuspend the peptide pellet in buffer A to a concentration of 1 mg peptides per ml, based on the protein amount estimated using the BCA assay at **step 1**.

3.2 SRM Assay Design

A protein SRM assay includes SRM transitions for at least one unique peptide generated upon digestion of the target protein, and ideally comprises transitions for multiple peptides from that protein to improve reliability in the protein detection and quantification steps. An SRM assay for a peptide in turn minimally includes the mass-to-charge ratio (m/z) of a peptide precursor ion and those of a set of fragments generated upon collision-induced dissociation of the precursor. The design of protein SRM assays involves selection of optimal peptide and fragment ion coordinates and is followed by refinement and validation of the assays, after which the assays can be applied to the quantification of the target proteins. The information required to design SRM assays can be obtained from in silico analyses, databases, or shotgun proteomics data, as described below (Fig. 1).

In silico design: Ideal peptides for SRM uniquely map to the target protein and show good ionization and fragmentation properties, resulting in intense SRM peaks. If the number of target proteins is limited (for example, below 20), ideal peptides and their fragments can simply be selected by testing the MS performance of all unique peptides generated upon tryptic digestion of each protein [7]. This involves retrieval of the target

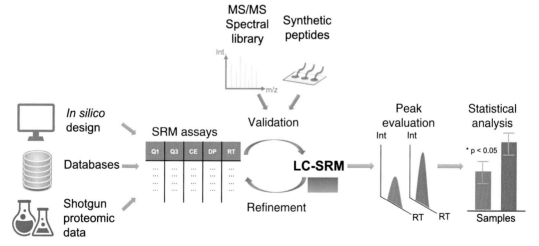

Fig. 1 Workflow of SRM-based targeted proteomic analyses. Coordinates of protein SRM assays are generated by in silico design, retrieved from Web-based repositories or derived from shotgun proteomics experiments. The assay coordinates are validated using MS/MS spectra filtered to an acceptable FDR or synthetic peptides. The assay coordinates are measured by SRM for assay refinement and the final SRM transitions are applied to perform quantitative SRM measurements across the samples of interest. The SRM peak data are scrutinized using tools such as Skyline and statistically evaluated

protein sequences from protein databases, their in silico tryptic digestion, prediction of all possible (b- and y-) ion fragments for each peptide precursor, and analysis of the intensity and specificity of the generated transitions by SRM, using samples that contain the protein of interest. Testing large numbers of transitions, even for a small set of proteins, might require considerable instrument time. Therefore, prioritization criteria are typically applied to analyze only transitions with the highest likelihood of resulting in detectable and reliable SRM peaks. For example, peptides which have been previously detected in proteomics databases with a large number of observations across different experiments should be prioritized (*see* below). Further prioritization criteria are in **Note 1**. When the number of target proteins is large, the number of peptides to test by SRM can be further reduced by computational prediction of the best responding peptides (those that can be detected with intense MS signals), using tools such as PeptideSieve [8], the ESP predictor [9], PeptideRank [10], or PeptidePicker [11]. The assays generated by in silico design need to be additionally validated (*see* below).

Databases: Web-based repositories of SRM assays, such as the SRMAtlas [12] (www.srmatlas.org), the CPTAC (Clinical proteomic Tumor Analysis Consortium) assay portal [13] (assays.cancer.gov), or Panorama [14] (panoramaweb.org/labkey/project/home/begin.view), have recently emerged (*see* **Note 2**). Coordinates for SRM assays can be retrieved from these databases, paying attention to the considerations reported in **Note 3**. Typically, the assays stored in such databases have already been validated. However, to grant reliability of the resulting SRM assay, care should be taken to ensure that the detected peaks match reported retention time constraints (upon retention time realignment to the local system setup, *see* **Note 4**) and reported fragment ion relative intensities.

Shotgun proteomics experiments: SRM coordinates can be selected from existing shotgun (LC-MS/MS) proteomics measurements performed on the same proteome and filtered to a reasonable (e.g., 1 %) protein false discovery rate (FDR). Precursor ions from unique peptides with the highest MS signal or the largest number of spectral counts should be prioritized. Fragment ions can be selected from shotgun MS/MS spectra and then tested by SRM analysis (*see* **Note 5**). We recommend testing the ten most intense b- and y-fragment ions selected from MS/MS spectra and depending on the combination of instruments used, prioritizing y-ions might be advisable. Assays generated from shotgun data are automatically validated by the existence of the corresponding MS/MS spectra, when these have been assigned to peptide sequences using broadly accepted confidence thresholds. However, care should be taken in ensuring that the detected SRM peaks match the retention times at which the MS/MS spectra were acquired, upon RT realignment to the applied LC-SRM setup.

We will now exemplify SRM assay design using the open-source software tool Skyline [5] (http://skyline.maccosslab.org), assuming that no assays are available in public databases for the proteins of interest. Here, we only provide a simplistic guide to the use of Skyline and recommend consulting the tutorials and videos on the Skyline website for more detailed instructions.

1. Import the amino acid sequences of the target proteins in FASTA format into Skyline. In silico digestion is performed by Skyline automatically according to the conditions specified under Peptide and Transition Settings.

2. Under Peptide Settings choose trypsin KR/P as the digestion protease, allow for no missed cleavages, and build a background proteome database from the FASTA file for your species of interest to check for peptide unicity in the proteome (proteotypicity). Choose carbamidomethylation of cysteines as a static modification and keep the other default settings. If shotgun datasets are available, build the corresponding spectral library and include it in the document using the Library tab, to guide the peptide and fragment selection.

3. Under Transition Settings, choose calculation of collision energies according to your instrument model, optionally apply the fragment prioritization criteria described in **Note 1**, using the Filter and Instrument tabs, and choose to pick ten fragment ions from library spectra, if available.

4. Skyline will generate a list of SRM assays based on the defined criteria. Instrument-dependent collision energies (CE) are calculated based on the peptide precursor m/z value. If needed, CEs can be optimized in CE ramping experiments (*see* **Note 6**).

5. Export the assay coordinates as a transition list which essentially contains the Q1 and Q3 m/z values, the CE, and the transition identifier (ID, i.e., peptide sequence, precursor charge, and fragment ion information). Measure the transition list by SRM in one or multiple runs, depending on its length (*see* below), and use biological samples or synthetic peptides to evaluate transition performance. It is advisable not to monitor more than 100 transitions within the same unscheduled SRM measurement, due to sensitivity considerations. Analyze the data as described in Subheading 3.3.

3.3 SRM Assay Validation and Refinement

The coordinates of an SRM assay require validation [2]; that is, it should be confirmed that the chosen transitions and associated retention times detect the target peptide and not a false-positive peak, based on an acceptable error rate. SRM assays from databases or shotgun data are typically already validated by statistically filtered LC-MS/MS spectra or other approaches, and can in theory be directly applied to protein quantification with the precautions described above. SRM assays from in silico design and computational prediction require validation by alternative approaches. One option is the acquisition of MS/MS spectra for each target peptide from the biological sample containing the protein of interest, their assignment to peptide sequences via database search, and statistical filtering to the chosen error rate. Another simple and cost-efficient possibility is the (MS/)MS analysis of either crude synthetic peptide analogues [15] of the targeted peptides (*see* **Note 7**) or tryptic digests of recombinant proteins [16]. These options are particularly useful when the targeted proteins are of low abundance and thus difficult to detect in the available biological samples or if the samples of interest are precious (e.g., from patients) and can only be used in the final quantification step.

For the refinement of SRM assays, the SRM coordinates derived from Subheading 3.2 are measured in SRM mode using the biological samples of interest. The resulting peaks are inspected to select a minimal set of suitable transitions and associated retention time (for scheduled SRM measurements) that defines the final quantitative assay for the protein. The three to five most intense transitions for the best ionizing peptides are typically chosen in this step. Shouldered peaks, likely resulting from multiple co-elutingpeptides sharing the same transition(s), should be discarded. Retention times of SRM peaks at the peak apex are annotated, if the aim is a scheduled SRM experiment. Note that the extracted retention times are valid only if the chromatographic settings remain the same in the quantification step (*see* **Note 4**).

Once the assays are validated as described above, SRM assay refinement can be performed with the tool Skyline.

1. First, open the Skyline (.sky) file from which the assays were derived, to enable matching of the measured transitions to the associated identifiers.

2. Import the raw SRM data as Results.

3. Inspect the peaks associated to each peptide transition group. Important criteria to evaluate if a peptide was identified and where in the chromatogram include (1) co-elution and shape similarity of all transitions for that peptide; (2) retention time of the selected peak matching the RT extracted during the validation step; and (3) relative intensities of fragment ions matching those extracted from the reference spectral library

(evaluated by the "dot product" score in Skyline) or measured from synthetic peptides.

4. To facilitate the peak inspection process, use the Skyline View option and display normalized peak areas (normalized to total area). The relative intensities of different transitions will be visualized by bar plots. In case you judge that a wrong peak has been selected by the software, correction can be applied by re-selecting the retention time frame (clicking and dragging the cursor beneath the x(RT)-axis) where the correct peak is located.

5. Transitions associated to shouldered peaks can be manually discarded by removing the corresponding fragment ions from the drop-down peptide/transition list of the Skyline interface. Similarly, you can here select to retain only the most intense three to five transitions per peptide by removing all the others in the same way, since Skyline displays peak area rankings in brackets (e.g., [1], [2], [3] …). Selection of most intense transition peaks may also be performed in bulk, using the Edit > Refine > Advanced option. Export the transition list with measured retention times for each peptide from Skyline as above.

3.4 LC-SRM/ MS Setup

A QqQ mass spectrometer operated with a nano-electrospray ion source and interfaced with a nano-LC system is required for SRM analysis, according to the setup described in Subheading 2.

1. Separate the peptide mixture using a linear gradient of 30 min, from 5 to 35 % buffer B.

2. We recommended the use of an MS cycle time of 2–2.5 s, to ensure acquisition of a sufficient number of data points per peak, with the described chromatographic setup, while maximizing the available dwell time per transition. For unscheduled SRM measurements, in one such LC-SRM the total number of transitions measured should be below 100, to ensure a dwell time of at least 20 ms per transition (*see* **Note 8**). For scheduled SRM measurements, we recommend a retention time window of 3–5 min and measurement of a maximum number of 1000–1200 SRM transitions.

3.5 Final Quantitative Analysis

Validated and refined SRM coordinates constitute the final SRM assay for a protein. Final SRM assays are directly applied to measure the target proteins using scheduled or unscheduled SRM acquisition. The resulting data are manually inspected or processed automatically using software tools such as the mProphet [6], to evaluate detection and (relative) amount of the protein in each sample. Manual data evaluation should include application of the constraints mentioned above in Subheading 3.3 to minimize the likelihood of false positives. In addition, if heavy isotope-labeled peptides are spiked into the sample as quantitation standards, elution time and relative fragment ion intensities of the heavy and light transitions should correspond. Tools for the automated analysis of SRM data such as the mProphet apply most of the constraints described above to SRM data and combine the resulting scores into a statistical model, enabling also calculation of an error rate for the SRM dataset. Manual inspection of such data should still be performed when statistical significance estimates differ for the same peptide measured in different samples to avoid missing data.

Once peptide detection is confirmed, SRM peaks are integrated, the corresponding areas or heights are extracted, and the quantitative data from multiple peptides and multiple transitions are statistically evaluated using tools such as MSStats [17]. MSstats (www.msstats.org) is an open-access R-based tool enabling the application of linear mixed-effects models to the statistical analysis of quantitative SRM data.

The software Skyline can be used for the semiautomated analysis of quantitative SRM data. An automated peak-picking function is implemented into Skyline, after which peak assignments are manually validated, as described below. Recent updates of the software have enabled users to apply the statistically calibrated scoring model of mProphet (*see* step-by-step procedure to include mProphet in the Skyline-based data analysis in one of the Skyline tutorials available online).

1. To semiautomatically evaluate quantitative SRM data with Skyline, open first the .sky file from which the final validated assays were derived and import the raw SRM data.

2. Skyline will automatically apply its default peak-picking algorithm to all imported measurements.

3. Manually inspect SRM peaks, using the validated peaks from Subheading 3.3 as reference to judge the correctness of the peak-picking step. Apply the same procedure as described in Subheading 3.3 to make corrections where needed.

4. Results can be exported at this stage in .csv format using the "Export Report" function. Skyline offers the possibility to customize the type of information that will be included in the exported data sheet (e.g., peptide sequence, length, transition area or height, retention time, and more—a full reference is available at the end of the Live Reports tutorial on the Skyline website). Data can be further processed using spreadsheet applications such as Microsoft Excel and statistical tools such as R. For example, a normalization factor can be applied that accounts for the variability in the total peptide amount across the sample set (*see* **Note 9**).

5. To perform further statistical analysis of the quantitative dataset, install the MSstats tool using the Tools > Tool Store option in

Skyline. For extended processing with MSstats, choose the "MSstats Input" option when exporting a report. The exported file can be directly processed by MSstats, in an R programming environment, following the user manual provided in www.msstats.org. The MSstats tool includes also a total-peptide amount normalization step (*see* **Note 10**).

4   Notes

1. When selecting peptides and fragment ions for SRM measurements, the following considerations apply. Peptides with a length below 6 and above 25 amino acids should be discarded, due to specificity and MS detectability issues, respectively. Avoiding methionine-containing peptides excludes the possibility of quantification artifacts due to methionine oxidation. Half-tryptic peptides and peptides embedding missed-cleavage sites should be avoided, especially in experiments aiming at the absolute quantification of the target protein. Focusing only on doubly and triply charged peptide precursors for peptides of average length, and only on doubly charged precursor ions if peptides do not contain histidines, -KP-, or -RP- bonds, is typically reasonable. Similarly, singly charged y-ions should be preferred, due to their high likelihood of detection. Fragment or precursor ions with m/z below 350–400 should be avoided, as in this m/z region typical environmental contaminant ions are detected. To further maximize specificity, fragment ions with low m/z, for example below that of the precursor, could be avoided. M/z values of a precursor and its fragment ion used in an SRM transition should be at least 5 Da apart, to avoid spurious signals from unfragmented precursors.

2. Currently, the SRMAtlas contains assays with high proteome coverage from four different organisms (*Mycobacterium tuberculosis, Saccharomyces cerevisiae, Mus musculus, and Homo sapiens*). The CPTAC database instead focuses specifically on the collection of assays for human cancer-related proteins.

3. The SRM assays deposited in publicly accessible databases have been generated using different types of instruments and assay transferability may vary depending on the system used. For example, the geometry of the collision cell may vary across MS systems, affecting the conservation of peptide fragmentation patterns. In addition, different equations to calculate peptide collision energies are suggested by the different instrument vendors (*see* MS or acquisition software user manual).

4. Peptide retention times depend on the specific chromatographic setup used, which includes parameters such as gradient coordinates, column dimension, and packing material or length of the transport capillaries, and may change slightly

over time with aging of the chromatographic column. Thus, RTs should be updated when changing chromatographic setup or upon fluctuating environmental conditions, e.g., when reproducing SRM assays from Web-based databases, or between the SRM assay validationphase and the final quantitative analyses, when these are conducted several days apart. This can be achieved using sets of synthetic peptides eluting over the whole chromatographic space and spiked into the sample (e.g., the iRT peptides [18]) and serving as normalization anchors. An equation that linearly correlates "old" and "new" retention times is typically extracted based on the anchor peptides and can be applied to predict the new RTs of the target peptides. Accurate retention times are also essential in scheduled SRM measurement, conducted typically with narrow RT windows to increase the throughput (e.g., 3–5 min). Therefore, even a slight deviation in RT could lead to a missing peak, resulting in misinterpretation of the data.

5. Shotgun proteomics instruments with q2-fragmentation (e.g., Q-TOFs) or employing higher energy collisional dissociation (HCD) [19] are preferable to guide selection of optimal fragment ions for SRM, as fragmentation patterns typically resemble those produced during SRM. However, also MS/MS spectra from instruments such as linear ion traps are a reasonable starting point to guide the selection of fragment ions for SRM, provided that several intense fragment y-ions (the intensity of b-ions is less conserved between QqQ and trapping instruments) are selected from those spectra and tested using SRM.

6. The sensitivity of an SRM assay can be further increased by optimizing specific peptide and MS-specific parameters, such as, most commonly, the collision energy. Experimental procedures for CE optimization, optionally using the Skyline platform, have been described [5, 20]. Using optimized collision energies instead of CEs calculated using equations provided by instrument vendors results typically in a signal increase below threefold [15]. We recommend performing CE optimization only when the target proteins are difficult to detect.

7. Synthetic peptides (optionally heavy labeled and spiked into the sample of interest) can also be measured in SRM mode to confirm the validity of the chosen transitions, but this option does not allow assignment of an error rate to the analysis.

8. The dwell time is the time the mass spectrometer spends on measuring a given SRM transition. The sum of dwell times for all the transitions measured is the cycle time of an SRM measurement, which in turn determines the number of data points collected along the chromatographic elution profile of a peptide. For example 100 SRM transitions measured with a dwell time of 20 ms each will result in a cycle time of about 2 s,

translating in the acquisition of a data point every 2 s for a given transition peak. Longer dwell times result in a higher signal-to-noise and thus sensitivity for the target analyte. However, increasing the number of transitions leads to a longer cycle time and less data points collected along a chromatographic peak, which can affect the quantitative precision.

9. Relative quantification results from SRM measurements will be affected by variations in the total peptide amount injected for the different samples measured. For example, let us assume that sample A is two times more concentrated than sample B due to sample processing issues and that the same volume is injected into the LC-SRM system for the two peptide samples. Albeit peptide P is present in the same amount in both samples, SRM measurements will report a twofold higher amount for peptide P in sample A. To address this issue, it is highly recommended to perform a data normalization step based on the total peptide amount in each sample (even if BCA assays are performed after protein extraction). One can measure by SRM a set of well-characterized (e.g., housekeeping) proteins whose abundance is assumed to remain constant across the different conditions and use their average abundance change as a normalization factor. Similar estimations of the relative total peptide amount can be extracted from absorbance measurements of the peptide mixtures at 280 nm before MS injection or from the total ion chromatogram (TIC) intensity of the samples measured in shotgun LC-MS/MS mode.

10. The MSstats package performs normalization of the total peptide amount loaded onto the LC-MS/MS system based on the comparison of the median intensity of all target analytes (or a reference set of analytes) between different runs. For samples where the majority of the measured targets change abundance, this normalization approach may not be suitable, and should be replaced by one of the methods mentioned in **Note 9**.

Acknowledgements

We thank Paul J. Boersema and Martin Soste (ETH Zurich) for insightful discussions. We thank Brendan MacLean (University of Washington) for critical reading of the manuscript. P.P. is supported by an EU-FP7 ERC Starting Grant (FP7-ERC-StG-337965), by a FP7-Reintegration grant (FP7-PEOPLE-2010-RG-277147), by a Professorship grant from the Swiss National Science Foundation (grant PP00P3_133670), and by a Promedica Stiftung (grant 2-70669-11). Y.F. is supported by an ETH Research Grant (grant 4412-1).

References

1. Aebersold R, Mann M (2003) Mass spectrometry-based proteomics. Nature 422:198–207

2. Picotti P, Aebersold R (2012) Selected reaction monitoring-based proteomics: workflows, potential, pitfalls and future directions. Nat Methods 9:555–566

3. Soste M, Hrabakova R, Wanka S et al (2014) A sentinel protein assay for simultaneously quantifying cellular processes. Nat Methods 11:1045–1048

4. Kusebauch U, Deutsch EW, Campbell DS et al (2014) Using PeptideAtlas, SRMAtlas, and PASSEL: comprehensive resources for discovery and targeted proteomics. Curr Protoc Bioinformatics 46:13.25.11–13.25.28

5. MacLean B, Tomazela DM, Shulman N et al (2010) Skyline: an open source document editor for creating and analyzing targeted proteomics experiments. Bioinformatics 26:966–968

6. Reiter L, Rinner O, Picotti P et al (2011) mProphet: automated data processing and statistical validation for large-scale SRM experiments. Nat Methods 8:430–435

7. Bereman MS, MacLean B, Tomazela DM et al (2012) The development of selected reaction monitoring methods for targeted proteomics via empirical refinement. Proteomics 12:1134–1141

8. Mallick P, Schirle M, Chen SS et al (2007) Computational prediction of proteotypic peptides for quantitative proteomics. Nat Biotechnol 25:125–131

9. Fusaro VA, Mani DR, Mesirov JP et al (2009) Prediction of high-responding peptides for targeted protein assays by mass spectrometry. Nat Biotechnol 27:190–198

10. Qeli E, Omasits U, Goetze S et al (2014) Improved prediction of peptide detectability for targeted proteomics using a rank-based algorithm and organism-specific data. J Proteomics 108:269–283

11. Mohammed Y, Domanski D, Jackson AM et al (2014) PeptidePicker: a scientific workflow with web interface for selecting appropriate peptides for targeted proteomics experiments. J Proteomics 106:151–161

12. Deutsch EW, Lam H, Aebersold R (2008) PeptideAtlas: a resource for target selection for emerging targeted proteomics workflows. EMBO Rep 9:429–434

13. Whiteaker JR, Halusa GN, Hoofnagle AN et al (2014) CPTAC assay portal: a repository of targeted proteomic assays. Nat Methods 11:703–704

14. Sharma V, Eckels J, Taylor GK et al (2014) Panorama: a targeted proteomics knowledge base. J Proteome Res 13:4505–4510

15. Picotti P, Rinner O, Stallmach R et al (2010) High-throughput generation of selected reaction-monitoring assays for proteins and proteomes. Nat Methods 7:43–46

16. Stergachis AB, MacLean B, Lee K et al (2011) Rapid empirical discovery of optimal peptides for targeted proteomics. Nat Methods 8:1041–1043

17. Choi M, Chang CY, Clough T et al (2014) MSstats: an R package for statistical analysis of quantitative mass spectrometry-based proteomic experiments. Bioinformatics 30:2524–2526

18. Escher C, Reiter L, MacLean B et al (2012) Using iRT, a normalized retention time for more targeted measurement of peptides. Proteomics 12:1111–1121

19. de Graaf EL, Altelaar AF, van Breukelen B et al (2011) Improving SRM assay development: a global comparison between triple quadrupole, ion trap, and higher energy CID peptide fragmentation spectra. J Proteome Res 10:4334–4341

20. Maclean B, Tomazella DM, Abatiello SE et al (2010) Effect of collision energy optimization on the measurement of peptides by selected reaction monitoring (SRM) mass spectrometry. Anal Chem 82:10116–10124

Chapter 5

Monitoring PPARG-Induced Changes in Glycolysis by Selected Reaction Monitoring Mass Spectrometry

Andreas Hentschel and Robert Ahrends

Abstract

As cells develop and differentiate, they change in function and morphology, which often precede earlier changes in signaling and metabolic control. Here we present a selected reaction monitoring (SRM) approach which allows for the parallel quantification of metabolic regulators and their downstream targets.

In particular we explain and describe how to monitor abundance changes of glycolytic enzymes upon PPARγ activation by using a label-free or a stable isotope-labeled standard peptide (SIS peptides) approach applying triple-quadrupole mass spectrometry. We further outline how to fractionate the cell lysate into cytosolic and nuclear fractions to enhance the sensitivity of the measurements and to investigate the dynamic concentration changes in those compartments.

Key words Metabolic control, Adipocytes, Selected reaction monitoring, SIS peptides, Quantification, Mass spectrometry, Proteomics

1 Introduction

In the past adipose tissue was just seen as a storage depot for free fatty acids. Recent studies and models replaced this with the idea that adipose tissue is a complex, essential, and highly active metabolic and endocrine organ [1, 2]. Adipocytes also play a central role in lipid and glucose metabolism and produce a large number of hormones and cytokines, whose dysregulation is known to be involved in metabolic syndrome, diabetes mellitus, and vascular diseases. Therefore it affects the entire set of organ systems in our body [3, 4].

The increasing rate of obesity in our society has led to intense interest in understanding the mechanisms underlying the formation of adipose tissue and its capacity to store fat [5–7]. The nuclear receptor and lipid-binding protein PPARγ is a master regulator of the formation and function of mature fat cells [8]. PPARγ is expressed and activated during adipocyte differentiation, and artificial induction of PPARγ in cells with adipogenic differentiation

Jörg Reinders (ed.), *Proteomics in Systems Biology: Methods and Protocols*, Methods in Molecular Biology, vol. 1394, DOI 10.1007/978-1-4939-3341-9_5, © Springer Science+Business Media New York 2016

potential can convert them into mature adipocytes [9]. In vitro studies suggest that PPARγ is the ultimate regulator of adipogenesis in a transcriptional cascade that also involves members of the C/EBP transcription factor family and an interconnected feedback loop system [10–14]. Additionally, PPARγ knockout mice fail to develop adipose tissue [10, 15, 16]. Consistent with these findings, humans with dominant-negative mutations in a single allele of PPARγ (the gene encoding PPARγ) have partial lipodystrophy and insulin resistance [17–19]. Adipose tissue also has a key role in directing whole-body glucose homeostasis. The realization that a fatty acid sensor like PPARγ might be an important regulator of glucose metabolism arose from the discovery that the insulin-sensitizing thiazolidinediones (TZDs) such as rosiglitazone are potent agonists for PPARγ [20, 21] and therefore activators of the glucose transporter GLUT4. In addition to glucose uptake regulation, PPARγ regulates many glycolytic enzymes by binding to their transcriptional promoters [22]. The activity of the pathway can be regulated at different key steps to ensure that glucose consumption and energy production match the needs of the cell. The key regulatory enzyme in mammalian glycolysis is phosphofructokinase, the rate-limiting catalyst.

The glycolytic pathway is extremely ancient in evolution and is common to nearly all living organisms. This pathway can be broken down into three stages: (1) the conversion of glucose into fructose 1,6-bisphosphate, (2) the cleavage of fructose 1,6-bisphosphate into two three-carbon fragments, and (3) the generating of ATP after the three-carbon fragments are oxidized to pyruvate [23]. Though the glycolytic pathway was one of the first to be discovered, its importance in modern biology is still being elucidated. Its metabolites and intermediates serve as entry points for many other important pathways and dysregulation is involved in multiple diseases. Because of the regulatory function of PPARγ in glucose metabolism, it is of great interest to monitor the changes in glycolysis upon its activation.

In this protocol we describe how to induce and analyze abundance changes of PPARγ and its downstream targets with a particular focus on glycolysis enzymes.

To estimate protein abundance of single proteins over time and/or activation ranges, immunological methods such as the enzyme-linked immunosorbent assay (ELISA) or Western blot analysis can be carried out. But those approaches are not suited to analyze entire pathways since they are expensive and difficult to assess regarding their specificity and their linear dynamic range of detection which is often very limited.

Mass spectrometry has significant advantages that overcome the limitations of antibody-based assays. It allows direct detection and quantification of peptides unique to a protein of interest. Quantification can be improved by using SIS peptides, allowing

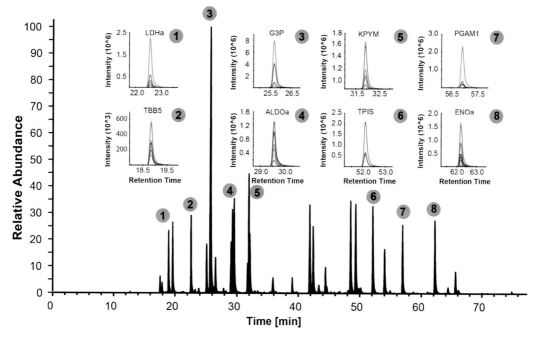

Fig. 1 Selected reaction monitoring (SRM) total ion chromatogram of peptides derived from glycolysis proteins out of an OP9 cytosolic protein extraction. The total ion chromatogram showing measurements of 50 proteins (150 peptides and 900 transitions total) during an 80-min experiment on a single sample consisting of 1 μg of total cytosolic protein digest. The *inset panels* display chromatograms of individual monitored peptides derived from 1 to 8: L-lactate dehydrogenase A, tubulin beta 5 (control), glyceraldehyde-3-phosphate dehydrogenase, fructose-bisphosphate aldolase A, pyruvate kinase, triosephosphate isomerase, phosphoglycerate mutase and alpha-enolase

the absolute quantification of targeted proteins. Another advantage of using mass spectrometry is the multiplexing capability, which allows for hundreds of proteins to be measured in parallel (Fig. 1). Further, this method is easy to automate, allowing the analysis of many samples in a row without loading and detecting each sample manually.

By using an LC/MS-based method in conjunction with a triple-quadrupole mass spectrometer, a selected reaction monitoring (SRM) experiment can be carried out. SRM mode works like a double-mass filter, which drastically reduces noise and increases selectivity.

To apply this approach, first, the proteins are digested into short peptides (5–50 amino acids) with a hydrolytic enzyme such as trypsin. After sample preparation the digest is subjected to nano reverse-phase liquid chromatography and ionized by nano electro-spray ionization (NSI).

The ionized peptides enter the mass spectrometer and the desired precursor peptide ion is selected by mass-to-charge ratio at the first quadrupole according to the developed method. Other ions generated in the ion source that have a different m/z ratio cannot pass the quadrupole (Q1) at this time. The selected precursor ion enters the second quadrupole (Q2) and is fragmented by collision-induced dissociation (CID). The generated fragments are transferred to the third quadrupole and one specific fragment is selected again and transferred to the detector where the abundance over the peptide retention time is measured.

To obtain the best sensitivity usually fragment ions are selected that have been found to give the best signal-to-noise ratio in previous experiments. The precursor ion and one of its corresponding fragment ions are known as a transition. To precisely monitor a peptide, generally two to three transitions are required. Furthermore at least two to three unique peptides per protein are needed, making it necessary to monitor six to nine transitions per protein. If SIS peptides are used for validation and quantification up to 18 transitions should be monitored per protein.

There are also challenges of using SRM mass spectrometry on triple-quadrupole instruments. Since the precision of mass measurement is limited to a mass accuracy of ± 0.3 Da and a resolution of 7500 (at m/z of 508) the triple quadrupole is not among the instruments with the highest mass accuracy and resolution. This may lead to false positives when similar MS/MS fragmentation patterns are detected simultaneously or at a different time point during analysis. To solve this issue the use of SIS is highly recommended. Another handicap is the limited number of total fragment ions that can be detected in each sample. This can be overcome by performing a scheduled analysis (each peptide, separate analysis window) to enhance the limited number of fragment ions significantly.

For validation and quantification of the chosen proteins SIS peptides should be used. A suitable reference peptide has to be selected based on the peptide sequence, retention time behavior, ionization efficiency, and its fragmentation pattern. The chosen peptide is then synthesized by solid-phase synthesis using light amino acids and one isotopically coded "heavy" amino acid. The result is a chemically identical peptide homolog that only differs in mass (8–10 Da) and can therefore be easily distinguished by SRM. SIS-peptides can be spiked in to the tryptic digest of the sample at a known concentration. Since the reference peptides elute at the same time and have the same fragmentation pattern as the endogenous peptides, the concentration of endogenous peptides can be directly measured as a ratio between the added isotopically coded SIS and the endogenous peptides (Fig. 2).

Here we describe a method and SRM assay to analyze and quantify specific glycolytic proteins and their regulator PPARγ after its activation. This method is also easily applicable to other signaling or metabolic pathways where dynamics occur.

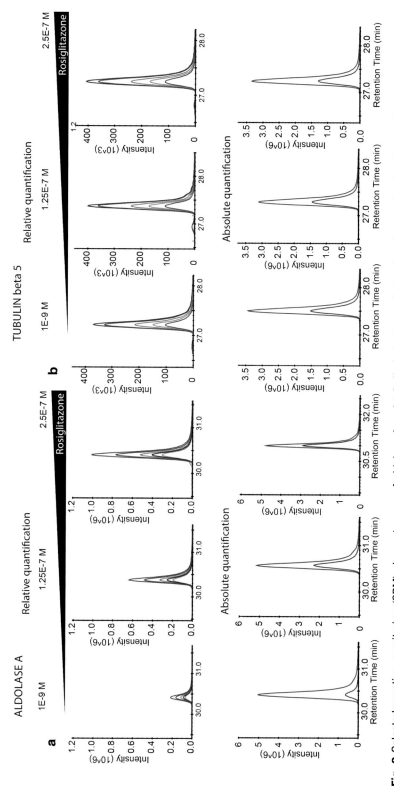

Fig. 2 Selected reaction monitoring (SRM) chromatogram of aldolase A and tubulin beta 5, monitoring the changes in abundance, after PPARγ activation. (**a**) Quantitative analysis of aldolase A (*upper panel*) using a label-free approach for relative quantification, *lower panel* displays the same peptide in an absolute quantification approach using internal isotope-coded standard peptides (SIS) after treating the OP9 cells with three different concentrations of rosiglitazone (1 nM, 112 nM, and 250 nM). (**b**) Transitions of the control protein tubulin beta 5 derived from same sample and analysis. The SIS peptides for absolute quantification were added in an amount of 500 fmol each, whereby every sample contained 1 μg of total cytosolic protein digest

2   Materials

All solutions have to be prepared with ultrapure water (prepared by purifying deionized water to attain a sensitivity of 18 MΩ cm at 25 °C) and analytical grade reagents. Prepare and store all reagents at 4 °C (unless indicated otherwise). All % values are given in v/v if not otherwise indicated.

2.1   Cell Culture of OP9 Cells

1. Cells: OP9, a stromal cell line from mouse bone marrow, was purchased from the Tokyo Metropolitan Institute of Medical Science.

2. High-serum culture media: MEMα containing 20 % fetal bovine serum (FBS) (Biochrome), 1 % penicillin/streptomycin (PSG) (GIBCO).

3. Low-serum culture media: MEMα containing 10 % fetal bovine serum (FBS) (Biochrome), 1 % penicillin/streptomycin (PSG) (GIBCO).

4. Cell culture dishes: Cell culture flasks (T75) (Sarstedt), disposable filter units, pore size 0.2 μm (Thermo) sterile serological pipettes 5, 10, 25 ml (VWR).

5. Rosiglitazone stock solution: 1 mg/ml Rosiglitazone in DMSO (Sigma), store at –80 °C.

6. Differentiation: Low-serum media containing 1 μM rosiglitazone.

2.2   Cell Harvesting and Lysis

1. Buffer: 1× Phosphate-buffered saline (PBS) pH 7.4 (GIBCO), warm up to 37 °C (water bath).

2. Trypsin: 0.05 % Trypsin-EDTA (GIBCO) with phenol-red indicator.

3. Trypsin stopping solution: High-serum culture media.

4. Cell culture dishes: 15 ml Falcon tubes (Sarstedt), sterile serological pipettes 5, 10, 25 ml (VWR), Neubauer improved counting chamber (Marienfeld).

5. Centrifuge: Eppendorf centrifuge 5804.

2.3   Protein Extraction

1. Buffer A: 10 mM HEPES pH 7.9 (Sigma, cell culture tested), 1.5 mM $MgCl_2$ (Sigma), 10 mM KCl (Sigma), 1× complete protease inhibitor cocktail (Roche), 1 mM DTT, 0.1 % digitonin (Sigma), please check for purity and protein contamination.

2. Buffer B: Same as buffer A but without digitonin.

3. Buffer C: 20 mM HEPES pH 7.9 (Sigma, cell culture tested), 1.5 mM $MgCl_2$ (Sigma), 450 mM KCl (Sigma), 25 % v/v glycerol, 0.2 EDTA (titriplex II Sigma).

4. BCA protein assay kit (Pierce).

5. Additional chemicals: Acetone (Sigma, Chromasolv plus for HPLC) ice-cold, 1 M NH_4HPO_3 (Sigma) pH 8, 6 M urea, (Sigma), 0.5 M TCEP (Sigma, pH 7), 0.5 M Iodoacetamide (Sigma), trypsin (Promega, sequencing grade), 99 % formic acid (FA) (Biosolve ULC/MS grade).

6. Centrifuge: Eppendorf centrifuge 5810R.

7. Additional equipment: 25, 30 gauge needle + syringe (luer-lock system), Thermomixer (Eppendorf) for 1.5 ml Eppendorf tubes.

2.4 Peptide Cleanup Procedure

1. Buffer A: 0.05 % Trifluoroacetic acid (TFA) (pH 3.0); can also use 5 % acetonitrile, 0.2 % formic acid, but can lose small peptides.

2. Buffer B: 80 % Acetonitrile (ACN), 0.2 % formic acid (FA) (pH below 4.0).

3. 100 % Methanol (Biosolve, ULC/MS).

4. Filters: SepPak C18 (100 mg, 1 cc), Waters #WAT023590, 1 mg capacity.

2.5 HPLC/MS Conditions and Materials

1. Standard peptides: Heavy arginine ($[^{13}C_6]$ $[^{15}N_4]$)- and lysine ($[^{13}C_6]$)-labeled peptides can be purchased from multiple vendors (*see* **Note 6**). We use our own synthesized heavy peptides and store them in 30 % ACN and 0.1 % TFA solution. Combine all the heavy peptides needed for the experiment to a final concentration of 1.6 µM and store the peptide mix, as well as the remainder of the concentrated individual peptide solutions, aliquoted at −80 °C.

2. HPLC buffers: A: 0.1 % FA in HPLC-grade water, B: 84 % ACN in 0.1 % FA.

3. Nano-HPLC instrument: Dionex UltiMate 3000 UHPLC equipped with an Acclaim PepMap RSLC c18 reversed-phase main column (Thermo) with 75 µm × 25 cm and 2 µm, 100 Å particles, and an Acclaim PepMap 100 c18 reversed-phase pre column (Thermo) with 100 µm × 2 cm and 5 µm, 100 Å particles.

4. MS instrument: TSQ Vantage (Thermo Fisher Scientific) triple-quadrupole mass spectrometer.

2.6 Data Analysis Software

1. Xcalibur 2.2.44, data analysis and instrument control.

2. Skyline 2.5 software suite—available for free from the MacCoss Lab website: https://brendanx-uw1.gs.washington.edu/labkey/project/home/software/Skyline/begin.view.

3. Microsoft Excel.

4. Origin 9.1.

3   Methods

3.1   Cell Culture

Each step must be carried out under sterile conditions.

1. Culture OP9 cells in high-serum culture media in cell culture flasks as needed. Seed them out at 10 % of confluence (1E + 5 in T15, 5E + 5 in T75, 1E + 6 in T175). Change the media every 2 days. After 4 days the cells are almost confluent (*see* **Note 1**).

2. For the experiment prepare 6 × T75 cell culture flasks with 10 % confluence and incubate them for 3 days (should contain three to four million cells after this period). Change the media on third day to low-serum culture media and add rosiglitazone to three of the flasks (end concentration 0.25 μM). Treat the other ones just with DMSO with the same concentration. Incubate all flasks for 48 h at 37 °C with 5 % CO_2 in humidified atmosphere.

3.2   Harvesting Cells

1. Aspirate the media from the flasks.

2. Add 7.5 ml of PBS (37 °C) to the cells. Move the flasks a bit to wash the remaining media and aspirate the PBS.

3. Add 7.5 ml of trypsin (RT) to the flasks and incubate them for 5 min in the incubator at 37 °C, 5 % CO_2.

4. Check the flasks under the microscope to ensure that all the cells are detached from the surface. If this is not the case, knock the flask carefully against your hand and check again.

5. Add 7.5 ml of high-serum culture media to the flasks to stop the trypsin reaction.

6. Use a 5 or 10 ml serological pipette to transfer everything of one flask into one Falcon tube. Rinse the surface with the suspension thoroughly to collect as much cells as possible.

7. Centrifuge the Falcon tubes at $200 \times g$ for 5 min (RT) to pellet the cells.

8. Aspirate the media and resuspend the pellet in 10 ml of PBS (37 °C).

9. Use a small amount (10 μl) of the cell suspension to count the cells with the Neubauer counting chamber.

10. Centrifuge again at $200 \times g$ for 5 min (RT) to pellet the cells and aspirate the PBS.

11. You can now store the cells by snap freezing them in liquid nitrogen and store them at −80 °C or go on with the cell lysis.

3.3   Cell Lysis and Fractionated Extraction

1. Resuspend the pellets in 100 μl lysis buffer (buffer A) per million cells; incubate on ice for 10 min, flicking the tube every minute. Can add more lysis buffer if needed.

2. Transfer the suspension into protein low-bind Eppendorf tube. Use low-bind Eppendorf tube for every following step.

3. Dounce cells by passing the suspension through a 25-gauge needle/syringe 5× and through a 30 g needle 3× (avoid bubbles while douncing and make sure to check for breakage of membranes by looking under a microscope; should only see nuclei and membrane remnants, not whole cells anymore) (*see* **Note 2**).

4. Centrifuge suspension for 10 min at 2300×*g* and 4 °C to pellet nuclei. Collect the supernatant.

5. Resuspend pellet in 30 µl per million cells of buffer A (no digitonin). Centrifuge suspension for 10 min at 2300×*g* and 4 °C.

6. Add supernatant to that collected in the previous step. The collected supernatants are the cytosol/membrane fraction. The pellet from this step should contain an enriched preparation of nuclei. Call this pellet #1 (*see* **Note 3**).

7. Resuspend pellet #1 (the nuclei) in high-salt solution (buffer C) to destroy the nuclei and strip most of the proteins from DNA. Use 50 µl of buffer C for every one million OP9 cells.

8. Shake vigorously to resuspend the pellet. Incubate the suspension at 4 °C for 15 min, and then at 5-min intervals for another 15 min, flick to resuspend, and place back on ice. Under these conditions most nuclear proteins will be extracted (*see* **Note 4**).

9. Centrifuge for 10 min at 4 °C at highest speed (20,000×*g*).

10. Transfer the supernatant into clean Eppendorf tubes. This is your nuclear extract and the pellet is your histone fraction (*see* **Note 5**).

11. Do an acetone precipitation on the nuclear and cytosolic fractions: add at least 3 volumes of ice-cold acetone (precooled to –20 °C). Incubate proteins overnight at –20 °C.

3.4 Protein Digest

1. Centrifuge all fractions at max speed with the centrifuge (20,000×*g*) for 20 min at 4 °C. Carefully remove the supernatant leaving the protein pellet intact.

2. Allow the acetone to evaporate from the uncapped tube at room temperature for 20 min. Do not over-dry pellet, or it may not dissolve properly.

3. Resuspend the nuclear and cytosolic pellet in 6 M urea to solubilize (10 µl per million cells at room temperature). Leave for at least 1 h in urea, vortexing often. If the pellet is not dissolving you can use sonication in a water bath for 3–5 min.

4. Dilute samples to less than or equal to 2 M urea using 10 mM NH_4HCO_3. Measure protein concentration with BCA kit. Make a duplicate assay (standard and each sample). 1:5 dilution of each sample is recommended to stay within the standard range.

5. Add the internal isotope-labeled standard peptide mix (final concentration should be 50–500 fmol/1 μg total proteins) (*see* **Note 6**).

6. Add TCEP (final concentration 10 mM). Shake for 30 min at 37 °C (Thermomixer).

7. Cool sample to room temperature. Add 0.5 M iodoacetamide (30 μl per 1 ml) and shake for 30 min in the dark at room temperature (Thermomixer with aluminum foil). Iodoacetamide has to be freshly made before use.

8. Add trypsin in a ratio of trypsin to protein (1:50).

9. Incubate shaking for over 12 h at 37 °C (Thermomixer).

10. After digestion stop the reaction by adding about 10 μl of 99 % FA (total volume 500 μl) to lower the pH to 3 or lower. Check with pH indicator paper.

3.5   Peptide Cleanup Procedure

1. Wash cartridges (SepPak C18) with 3× volumes (1 volume is equivalent of 1 ml) of methanol (100 %), fast flow (with vacuum applied). This gets rid of contaminants in the cartridge. Make sure not to run the filter dry. Stop when meniscus is just above it.

2. Equilibrate with 3× volumes of buffer A, fast flow (with vacuum applied).

3. Load sample (pH should be <3), slow flow, best by gravity.

4. Wash and desalt with 3× volumes of buffer A, fast flow (with vacuum applied).

5. Elute peptides with 1 ml buffer B, slow flow (gravity). Elute the remaining fluid by vacuum past filter (after gravity elution).

6. Vacuum centrifuge sample to dryness and resuspend in 2 % acetonitrile and 0.1 % FA (want a concentration of 1 μg/μl as stock). This can take up to 5 h, depending on the vacuum centrifuge. Best to use a high-performance solvent-resistant vacuum centrifuge connected to a vacuum pump of sufficient power.

3.6   Setting Up the SRM Method in Skyline

Before starting with the measurements you have to set up the right method containing the proteins and peptides of interest including the transitions for the SIS peptides.

1. First, set up the Skyline document with the following preferences:

Peptide settings

(a) Digestion: Trypsin [KR|P], missed cleavages 0

(b) Filter: Min length 8, max length 25, exclude N-terminal AAs 25, you can also exclude all peptides that contain specific AAs like Cys or Met, as needed

(c) Modifications: Carbamidomethyl (C)

Transition settings

(a) Filter: Precursor charges 2 and 3, ion charges 1, ion types y

(b) Product ions: 6 ions

2. After those preparations are done, load a spectral library containing the proteome of the species of interest (*see* **Note 7**).

3. Copy the amino acid sequences of each protein from Table 1 and paste it directly into the new Skyline document. The program will automatically rank the best transitions for each peptide according to the loaded spectral library (*see* **Note 8**).

4. Export the method (File/Export/Transition list) and choose multiple methods. For max concurrent transitions choose 100 or 120 and method type: standard.

5. You can now import all the method files into the TSQ Vantage program to search for all the peptides with their transitions and get the right retention times.

3.7 Selected Reaction Monitoring Mass Spectrometry

1. Set up the following program for the separation/quantitation of peptides:

   Separation: Linear gradient from 3 to 35 % ACN over 60 min with a flow rate of 350 nl/min for separation and a 20 μl/min flow rate for sample loading.

   Polarity: Positive (tune file).

   Resolution for Q1 and Q3: 0.7u FWHM (method file).

   Emitter voltage: 1200–1500 V (tune file).

   Temperature of the transfer capillary: 270 °C (tune file).

   Declustering voltage: 10.

2. Unscheduled SRM: To obtain the right retention times for scheduled SRM you first have to run an unscheduled method with the internal standard peptides. Retention times for specific transitions may vary slightly from run to run depending on the composition of the sample and the performance of the instrument.

3. Scheduled SRM: Maximum time window 5 min, cycle time 1 s, average dwell time 20 ms.

4. Inject a volume of 5–15 μl with a total peptide amount of 1 μg of the sample and 50–500 fmol of total amount of your SIS peptides and start the measurement. Repeat each analysis at least twice to account for run-to-run variability of the system. Be sure that at least one blank is in between each sample measurement.

5. To verify proper operating performance of the system and to determine if the LC/MS needs cleaning or calibration, perform SRM on a standard peptide mix at regular intervals between the sample measurements and at least once per day.

Table 1
List of proteins, peptides, and transitions for monitoring PPARγ-induced glycolytic changes. The transition list for PPARγ and other nuclear proteins can be found in Ahrends et al. (2014, Science, DOI: 10.1126/science.1252079)

Accession	Protein	Gene	Peptide			Fragment			
Q8VDL4	ADP-dependent glucokinase	Adpgk	LGPAPVPVGPLSPESR	786.9408++	791.9450++	y12	1234.68	1244.69	1
						y10	1038.56	1048.57	1
						y8	842.44	852.44	1
			VAGTQACATETIDTNR	854.4020++	859.4061++	y10	1180.53	1190.53	1
						y9	1020.50	1030.50	1
						y8	949.46	959.47	1
P05064	Fructose-bisphosphate aldolase A	Aldoa	GILAADESTGSIAK	666.8539++	670.8610++	y11	1049.51	1057.53	1
						y10	978.47	986.49	1
						y7	663.37	671.38	1
			QLLLTADDR	522.7878++	527.7920++	y7	803.43	813.43	1
						y6	690.34	700.35	1
						y5	577.26	587.27	1
			ADDGRPFPQVIK	671.8593++	675.8664++	y8	984.60	992.61	1
						y7	828.50	836.51	1
						y5	584.38	592.39	1
P17182	Alpha-enolase	Eno1	TIAPALVSK	450.2817++	454.2888++	y7	685.42	693.44	1
						y6	614.39	622.40	1
						y5	517.33	525.35	1
			YITPDQLADLYK	720.3745++	724.3816++	y10	1163.59	1171.61	1
						y9	1062.55	1070.56	1
						y5	609.32	617.34	1
			SCNCLLLK	504.2543++	508.2614++	y6	760.44	768.45	1
						y5	646.40	654.41	1
						y4	486.36	494.38	1
P21550	Beta-enolase	Eno3	SGETEDTFIADLVVGLCTGQIK	1177.0832++	1181.0903++	y13	1373.75	1381.76	1
						y12	1302.71	1310.72	1
						y11	1187.68	1195.70	1
			AAVPSGASTGIYEALELR	902.9756++	907.9797++	y10	1164.63	1174.63	1
						y9	1063.58	1073.59	1
						y8	1006.56	1016.57	1

P16858	Glyceraldehyde-3-phosphate dehydrogenase	Gapdh	IVSNASCTTNCLAPLAK	910.4557++	914.4628++	y12	1335.64	1343.65	1
						y10	1088.58	1096.59	1
						y5	499.32	507.34	1
			GAAQNIIPASTGAAK	685.3753++	689.3824++	y11	1042.59	1050.60	1
						y9	815.46	823.48	1
						y8	702.38	710.39	1
			LISWYDNEYGYSNR	890.4023++	895.4064++	y9	1117.45	1127.46	1
						y6	759.34	769.35	1
						y5	596.28	606.29	1
P06745	Glucose-6-phosphate isomerase	Gpi	ELFEADPER	553.2617++	558.2658++	y6	716.32	726.32	1
						y5	587.28	597.28	1
						y3	401.21	411.22	1
			TFTTQETITNAETAK	828.4098++	832.4169++	y10	1077.54	1085.55	1
						y9	948.50	956.51	1
						y7	734.37	742.38	1
			HFVALSTNTAK	594.8222++	598.8293++	y10	1051.58	1059.59	1
						y9	904.51	912.52	1
						y8	805.44	813.45	1
Q9WUA3	6-Phosphofructokinase type C	Pfkp	LGITNLCVIGGDGSLTGANLFR	1124.5914++	1129.5955++	y14	1377.71	1387.72	1
						y13	1264.63	1274.63	1
						y10	1035.56	1045.56	1
			VTILGHVQR	511.8089++	516.8130++	y6	709.41	719.41	1
						y5	596.33	606.33	1
						y4	539.30	549.31	1
			EIGWADVGGWTGQGGSILGTK	1045.0211++	1049.0282++	y14	1318.67	1326.68	1
						y11	1018.55	1026.56	1
						y8	732.43	740.43	1
P52480	Pyruvate kinase	Pkm	NTGIICTIGPASR	680.3561++	685.3602++	y9	974.51	984.51	1
						y8	861.42	871.43	1
						y7	701.39	711.40	1
			GIFPVLCK	467.2650++	471.2721++	y6	763.42	771.43	1
						y5	646.35	624.36	1
						y4	519.30	527.31	1
			DAVLNAWAEDVDLR	793.8941++	798.8982++	y10	1188.56	1198.57	1
						y9	1074.52	1084.52	1
						y8	1003.48	1013.49	1

(continued)

Table 1
(continued)

P09411	Phosphoglycerate kinase 1	Pgk1	YSLEPVAAELK	610.3321++	614.3392++	y8	856.48	864.49	1
						y7	727.43	735.44	1
						y6	630.38	638.39	1
			ITLPVDFVTADK	659.8663++	663.8734++	y9	991.51	999.52	1
						y7	795.39	803.40	1
						y6	680.36	688.37	1
			GCITIIGGGDTATCCAK	877.8971++	881.9042++	y12	1210.52	1218.53	1
						y11	1097.44	1105.44	1
						y5	639.26	647.27	1
P06151	L-Lactate dehydrogenase	Ldha	LVIITAGAR	457.2951++	462.2992++	y7	701.43	711.43	1
						y6	588.35	598.35	1
						y5	475.26	485.27	1
			FIIPNIVK	472.3024++	476.3095++	y7	796.53	804.54	1
						y6	683.45	691.45	1
						y5	570.36	578.37	1
			VTLTPEEEAR	572.7959++	577.8000++	y8	944.47	954.47	1
						y7	831.38	841.39	1
						y6	730.34	740.34	1
P17751	Triosephosphate isomerase	Tpi1	IAVAAQNCYK	569.2897++	573.2968++	y8	953.45	961.46	1
						y7	854.38	862.39	1
						y6	783.35	791.35	1
			DLGATWVVLGHSER	770.3993++	775.4035++	y8	896.49	906.50	1
						y7	797.43	807.43	1
						y5	585.27	595.28	1
			VVLAYEPVWAIGTGK	801.9481++	805.9552++	y9	928.53	936.53	1
						y7	732.40	740.41	1
						y6	546.32	554.33	1
O08528	Hexokinase-2	Hk2	LPLGFTFSFPCHQTK	890.4480++	894.4551++	y8	1004.46	1012.47	1
						y6	770.36	778.37	1
						y5	673.31	681.32	1
			NILIDFTK	482.2791++	486.2862++	y6	736.42	744.43	1
						y5	623.34	631.35	1
						y4	510.26	518.27	1

Accession	Protein	Gene	Peptide			Fragment			
P17710	Hexokinase-1	Hk1	LSDEILIDILTR	700.9034++	705.9075++	y7	843.53	853.53	1
						y6	730.45	740.45	1
						y5	617.36	627.37	1
			GDFIALDLGGSSFR	727.8673++	732.8715++	y10	1022.53	1032.53	1
						y9	951.49	961.49	1
						y6	610.29	620.30	1
			SANLVAATLGAILNR	742.4332++	747.4373++	y11	1098.66	1108.67	1
						y10	999.59	1009.60	1
						y6	643.39	653.39	1
Q9DBJ1	Phosphoglycerate mutase 1	Pgam1	VLIAAHGNSLR	575.8382++	580.8423++	y8	825.43	835.44	1
						y7	754.3955	764.40	1
						y6	683.35	693.36	1
			YADLTEDQLPSCESLK	934.9328++	938.9399++	y10	1176.56	1184.57	1
						y8	933.47	941.48	1
						y7	820.39	828.40	1
			ALPFWNEEIVPQIK	842.4589++	846.4660++	y9	1069.59	1077.60	1
						y5	584.38	592.39	1
						y4	485.31	493.32	1
Q9QXG4	Acetyl-coenzyme A synthetase	Acss2	VAFYWEGNEPGETTK	864.3992++	868.4063++	y10	1061.47	1069.48	1
						y9	932.43	940.44	1
						y6	632.32	640.33	1
			AELGMNDSPSQSPPVK	828.8985++	832.9056++	y11	1155.56	1163.57	1
						y9	926.49	934.50	1
						y8	839.46	847.47	1
			IGPIATPDYIQNAPGLPK	933.0120++	937.0191++	y12	1312.69	1320.70	1
						y7	696.40	704.41	1
						y6	582.36	590.37	1
P99024	Tubulin beta-5 chain	Tubb5	ISVYYNEATGGK	651.3222++	655.3293++	y9	1002.45	1010.46	1
						y8	839.39	847.40	1
						y7	676.33	684.34	1
			YLTVAAVFR	520.3004++	525.3045++	y7	763.45	773.45	1
						y6	662.40	672.40	1
						y5	563.33	573.33	1
			NSSYFVEWIPNNVK	848.9201++	852.9272++	y8	999.53	1007.54	1
						y6	684.40	692.41	1
						y5	571.32	579.33	1

3.8 *Data Analysis*

1. Import your raw files into Skyline.

2. Review the chromatograms of each peptide and check for selection of the correct peaks. After your unscheduled/scheduled SRM analysis and data refinement your dataset should contain at least three unique peptides for each protein with three transitions each.

3. Use the export function of Skyline to export a report containing transition results to create a .csv file containing the following data: PeptideSequence, ProteinName, ReplicateName, Peptide RetentionTime, and Total Area (for label free) or Ratio ToStandard (for SIS).

4. To compare the different samples (OP9 control with DMSO against PPARγ activated cells) and calculate the fold change of each peptide do the following:

    (a) Label free: Divide every total area by the control (DMSO) to obtain the fold change over the treatment experiment.

    (b) SIS: The ratio of the control sample is set to one and all other ratios to standard values are divided by the ratio of the control.

    (c) If the exact amount of the injected SIS peptide is known also the absolute amount of the endogenous protein can be estimated by the peptide-to-standard ratio.

4   Notes

1. Do not allow the cells to reach full confluence until right before the experiment. It is best to split the cells every 2 days to keep them between 10 and 70 % confluence. OP9 cells are prone to spontaneous differentiation if they grow too confluent.

2. At this point you can stain with trypan blue. Pipette approximately 2–5 μl from your sample onto an object slide, add 1–2 μl of the trypan blue solution, and mix gently with the pipette. Check under the microscope for the destroyed cells. If the mechanical lysis was successful, you should only see blue stained cell remnants. Everything that is not stained was not lysed.

3. This is a time-critical step because you can lose nuclear proteins by diffusion if you wait too long to take the supernatant.

4. Do not exceed 500 mM KCl or you will extract histones.

5. For the further steps the histone fraction is not needed.

6. SIS peptides that are used in this protocol were synthesized in-house without any modifications such as carbamidomethylation. SIS peptides without those modifications have to be added to the digest before adding TCEP and IAA treatment. If it is not possible to synthesize the peptides on your own

there it is possible to buy them from different vendors (Thermo Scientific, JPT, and Sigma Aldrich). SIS peptides that are already carbamidomethylated can be added to the digest after the sample has been treated with IAA and TCEP.

7. Spectral libraries for SRM method design and for data analysis can be either directly added to a Skyline document or a custom library can be built within Skyline. For the latter, data-dependent acquisition (shotgun) as well as SRM-triggered MS 2 measurements searched with one of the common search engines can be used. Here we used our own spectral libraries obtained from data-dependent acquisition measurements.

8. Even if Skyline chooses the transitions according to the transition settings and the spectral library you should check every peptide manually for the right transitions. At this optimization stage using a triple-quadrupole instrument, it is recommended to try five to six transitions for each peptide ranging from the m/z value of around 500–1200 for the unscheduled measurements. The lower the m/z value gets, the more unspecific the transition is. After the unscheduled measurements you can refine the transitions by choosing the three most intense transitions per peptide for the actual experiments in order to maximize the dwell time and to increase the sensitivity. Since the heavy-labeled standard peptides are very expensive, they should only be ordered once the best precursor peptides and transitions have been selected based on the sample of interest. If you are able to produce your own standard peptides measuring the precursors of them directly with an unscheduled method to get the right retention times is possible. If the retention time or fragmentation pattern differs between the heavy and the light peptide, you are analyzing a false-positive signal. In such case, the design of the peptide transitions should be started from scratch.

References

1. Kershaw EE, Flier JS (2004) Adipose tissue as an endocrine organ. J Clin Endocrinol Metab 89(6):2548–2556. doi:10.1210/jc.2004-0395

2. Rosen ED, Spiegelman BM (2006) Adipocytes as regulators of energy balance and glucose homeostasis. Nature 444(7121):847–853. doi:10.1038/nature05483

3. Tilg H, Moschen AR (2008) Inflammatory mechanisms in the regulation of insulin resistance. Mol Med 14(3–4):222–231. doi:10.2119/2007-00119.Tilg

4. Hajer GR, van Haeften TW, Visseren FL (2008) Adipose tissue dysfunction in obesity, diabetes, and vascular diseases. Eur Heart J 29(24):2959–2971. doi:10.1093/eurheartj/ehn387

5. Van Gaal LF, Mertens IL, De Block CE (2006) Mechanisms linking obesity with cardiovascular disease. Nature 444(7121):875–880. doi:10.1038/nature05487

6. Kahn SE, Hull RL, Utzschneider KM (2006) Mechanisms linking obesity to insulin resistance and type 2 diabetes. Nature 444(7121):840–846. doi:10.1038/nature05482

7. Olshansky SJ, Passaro DJ, Hershow RC et al (2005) A potential decline in life expectancy in the United States in the 21st century. N Engl J Med 352(11):1138–1145. doi:10.1056/NEJMsr043743

8. Rosen ED, Walkey CJ, Puigserver P et al (2000) Transcriptional regulation of adipogenesis. Genes Dev 14(11):1293–1307

9. Tontonoz P, Hu E, Spiegelman BM (1994) Stimulation of adipogenesis in fibroblasts by PPAR gamma 2, a lipid-activated transcription factor. Cell 79(7):1147–1156

10. Rosen ED, Sarraf P, Troy AE et al (1999) PPAR gamma is required for the differentiation of adipose tissue in vivo and in vitro. Mol Cell 4(4):611–617

11. Wu Z, Rosen ED, Brun R et al (1999) Cross-regulation of C/EBP alpha and PPAR gamma controls the transcriptional pathway of adipogenesis and insulin sensitivity. Mol Cell 3(2):151–158

12. Rosen ED, Hsu CH, Wang X et al (2002) C/EBPalpha induces adipogenesis through PPARgamma: a unified pathway. Genes Dev 16(1):22–26. doi:10.1101/gad.948702

13. Park BO, Ahrends R, Teruel MN (2012) Consecutive positive feedback loops create a bistable switch that controls preadipocyte-to-adipocyte conversion. Cell Rep 2(4):976–990. doi:10.1016/j.celrep.2012. 08.038

14. Ahrends R, Ota A, Kovary KM et al (2014) Controlling low rates of cell differentiation through noise and ultrahigh feedback. Science 344(6190):1384–1389. doi:10.1126/science. 1252079

15. Barak Y, Nelson MC, Ong ES et al (1999) PPAR gamma is required for placental, cardiac, and adipose tissue development. Mol Cell 4(4):585–595

16. Kubota N, Terauchi Y, Miki H et al (1999) PPAR gamma mediates high-fat diet-induced adipocyte hypertrophy and insulin resistance. Mol Cell 4(4):597–609

17. Agarwal AK, Garg A (2002) A novel heterozygous mutation in peroxisome proliferator-activated receptor-gamma gene in a patient with familial partial lipodystrophy. J Clin Endocrinol Metab 87(1):408–411. doi:10.1210/jcem.87.1.8290

18. Hegele RA, Cao H, Frankowski C et al (2002) PPARG F388L, a transactivation-deficient mutant, in familial partial lipodystrophy. Diabetes 51(12):3586–3590

19. Savage DB, Tan GD, Acerini CL et al (2003) Human metabolic syndrome resulting from dominant-negative mutations in the nuclear receptor peroxisome proliferator-activated receptor-gamma. Diabetes 52(4):910–917

20. Forman BM, Tontonoz P, Chen J et al (1995) 15-deoxy-delta 12, 14-prostaglandin J2 is a ligand for the adipocyte determination factor PPAR gamma. Cell 83(5):803–812

21. Lehmann JM, Moore LB, Smith-Oliver TA et al (1995) An antidiabetic thiazolidinedione is a high affinity ligand for peroxisome proliferator-activated receptor gamma (PPAR gamma). J Biol Chem 270(22):12953–12956

22. Mikkelsen TS, Xu Z, Zhang X et al (2010) Comparative epigenomic analysis of murine and human adipogenesis. Cell 143(1):156–169. doi:10.1016/j.cell.2010.09.006

23. Masters C, Reid S, Don M (1987) Glycolysis—new concepts in an old pathway. Mol Cell Biochem 76(1):3–14. doi:10.1007/BF00219393

Chapter 6

A Targeted MRM Approach for Tempo-Spatial Proteomics Analyses

Annie Moradian, Tanya R. Porras-Yakushi, Michael J. Sweredoski, and Sonja Hess

Abstract

When deciding to perform a quantitative proteomics analysis, selectivity, sensitivity, and reproducibility are important criteria to consider. The use of multiple reaction monitoring (MRM) has emerged as a powerful proteomics technique in that regard since it avoids many of the problems typically observed in discovery-based analyses. A prerequisite for such a targeted approach is that the protein targets are known, either as a result of previous global proteomics experiments or because a specific hypothesis is to be tested. When guidelines that have been established in the pharmaceutical industry many decades ago are taken into account, setting up an MRM assay is relatively straightforward. Typically, proteotypic peptides with favorable mass spectrometric properties are synthesized with a heavy isotope for each protein that is to be monitored. Retention times and calibration curves are determined using triple-quadrupole mass spectrometers. The use of iRT peptide standards is both recommended and fully integrated into the bioinformatics pipeline. Digested biological samples are mixed with the heavy and iRT standards and quantified. Here we present a generic protocol for the development of an MRM assay.

Key words MRM, Quadrupole mass spectrometry, Quantitation

1 Introduction

The ultimate goal of a quantitative proteomics experiment is to identify and quantify protein changes in time and/or space. Particularly, when monitoring the changes of specific proteins over many time and/or space points, the use of multiple reaction monitoring (MRM) has emerged as a powerful technique because of its high quantitative precision and sensitivity resulting in low detection limits [1–5]. A prerequisite for such a targeted approach is that the protein targets are known, either as a result of previous global proteomics experiments or because a specific hypothesis is to be tested.

Typically, proteotypic peptides with favorable mass spectrometric properties are selected for each protein and synthesized

Jörg Reinders (ed.), *Proteomics in Systems Biology: Methods and Protocols*, Methods in Molecular Biology, vol. 1394, DOI 10.1007/978-1-4939-3341-9_6, © Springer Science+Business Media New York 2016

a   QTRAP mode

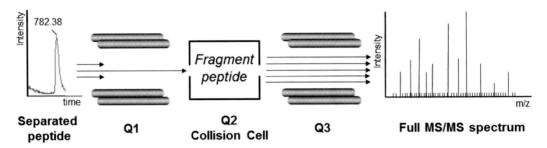

b   QQQ mode

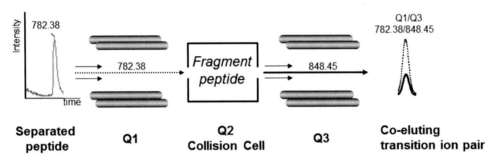

Fig. 1 The two operational modes of a QTRAP instrument. (**a**) QTRAP mode. Selected ions of separated peptides pass through Q1 and are fragmented in Q2, all fragment ions pass through Q3, and a full MS/MS spectrum is recorded. This mode is particularly useful in the early stages of assay development when identities of peptides need to be confirmed. (**b**) QQQ mode. Selected ions of separated peptides pass through Q1 and are fragmented in Q2, and all selected fragment ions pass through Q3. Co-eluting transition ion pairs are recorded. This mode is particularly useful in maximizing sensitivity of known peptides

with a heavy isotope. Retention times and calibration curves are determined using predominantly triple-quadrupole mass spectrometers (Fig. 1), although quadrupole time-of-flight and Orbitrap mass spectrometers can also be used. We preferentially use the QTRAP in the trap mode (Fig. 1a) during the initial stages of assay development, to establish retention time of the peptides, select the best suited three to five fragment ions, and optimize scheduling. Once this is accomplished, scheduled analysis is performed in the more sensitive QQQ mode (Fig. 1b). To determine realistic limits of detection (LOD) and quantitation (LOQ), it is recommended to analyze the heavy standards in a complex biological background. This prevents unwanted absorption of the peptides and helps to identify interferences of the peptides with the background at an early stage of the method development. Thus, digested biological samples are mixed with the heavy and retention time iRT standards and quantified over a dynamic range [6].

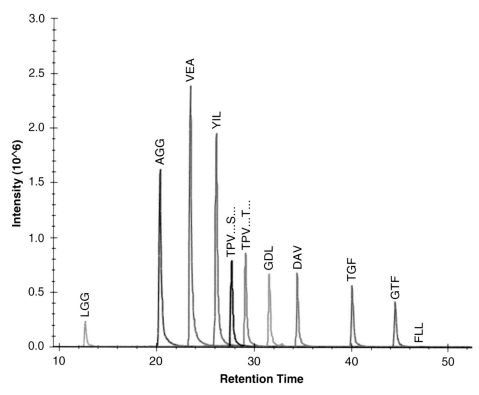

Fig. 2 Elution profile of 11 iRT peptides in a 45-min gradient. Peptides are LGGNETQVR (LGG), AGGSSEPVTGLADK (AGG), VEATFGVDESANK (VEA), YILAGVESNK (YIL), TPVISGGPYYER (TPV…S…), TPVITGAPYYER (TPV…T…), GDLD AASYYAPVR (GLD), DAVTPADFSEWSK (DAV), TGFIIDPGGVIR (TGF), GTFIIDPAAIVR (GTF), FLLQFGAQGSPLFK (FLL)

A typical elution profile of the iRT peptides is shown in Fig. 2. The use of iRT peptide standards is recommended to monitor and ensure consistent chromatographic performance throughout the assay [6]. Additionally, the iRT peptide standards can be fully integrated into the bioinformatics pipeline using Skyline [7]. Figure 3 shows an example of a calibration curve of standard peptides and their transition ion responses from 30 to 500,000 attomol, measured in a complex background, e.g., HeLa lysate if a HeLa cell culture is to be investigated. The calibration curves of the heavy standards can be used to determine targeted peptide amounts. As shown in Fig. 3, a response of 100,000 AU for a targeted peptide could be associated with an amount of 550 amol, in comparison to the calibration curve of the heavy standard shown (*see* **Note 1**). The most accurate quantitative results are usually achieved when the heavy standards are in the same range as the expected light samples. For this, it might be necessary to do a preliminary analysis first, followed by a full analysis with heavy peptides spiked in according to the preliminary results. Here, we provide a generic guideline for the development of an MRM assay.

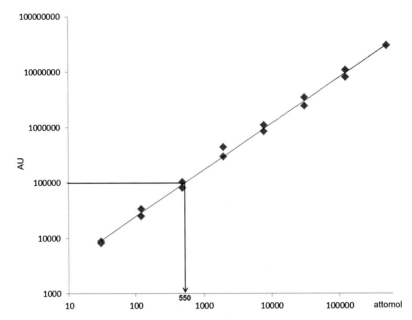

Fig. 3 Example of a calibration curve for a standard peptide covering a dynamic range from 30 to 500,000 attomol. This curve can be used to quantify unknown samples. If an unknown sample has a response of 100,000 AU, the sample contains 550 attomol of that particular peptide

2   Materials

2.1   Chemicals

2.1.1   LC and LC Samples

1. Solvent A: 0.2 % Formic acid. Add 2 mL of formic acid (for mass spectrometry; 98 %) to 998 mL of water (HPLC gradient grade quality).

2. Solvent B: Acetonitrile containing 0.2 % formic acid. Add 2 mL of formic acid (for mass spectrometry; 98 %) to 998 mL of acetonitrile (HPLC gradient grade quality).

3. Synthetic heavy standard peptides (either prepared in-house or commercially).

4. Synthetic iRT peptides (either prepared in-house or commercially).

5. Standard solvent (30 % acetonitrile, 70 % of 0.2 % formic acid). Add 30 mL of acetonitrile to a 69.8 mL of water and 0.2 mL of formic acid.

6. All reagents should be of the highest purity available.

2.1.2   FASP

1. *Lysis buffer.* Add 5 g sodium dodecyl sulfate (SDS), 1.58 g (100 mM) Tris/HCl (pH 7.6; adjust with HCl as necessary), 1.54 g (100 mM) D,L-dithiothreitol (DTT), supplemented with 1× protease inhibitor cocktail (complete, EDTA free, Roche) and 17.42 mg (1 mM) phenylmethanesulfonyl fluoride (PMSF) brought up to 100 mL with ddH$_2$O.

2. *Urea A (UA) buffer.* Add 48.05 g (8 M) urea and 1.58 g (100 mM) Tris/HCl (pH 8.5, adjust with HCl as necessary) to a total volume of 100 mL of ddH$_2$O; prepare fresh on the day of the digestion.

3. *Urea B (UB) buffer.* Add 48.05 g (8 M) urea and 1.58 g (100 mM) Tris/HCl (pH 8.0, adjust with HCl as necessary) to a total volume of 100 mL of ddH$_2$O; prepare fresh on the day of the digestion.

4. *27 mM Tris(2-carboxyethyl)phosphine hydrochloride (TCEP) solution.* Add 14.33 g (500 mM) TCEP to 1.58 g (100 mM) Tris/HCl (pH 8.5, adjust with HCl as necessary) in a total volume of 100 mL of ddH$_2$O and store at –20 °C. Right before reduction step, thaw aliquot and dilute 5.4 µL of 500 mM TCEP by adding 94.6 µL of UA buffer.

5. *Alkylating solution.* Prepare fresh using 0.925 mg (50 mM) iodoacetamide (IAA) in 1 mL of UA buffer.

6. *50 mM Ammonium bicarbonate (NH4HCO3) solution.* Add 0.40 g of NH$_4$HCO$_3$ to 100 mL of H$_2$O.

7. *100 mM Calcium chloride (CaCl2) solution.* Add 14.69 mg of calcium chloride dehydrate to 1 mL of water.

2.1.3 Preparation of Heavy Standard Peptide Solutions

1. Heavy standard proteotypic peptides of target proteins are synthesized, either commercially or if available in-house (*see* **Note 2**).

2. Prepare solutions of peptide standards at a nominal concentration of 1 pmol/µL using 30 % acetonitrile and 70 % of 0.2 % formic acid (*see* **Note 3**).

3. Eppendorf tubes; pipettes.

2.1.4 Preparation of iRT Peptide Solutions

1. Eleven iRT peptides (LGGNETQVR (LGG), AGGSSEPVT GLADK (AGG), VEATFGVDESANK (VEA), YILAGVESNK (YIL), TPVISGGPYYER (TPV…S), TPVITGAPYYER (TPV…T), GDLDAASYYAPVR (GLD)) are synthesized, either commercially or if available in-house (*see* **Note 4**).

2. Prepare a solution containing all iRT peptide standards at a nominal concentration of 1 pmol/µL using 30 % acetonitrile and 70 % of 0.2 % formic acid.

3. Eppendorf tubes; pipettes.

2.2 Liquid Chromatography-Mass Spectrometry (LC-MS/MS)

1. QTRAP 6500 (AB Sciex, Framingham, MA) with an Eksigent ekspert nanoLC 425 pump, ekspert nanoLC400 autosampler, ekspert cHiPLC, and Analyst software. The LC system is coupled to a NanoSpray III Source and Heated Interface (AB/Sciex).

2. CHiPLC Chrom XP C18-CL 3 µm trap column, 120 Å (200 µm × 0.5 mm).

3. CHiPLC Chrom XP C18-CL 3 μm column, 120 Å (75 μm × 150 mm).

4. Sonicate solvent A and B prior to use (*see* **Note 5**).

2.3   Data Analysis   Skyline software (*see* **Note 6**).

3   Methods

3.1   Preparation of HeLa Tryptic Peptide Solutions Using a Modified Filter-Assisted Sample Preparation (FASP) Procedure [8]

1. Resuspend cells or tissue in a 1:10 sample-to-buffer ratio and incubate at 95 °C for 5 min.

2. Lyse cells or homogenize tissue using standard methods. Clarify lysate by centrifugation at 16,000 × g for 5 min at 20 °C and determine protein concentration (*see* **Note 7**).

3.1.1   Lysate Preparation

3.1.2   FASP (Double Digestion)

1. Combine up to 200 μg of lysate with 200 μL of UA buffer in a 30 K microcon filter unit and centrifuge at 14,000 × g for 15 min.

2. Discard flow-through from the collection tube, add an additional 200 μL of UA buffer to the filter unit, and centrifuge at 14,000 × g for 15 min.

3. Repeat **step 2** at least two additional times (*see* **Note 8**).

4. Discard the flow-through, add 100 μL of UA buffer containing 27 mM TCEP to the filter unit, and mix at 600 rpm in a thermo-mixer for 1 min. Then incubate for 20 min at room temperature without shaking (*see* **Note 9**).

5. After the incubation with TCEP is complete, centrifuge the filter unit at 14,000 × g for 10 min.

6. Discard the flow-through, add an additional 100 μL of UA buffer to the filter unit, and then centrifuge at 14,000 × g for 15 min.

7. Discard the flow-through, add 100 μL of alkylating solution, and mix at 600 rpm in a thermo-mixer for 1 min. Then incubate for 20 min at room temperature without shaking, in the dark (*see* **Note 10**).

8. After the incubation with the alkylation solution is complete, centrifuge the filter unit at 14,000 × g for 10 min.

9. Discard the flow-through, add 100 μL of UB buffer to the filter unit, and centrifuge at 14,000 × g for 15 min.

10. Repeat **step 9** two additional times.

11. Transfer the filter unit to a new collection tube, add 40 μL of UB buffer containing Lys-C (enzyme-to-protein ratio 1:50),

and mix at 600 rpm for 1 min in the thermo-mixer, followed by a 4-h incubation at room temperature without shaking in the dark.

12. Then add 120 µL of trypsin dissolved in 50 mM NH_4HCO_3 (enzyme-to-protein ratio 1:100), along with 1.6 µL of 100 mM $CaCl_2$ and mix for 1 min at 600 rpm in a thermo-mixer (*see* **Note 11**).

13. Incubate the units at room temperature for 14 h, in the dark.

14. Centrifuge the filter unit at $14,000 \times g$ for 15 min (*see* **Note 12**).

15. To increase recovery of peptides, add 100 µL of 50 mM NH_4HCO_3 to the filter unit and centrifuge at $14,000 \times g$ for 15 min (*see* **Note 12**).

16. Repeat **step 15** two additional times and combine all flow-through (*see* **Note 12**).

17. Add enough formic acid to bring up the concentration to 5 %.

18. Lyophilize the peptide solution.

19. Resuspend the lyophilized peptides in 100 µL of 0.2 % formic acid and desalt by HPLC, e.g., with a C8 peptide macrotrap (3×8 mm) (200 µg maximum capacity) (*see* **Note 13**).

20. Lyophilize the desalted peptides and store at –20 °C until ready for mass spec analysis (*see* **Note 14**).

21. Prior to analysis, resuspend HeLa digest in 0.2 % formic acid at a nominal concentration of 4 µg/µL.

3.2 Preparation of Serial Dilutions of Heavy Standard Peptide and iRT Peptide Solutions in Complex Background

1. To 1.5 µL of trypsin-digested HeLa lysate (4 µg/µL), add 3 µL of heavy standard solution (1 pmol/µL), 0.6 µL of iRT peptide solution (1 pmol/µL), and 0.9 µL of 0.2 % formic acid: solution A, 500 fmol/µL of heavy standard.

2. Keep 1 µL of solution A (500 fmol/µL) for injection (*see* **Note 15**).

3. Dilute 1.5 µL of solution A (500 fmol/µL) in 1.125 µL trypsin-digested HeLa lysate, 0.45 µL iRT peptide solution (1 pmol/µL), and 2.925 µL of 0.2 % formic acid: solution B, 125 fmol/µL of heavy standard.

4. Keep 1 µL of solution B (125 fmol/µL) for injection.

5. Dilute 1.5 µL of solution B (125 fmol/µL) in 1.125 µL trypsin-digested HeLa lysate, 0.45 µL iRT peptide solution (1 pmol/µL), and 2.925 µL of 0.2 % formic acid: solution C, 31.3 fmol/µL of heavy standard.

6. Keep 1 µL of solution C (31.3 fmol/µL) for injection.

7. Dilute 1.5 µL of solution C (31.3 fmol/µL) in 1.125 µL trypsin-digested HeLa lysate, 0.45 µL iRT peptide solution (1 pmol/µL), and 2.925 µL of 0.2 % formic acid: solution D, 7.8 fmol/µL of heavy standard.

8. Keep 1 μL of solution D (7.8 fmol/μL) for injection.

9. Dilute 1.5 μL of solution D (7.8 fmol/μL) in 1.125 μL trypsin-digested HeLa lysate, 0.45 μL iRT peptide solution (1 pmol/μL), and 2.925 μL of 0.2 % formic acid: solution E, 1.95 fmol/μL of heavy standard.

10. Keep 1 μL of solution E (1.95 fmol/μL) for injection.

11. Dilute 1.5 μL of solution E (1.95 fmol/μL) in 1.125 μL trypsin-digested HeLa lysate, 0.45 μL iRT peptide solution (1 pmol/μL), and 2.925 μL of 0.2 % formic acid: solution F, 488 amol/μL of heavy standard.

12. Keep 1 μL of solution F (488 amol/μL) for injection.

13. Dilute 1.5 μL of solution F (488 amol/μL) in 1.125 μL trypsin-digested HeLa lysate, 0.45 μL iRT peptide solution (1 pmol/μL), and 2.925 μL of 0.2 % formic acid: solution G, 122 amol/μL of heavy standard.

14. Keep 1 μL of solution G (122 amol/μL) for injection.

15. Dilute 1.5 μL of solution G (122 amol/μL) in 1.125 μL trypsin-digested HeLa lysate, 0.45 μL iRT peptide solution (1 pmol/μL), and 2.925 μL of 0.2 % formic acid: solution H, 30.5 amol/μL of heavy standard.

16. Keep 1 μL of solution H (30.5 amol/μL) for injection.

17. Dilute 1.5 μL of solution H (30.5 amol/μL) in 1.125 μL trypsin-digested HeLa lysate, 0.45 μL iRT peptide solution (1 pmol/μL), and 2.925 μL of 0.2 % formic acid: solution I, 7.6 amol/μL of heavy standard.

18. Keep 1 μL of solution I (7.6 amol/μL) for injection (*see* **Note 16**).

19. Repeat **steps 1–17** two additional times so that three samples are created for each solution.

20. Calculate 2–3 transitions per iRT, heavy standard, and light sample peptide.

3.3 LC-MRM of Heavy Proteotypic Standard Peptide Solutions

1. For the separation and analysis of the samples, a nanoLC QTRAP 6500 (AB Sciex, Framingham, MA) can be used (*see* **Note 5**).

2. The peptides are separated using a CHiPLC Chrom XP C18-CL 3 μm column, 120 Å (75 μm × 150 mm) equipped with a CHiPLC Chrom XP C18-CL 3 μm trap column, 120 Å (200 μm × 0.5 mm) at a column temperature of 45 °C and a flow rate of 300 nL/min. Solvent A is 0.2 % formic acid and solvent B is 98.8 % acetonitrile containing 0.2 % formic acid. Linear gradients from 5 to 30 % B are applied within 45 min, 30–90 % B in 2 min, followed by 100 % B for 10 min. Preliminary mass spectra are recorded in positive ion mode acquiring data from the transition lists, initially in the QTRAP

mode (Fig. 1a) to confirm correct peak identification. Optimization of declustering potentials and collision energy is done automatically in Skyline. Subsequent analyses are performed using MRM scheduling in QQQ mode (Fig. 1b) (*see* **Note 17**).

3. Inject 1 µL of solution I through A (*see* **Note 18**).

4. Inject 1 µL of 0.2 % formic acid solution (blank).

5. Repeat **steps 5** and **6** two more times.

6. After an initial data analysis, spike in the heavy standard concentration to your sample that is most appropriate, i.e., at roughly the same concentration range.

7. Inject 1 µL of this adjusted solution.

8. Repeat **step 7** two additional times for triplicate measurements.

3.4 Data Analysis

Import the wiff (or appropriate raw) files of all measurements into Skyline software (*see* **Note 19**).

4   Notes

1. In addition to the calibration curves and dynamic range, analytical validation is achieved by testing repeatability (technical replicates, injecting the same sample ten times), reproducibility (biological replicates, injecting ten samples that have been prepared the same way ten times), limit of detection (best approximated by using ca. three times noise), and limit of quantitation (best approximated by using ca. ten times noise), similar to common practice in the pharmaceutical field [2, 9–12].

2. Generally, it is sufficient to have one heavy-labeled amino acid such as [$^{13}C_6$]Lys or [$^{13}C_6$]Arg at the C-terminal side of the tryptic peptide. It may be necessary to purify and/or desalt the peptides after synthesis. Peptides containing Cys, Met, His, N-terminal Glu or Gln, glycosylation site motifs (NXS/T), or Pro following Lys should be avoided if possible.

3. If the stock solutions are to be stored in the freezer, we recommend using higher concentrations (nmol/µL) for storage to avoid sample adsorption to the storage vials.

4. iRT peptides are used to assess chromatographic reproducibility. They also enable users to schedule their analysis in smaller windows and help verify transitions in case of interference. Finally, they aid in transferring methods from one lab to another lab.

5. The use of this setup is not mandatory. Other QQQ mass spectrometers and quadrupole-based mass spectrometers such as

Q-TOF or Q-Exactive from other vendors with nanoLC setup are equally suited for this purpose.

6. This software tool has been developed by the MacCoss lab and is freely available at proteome.gs.washington.edu/software/skyline/. Extensive documentation is provided on the Skyline website for data analysis and interpretation. Commercial packages may serve the same purpose.

7. Depending on the cell or tissue type used lysis methods will vary.

8. The purpose of **steps 1** and **2** is to wash away the SDS; since proteins are extended and denatured they will remain in the filter unit. It may be necessary to repeat this wash step multiple times. We have found five washes to be sufficient for most samples.

9. The final concentration of TCEP once the 100 μL is added to the 30 μL of filtrate that remains in the filter unit will be 20 mM. The purpose of this step is to ensure complete reduction of all disulfide bridges.

10. The purpose of this step is to alkylate the cysteine residues to prevent disulfide bridges from reforming.

11. Adding $CaCl_2$ will enhance the activity of trypsin.

12. The flow-through contains the trypsin-digested peptides. Keep!

13. The FASP procedure is optimized for the recovery of purified tryptic peptides from intact cells or tissue. Loss of sample will be greater than 50 %; therefore adjust the amount of starting material accordingly. A total of 100 μg will be sufficient for a comprehensive assay development.

14. Determine peptide concentration, e.g., through UV response during the desalting step.

15. The resulting 1 μL contains 500 fmol of each heavy standard and 100 fmol of each iRT peptide.

16. The serial dilutions are prepared to determine linearity, dynamic range, and LOD/LOQ values.

17. While theoretically all samples can be analyzed without MRM scheduling, it is highly recommended to take advantage of the MRM scheduling. Once retention times have been recorded in a preliminary analysis, transitions in the following analysis according to the predetermined retention times should be scheduled. This ensures higher sensitivity and more data points across the peaks for smoother peak shapes and better accuracy in quantification.

18. Start with the lowest concentration to avoid carryover.

19. The files are imported and show target masses, individual transitions per peptide, peak areas, and retention times over the multiple analysis. Considerable manual intervention is needed to confirm assignment of peaks.

Acknowledgement

This work was supported by the Beckman Institute, the Gordon and Betty Moore Foundation through Grant GBMF775, and the NIH through grant 1S10OD010788-01A1.

References

1. Picotti P, Aebersold R (2012) Selected reaction monitoring-based proteomics: workflows, potential, pitfalls and future directions. Nat Methods 9(6):555–566. doi:10.1038/nmeth.2015

2. Carr SA, Abbatiello SE, Ackermann BL et al (2014) Targeted peptide measurements in biology and medicine: best practices for mass spectrometry-based assay development using a fit-for-purpose approach. Mol Cell Proteomics 13(3):907–917. doi:10.1074/mcp.M113.036095

3. Huttenhain R, Soste M, Selevsek N et al (2012) Reproducible quantification of cancer-associated proteins in body fluids using targeted proteomics. Sci Transl Med 4(142):142ra194. doi:10.1126/scitranslmed.3003989

4. Kuzyk MA, Smith D, Yang J et al (2009) Multiple reaction monitoring-based, multiplexed, absolute quantitation of 45 proteins in human plasma. Mol Cell Proteomics 8(8):1860–1877. doi:10.1074/mcp.M800540-MCP200

5. Picotti P, Rinner O, Stallmach R et al (2010) High-throughput generation of selected reaction-monitoring assays for proteins and proteomes. Nat Methods 7(1):43–46. doi:10.1038/nmeth.1408

6. Escher C, Reiter L, MacLean B et al (2012) Using iRT, a normalized retention time for more targeted measurement of peptides. Proteomics 12(8):1111–1121. doi:10.1002/pmic.201100463

7. MacLean B, Tomazela DM, Shulman N et al (2010) Skyline: an open source document editor for creating and analyzing targeted proteomics experiments. Bioinformatics 26(7):966–968. doi:10.1093/bioinformatics/btq054

8. Wisniewski JR, Zougman A, Nagaraj N et al (2009) Universal sample preparation method for proteome analysis. Nat Methods 6(5):359–362. doi:10.1038/nmeth.1322

9. Hess S, Akermann M, Ropte D et al (2001) Rapid and sensitive LC separation of new impurities in trimethoprim. J Pharm Biomed Anal 25(3–4):531–538

10. Hess S, Dolker M, Haferburg D et al (2001) Separation, analyses and syntheses of trimethoprim impurities. Pharmazie 56(4):306–310

11. Hess S, Muller CE, Frobenius W et al (2000) 7-deazaadenines bearing polar substituents: structure-activity relationships of new A(1) and A(3) adenosine receptor antagonists. J Med Chem 43(24):4636–4646. doi:10.1021/Jm000967d

12. Hess S, Teubert U, Ortwein J et al (2001) Profiling indomethacin impurities using high-performance liquid chromatography and nuclear magnetic resonance. Eur J Pharm Sci 14(4):301–311

Chapter 7

Targeted Phosphoproteome Analysis Using Selected/Multiple Reaction Monitoring (SRM/MRM)

Jun Adachi, Ryohei Narumi, and Takeshi Tomonaga

Abstract

Mass spectrometry-based phosphoproteomics has been rapidly spread based on the advancement of mass spectrometry and development of efficient enrichment techniques for phosphorylated proteins or peptides. Non-targeted approach has been employed in most of the studies for phosphoproteome analysis. However, targeted approach using selected/multiple reaction monitoring (SRM/MRM) is an indispensible technique used for the quantitation of known targets especially when we have many samples to quantitate phosphorylation events on proteins in biological or clinical samples. We herein describe the application of a large-scale phosphoproteome analysis and SRM-based quantitation for the systematic discovery and validation of biomarkers.

Key words Phosphoproteomics, Selected/multiple reaction monitoring, Targeted proteomics, Biomarker, IMAC

1 Introduction

Mass spectrometry (MS) is a powerful tool to identify protein phosphorylation sites, and it is possible to identify over 50,000 phosphorylation sites using the state-of-the-art proteomic analysis platform [1]. Combined with a quantification method such as metabolic labeling method (e.g., SILAC), chemical labeling method (e.g., TMT, iTRAQ), or label-free quantification method, it is possible to perform systematic quantification of protein phosphorylation. These non-targeted analyses do not require previous knowledge of the target proteins; thus it is suitable to use at discovery phase in order to identify key phosphorylation events or biomarker candidates and create a new hypothesis. Chemical labeling techniques are particularly useful for quantitatively comparing proteomes between clinical samples such as tissues or plasma/serum. For example, a large-scale phosphoproteome analysis can be performed by combining chemical labeling with

Jörg Reinders (ed.), *Proteomics in Systems Biology: Methods and Protocols*, Methods in Molecular Biology, vol. 1394,
DOI 10.1007/978-1-4939-3341-9_7, © Springer Science+Business Media New York 2016

phosphopeptide-enrichment technique, and has recently been applied to the discovery of biomarker candidates using tissue samples [2].

Another approach for MS-based phosphopeptide quantification is selected reaction monitoring (SRM, also called as MRM, multiple reaction monitoring). SRM is a targeted approach to quantitate specific peptides by monitoring specific precursor-to-product ion transitions during liquid chromatography (LC)-MS/MS in a triple-quadrupole instrument. Thus, SRM is useful for an extensive validation for tens or hundreds of biomarker candidates identified by a phosphoproteome analysis at discovery phase. Antibody-based validation is also common; however, available number of phosphorylation-specific antibody is limited in the case of protein phosphorylation. Thus MS-based phosphoproteomic targeted analysis such as SRM has an importance at the validation phase.

Here, we describe the application of a large-scale phosphoproteome analysis and SRM-based quantification to develop a strategy for the systematic discovery and validation of biomarkers using tissue samples or cultured cells. At first we identify differentially modulated phosphopeptides using immobilized metal ion affinity chromatography (IMAC) coupled with non-targeted quantitative proteomic analysis. Identified phosphopeptide candidates are then validated by the SRM analysis. This systematic approach has enormous potential for the discovery of bona fide disease-specific biomarkers (Fig. 1).

2 Materials

2.1 Reagents and Equipment for Sample Preparation and Enzymatic Digestion (See Note 1)

1. Phase-transfer surfactants A (PTS-A) buffer: 50 mM ammonium bicarbonate.

2. Phase-transfer surfactants B (PTS-B) buffer: 50 mM ammonium bicarbonate, 12 mM sodium deoxycholate, 12 mM sodium N-lauroyl sarcosinate.

3. Lysis buffer: PhosSTOP phosphatase inhibitor cocktail (Roche Diagnostics, Manheim, Germany) dissolved in PTS-B buffer. Prepare the lysis buffer just before the use.

4. LysC stock solution: 0.004 unit/μL lysyl endopeptidase (LysC) (Wako Pure Chemical Industries, Osaka, Japan). Dissolve lyophilized LysC in water (see Note 2). Store at –80 °C.

5. Trypsin stock solution: 1 μg/μL Trypsin (proteomics grade; Roche Diagnostics). Dissolve lyophilized trypsin in 10 mM HCl (see Note 3). Store at –80 °C.

6. DTT solution (10×): 100 mM dithiothreitol (DTT). Add 65 μL of PTS-A buffer to 1 mg of DTT and dissolve immediately prior to use.

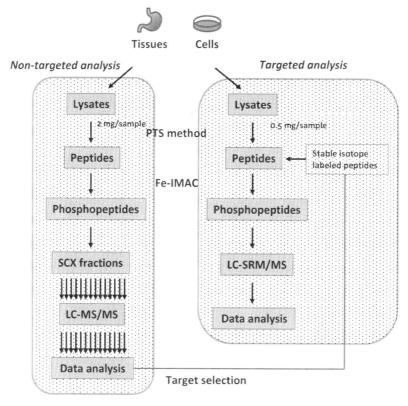

Fig. 1 Workflows of non-targeted analysis and targeted (SRM) analysis. In non-targeted analysis, proteins are denatured and digested using phase transfer surfactant (PTS) method. Phosphopeptides are enriched using Fe-IMAC, followed by SCX off-line fractionation and LC-MS/MS. In the SRM analysis, the individual sample is processed by PTS method and spiked with a mixture of the isotope-labeled peptides. The resulting sample is applied to Fe-IMAC to enrich the phosphopeptides and analyzed by SRM analysis

7. Iodoacetamide (IAA) solution (10×): 500 mM Iodoacetamide. Add 10.8 μL of PTS-A buffer to 1 mg of IAA and dissolve immediately prior to use.

8. DC Protein Assay Kit (Bio-Rad, Hercules, CA).

9. Tissue grinder.

10. Liquid nitrogen.

11. Phosphate-buffered saline (PBS) buffer.

12. Bovine serum albumin (BSA), e.g., Pierce BSA Protein Assay Standards (Thermo Scientific, Rockford, IL).

13. Benchtop centrifuge.

14. Sonicator, e.g., Bioruptor-UCD-250 (Cosmo Bio, Tokyo, Japan).

15. Speed Vac.

16. Ethylacetate.

17. Trifluoroacetic acid (TFA) (HPLC grade).

2.2 Reagent for Desalting Peptide Mixtures by C18-Stage Tip (See Note 4)

1. Empore™ C18 47 mm disk (3 M, St. Paul, MN).
2. 200 μL pipet tips.
3. Methanol (LC-MS grade)
4. 80 % Acetonitrile, 0.1 % TFA: Mix acetonitrile (LC-MS grade), distilled water (LC-MS grade), and TFA.
5. 2 % Acetonitrile, 0.1 % TFA.
6. 60 % Acetonitrile, 0.1 % TFA.

2.3 Reagents for IMAC

1. ProBond™ Nickel-Chelating Resin (Life Technology, Carlsbad, CA).
2. 50 mM EDTA.
3. 0.1 % Acetic acid.
4. 100 mM $FeCl_3$, 0.1 % acetic acid.
5. 2 % Acetonitrile, 0.1 % TFA.
6. 60 % Acetonitrile, 0.1 % TFA.
7. 1 % Phosphate: Dilute phosphoric acid (HPLC grade, 85 %) 85 times with water.

2.4 Strong Cation-Exchange Chromatography

1. Buffer A: 25 % Acetonitrile, 10 mM H_3PO_4 (pH 3). Mix 250 mL of acetonitrile, approximately 650 mL of water, and 685 μL of phosphoric acid. Adjust to pH 3.0 by adding KOH solution and to 1 L with water.
2. Buffer B: 25 % Acetonitrile, 10 mM H_3PO_4 (pH 3), 1 M KCl. Mix 250 mL of acetonitrile, approximately 650 mL of water, 685 μL of phosphoric acid, and 74.55 g of KCl. Adjust to pH 3.0 by adding KOH solution and to 1 L with water.
3. HPLC system, e.g., Prominence UFLC (Shimadzu, Kyoto, Japan).
4. Strong cation-exchange (SCX) column, e.g., 50 mm × 2.1 mm, 5 μm, 300 Å, ZORBAX 300SCX (Agilent Technology, Santa Clara, CA).

2.5 Non-targeted LC-MS/MS Analysis

1. Buffer-A: 0.1 % Formic acid, 2 % acetonitrile.
2. Buffer-B: 0.1 % Formic acid, 90 % acetonitrile.
3. Sample vial or sample plate for LC-MS/MS analysis.
4. Mass spectrometer for LC-MS/MS analysis, e.g., LTQ-Orbitrap Velos mass spectrometer (Thermo Scientific).
5. Nano-electrospray ion source.
6. Nano-LC system, e.g., paradigm system (Michrom Biosciences, Auburn, CA, USA).
7. Analytical column, e.g., a self-packed ESI column (*see* **Note 5**).

8. Trap column, e.g., L-column2 ODS (Chemicals Evaluation and Research Institute, Tokyo, Japan).

9. Software for non-targeted proteomic data analysis, e.g., Proteome Discoverer 1.3 (Thermo Scientific) connected to a search engine Mascot server 2.4 (Matrix Science) (*see* **Note 6**).

10. 2 % Acetonitrile, 0.1 % TFA.

2.6 Targeted (SRM/ MRM-Based) LC-MS/ MS Analysis	Basically same reagents and equipment are used except for those listed below.

1. Stable isotope-labeled peptides (SI peptides) (crude grade) (Thermo Fisher Scientific) (*see* **Note 7**).

2. Mass spectrometer for SRM analysis, e.g., TSQ Vantage triple-quadruple mass spectrometer (Thermo Scientific).

3. Software for targeted (SRM/MRM) proteomic analysis, e.g., Pinpoint 1.2 (Thermo Scientific), which is a software to quantitate the peak areas (quantitative data of targeted peptides) from the raw data of SRM analysis as well as to develop the SRM methods.

4. 2 % Acetonitrile, 0.1 % TFA.

5. 2 % Acetonitrile, 0.1 % TFA, 25 μg/mL EDTA.

6. 1 pmol/μL BSA digest solution.

3 Methods

3.1 Sample Preparation for Tissue Sample	1. Chill the stainless tissue pulverizer in liquid nitrogen. Place a piece of frozen tissue in the chilled device and pulverize the tissue by striking the device with a mallet several times.

2. Check the size of crushed particles. Keep pulverizing the particles until there are no large pieces left in the particles.

3. Transfer the grinded tissue into the chilled tube. Store at –80 °C.

4. Take the required amount of pulverized tissue sample into a microcentrifuge tube (*see* **Note 8**). If a high level of blood contamination is predicted, wash the sample by an appropriate volume of PBS.

5. Add cold lysis buffer, approximately 15 μL per 1 mg of tissue (*see* **Note 9**). Suspend the tissue by pipetting.

6. Immediately place the sample tubes into an ice-cold water bath in the Bioruptor-UCD-250 sonicator. Homogenize the sample by the sonication for 10 min (10 cycles of 30 s on/30 s off) with the amplitude set to 250 W (*see* **Note 10**).

7. Centrifuge the sample at $100,000 \times g$ for 30 min at 4 °C. Collect the supernatant into a new tube. Place a small amount of the

sample into another tube to determine the protein concentration. Store the remainder at −80 °C.

8. The protein concentration is determined by a DC protein assay kit using BSA as the standard.

3.2 Sample Preparation for Cultured Cells

1. Wash cells in the tissue culture dish by directly adding cold PBS and rocking gently. Aspirate PBS and repeat. Keep tissue culture dish on ice throughout.

2. Add appropriate volume of ice-cold lysis buffer to the dish, approximately 1 mL for a 100 mm tissue culture dish.

3. Scrape cells from the surface using a rubber spatula. Transfer the cells into the chilled tube.

4. Sonicate the samples in an ice-cold water bath using the Bioruptor-UCD-250 sonicator. Homogenize the sample by the sonication for 10 min (10 cycles of 30 s on/30 s off) with the amplitude set to 250 W. Repeat the sonication if the sample is still viscose.

5. Take a small amount of the sample into another tube to determine the protein concentration. Store the remainder at −80 °C.

6. The protein concentration is determined by a DC protein assay kit using BSA as the standard.

3.3 Protein Digestion

1. Add the homogenate to a new tube. Dilute the homogenate with the lysis buffer to a concentration that is constant across all samples (see Note 11).

2. Reduce cysteine residues with 10 mM dithiothreitol (DTT) for 30 min.

3. Alkylate the residues with 50 mM iodoacetamide (IAA) for 30 min in the dark.

4. Dilute the sample five times with PTS-A buffer.

5. Digest the sample by 1:100 (w/w) trypsin for 12 h at 37 °C.

6. Add an equal volume of ethyl acetate. Acidify the sample by adding 1/200 volume of trifluoroacetic acid (TFA) in order to transfer the detergents from the water layer into the ethyl acetate layer while most of the peptides remain in the water layer. Mix the ethyl acetate and water layers well by vortexing the tube. Centrifuge the tube at $10,000 \times g$ for 10 min at room temperature to separate the sample into two layers. Discard the upper ethyl acetate layer.

7. Dry the aqueous layer using Speed Vac and store at −80 °C.

3.4 Preparation of Fe-IMAC Resin

1. Transfer 2 mL slurry (1 mL resin) of Probond™ nickel-chelating resin to empty spin columns (see Note 12). Centrifuge the resin at $150 \times g$ for 2 min to discard the flow through.

2. In order to remove nickel ions from resin, wash the resin by 50 mM EDTA solution (3 mL of the solution per 1 mL of the resin) and centrifuge at $150 \times g$ for 2 min. Repeat this step until the color of the resin turns white (*see* **Note 13**).

3. Add water (3 mL of water per 1 mL of the resin) to the resin and centrifuge at $150 \times g$ for 2 min. Discard the flow through.

4. Add 1 % acetate solution (3 mL of the solution per 1 mL of the resin) to the resin in the column, centrifuge at $150 \times g$ for 2 min, and discard the flow through. Repeat this step once more.

5. Add 100 mM $FeCl_3$ in 0.1 % acetic acid (2 mL of the solution per 1 mL of the resin) to the resin and centrifuge at $150 \times g$ for 2 min to chelate iron ions to the resin. Repeat this step once more.

6. Wash the resin by 1 % acetate solution (3 mL of the solution per 1 mL of the resin) and centrifuge at $150 \times g$ for 2 min. Repeat this step twice more.

7. Add 60 % acetonitrile and 0.1 % TFA solution (3 mL of the solution per 1 mL of the resin) to the resin in the column, centrifuge at $150 \times g$ for 2 min, and discard the flow through. Repeat this step once more (*see* **Note 14**).

3.5 Phosphopeptide Enrichment by Fe-IMAC for Non-targeted Phosphoproteome Analysis

1. Dissolve the tryptic digests prepared from the tissue samples or cultured cells in 60 % acetonitrile and 0.1 % TFA solution.

2. Load the digests to the Fe-IMAC resin. Centrifuge at $150 \times g$ for 2 min and discard the flow through.

3. Wash the resin by 60 % acetonitrile and 0.1 % TFA solution (3 mL of the solution per 1 mL of the resin) and centrifuge at $150 \times g$ for 2 min to wash off the non-phosphopeptides. Repeat this step twice more.

4. Add 2 % acetonitrile and 0.1 % TFA solution (3 mL of the solution per 1 mL of the resin) to the resin. Centrifuge at $150 \times g$ for 2 min and discard the flow through.

5. Elute phosphopeptides by 1 % phosphate solution (1 mL of the solution per 1 mL of the resin) and centrifuge at $150 \times g$ for 2 min. Collect the elute into a tube. Repeat this step once more and then collect the second elute into the same tube.

6. Desalt the elute with a disposable solid-phase extraction (SPE) device such as Sep-Pak C18 cartridge.

7. Dry the sample using Speed Vac and store at −80 °C.

3.6 Strong Cation-Exchange Chromatography for Non-targeted Phosphoproteome Analysis

1. Dissolve the sample in mobile buffer A (*see* **Note 15**).

2. Apply the sample to an HPLC and separate on an SCX column using a linear gradient of mobile buffer A and B and sequentially collect eluate every 1 min (*see* **Note 16**).

3. Adjust the number of the fractions for the subsequent MS analysis by combining the fractions based on the peak intensity on the HPLC chromatogram (*see* **Note 17**).

4. Evaporate acetonitrile in the fractions using Speed Vac.

5. Desalt the combined fractions with a disposable SPE device such as C18-Stage Tip (*see* **Note 4**).

6. Elute the sample into a sample vial for MS analysis and then dry it using Speed Vac.

7. Store at −80 °C until MS analysis.

3.7 LC-MS/MS for Non-targeted Phosphoproteome Analysis

1. Add 10 μL of 2 % acetonitrile and 0.1 % TFA to each sample vial.

2. Vortex each vial for 1 min and then spin down.

3. Set the operating parameters of the mass spectrometer (*see* **Note 18**).

4. Analyze each sample by LC-MS/MS (*see* **Note 19**).

5. Apply the acquired raw file to data analysis software with search engine to identify and quantify the phosphopeptides (*see* **Note 20**)

6. Select the phosphopeptide that has to be validated in the subsequent SRM analysis.

3.8 Sample Preparation for Targeted Analysis

1. Prepare a homogenate as previously described in Subheading 3.1 or 3.2.

2. Digest the homogenate according to **steps 1–5** in Subheading 3.3.

3. Add all stable isotope-labeled (SI) peptides to each sample (*see* **Notes 21** and **22**).

4. Extract the peptides from the sample as described in **steps 6–7** in Subheading 3.3.

5. Desalt the resulting sample with C18-Stage Tip.

6. Prepare the micro-scale IMAC column using C18-Stage Tip. Pack two disks of C18 disk at the end of a 200 μL pipet tip and then load 50 μL of Fe-IMAC resin. Centrifugation at $800 \times g$ for 2 min.

7. Load the desalted sample in 60 % acetonitrile and 0.1 % TFA to the IMAC-C18-Stage Tip and then centrifuge at $600 \times g$ for 5 min.

8. Wash the column by 200 μL of 60 % acetonitrile and 0.1 % TFA and then centrifuge at $800 \times g$ for 2 min. Repeat this step twice more.

9. Add 200 μL of 0.1 % TFA to the IMAC-C18-Stage Tip and then centrifuge at $800 \times g$ for 2 min to equilibrate the C18 resin under the IMAC resin.

10. Elute phosphopeptides by 100 μL of 1 % phosphate and then centrifuge at $800 \times g$ for 2 min. Eluted peptides are trapped on the C18 disk. Repeat this step once more.

11. Add 200 μL of 0.1 % TFA to the IMAC-C18-Stage Tip and then centrifuge at $2300 \times g$ for 2 min to wash the C18 disk.

12. Elute the phosphopeptides bound to the C18 disk to a sample tube using 60 μL of 60 % acetonitrile and 0.1 % TFA.

13. Dry the sample using Speed Vac.

14. Store at −80 °C until MS analysis.

3.9 Targeted Analysis by LC-SRM/MS

1. In order to acquire target information (retention time, mass of target ion, and transition ions) prior to the analysis, prepare a mixture of the stable isotope-labeled peptide (SI peptides), which has the same sequence as the phosphopeptide selected from the results of non-targeted analysis. Analyze the mixture by LC-MS/MS using data-dependent mode (*see* **Note 23**).

2. Create a primary method for the subsequent SRM analysis by analyzing the acquired MS data and selecting the precursor ions of each target observed with a strong signal intensity (doubly, triply, or higher charged ions) and the product ions generated from the precursor ion with a strong signal intensity (*see* **Note 24**).

3. Optimize the parameters (m/z of product ions and collision energy (CE)) of the SRM method (*see* **Note 25**).

4. Add 10 μL of 2 % acetonitrile, 0.1 % TFA, and 25 μg/mL EDTA to each sample.

5. Vortex each vial for 1 min to dissolve the peptides and then spin down.

6. Set the optimized SRM method and other operating parameters for the SRM analysis (*see* **Note 26**).

7. Analyze each sample by LC-SRM/MS (*see* **Note 27**)

8. Analyze the acquired raw data by the software (*see* **Note 28**). Target peptides are compared across the samples by extracted ion chromatogram (XIC) intensity of each SRM transition and then normalize the values of the endogenous targeted peptides to those of the corresponding SI peptides.

4 Notes

1. The procedures used for homogenizing samples and enzymatic digestion are based on phase transfer surfactant (PTS)-aided trypsin digestion as described in a previous study [3].

2. Add 0.5 mL of water to a bottle containing 2.0 unit of lyophilized LysC.

3. Add 100 µL of 10 mM HCl to a bottle containing 100 µg of lyophilized trypsin.

4. Peptide mixtures are desalted using C18-Stage Tip or other solid-phase extraction (SPE) devices such as Oasis HLB. Desalting by C18-Stage Tip is performed as described in a previous study [4]. Briefly, a small 47 mm Empore™ C18 disk is stamped out using a blunt-ended syringe needle (16 G), and then the layers are placed in a 200 µL pipet tip by pushing them from the top of the tip using a plunger. C18-Stage Tip is preconditioned by methanol (for swelling), 80 % acetonitrile, 0.1 % TFA (for washing), and 2 % acetonitrile and 0.1 % TFA (for equilibrating). After the sample is applied to the C18-Stage Tip, the tip is washed by 2 % acetonitrile and 0.1 % TFA. Elution of the peptides is performed by 60 % acetonitrile and 0.1 % TFA. The volume of all solutions is 20 µL per layer of C18 disk.

5. We make use of 20 cm long nano HPLC columns with an inner diameter of 100 µm packed in-house with 3 µm C18 beads (L-column ODS, CERI, Tokyo, Japan).

6. We obtain protein and peptide lists and qualitative data including Mascot ion score and local probability of phosphorylation sites using Proteome Discoverer 1.3.

7. We mostly replace lysine or arginine at the C-terminal of target peptides with isotope-labeled lysine ($^{13}C_6$, $^{15}N_2$) or arginine ($^{13}C_6$, $^{15}N_4$) in order to make y-ion fragments heavier than those of endogenous peptides. When the amino acid at the C-terminal is not lysine or arginine (e.g., the C-terminal of a protein), we replace the other amino acids (e.g., alanine) at or near the C-terminal with the other isotope-labeled one (e.g., Alanine-$^{13}C_3$,$^{15}N_1$).

8. Two milligrams protein for non-targeted phosphoproteome analysis and 0.5 mg protein for SRM analysis are required in our study. To obtain enough amount of protein, we use more than 40 mg of tissue or a 15 cm dish (in the case of Hela cells) if possible.

9. By adding the buffer to samples at this ratio, we can generally obtain a solution containing 5–15 mg of proteins per mL.

10. After several rounds of sonication, we check whether the residual pieces of the tissue are left or not for the tissue sample. If the tissues are completely dissolved, we stop the sonication. If not, a few rounds of sonication are additionally performed until the samples are solved uniformly. At this stage, we consider the proteins to be sufficiently extracted from the tissue and stop the sonication.

11. We use 2 mg of protein for non-targeted phosphoproteome analysis and 0.5 mg protein per individual sample for SRM analysis. As shown in Fig. 1, we optimized these initial protein amounts to fit the loading capacity of our LC-MS system (2–3 µg peptides per injection).

12. The ProBond resin is initially provided as a 50 % slurry in 20 % ethanol. We use 1 mL of the resin (2 mL of the suspension) for up to 2 mg of proteins.

13. When nickel ions are released from the resin by EDTA, the color of the resins turns from blue to white.

14. We store the Fe-IMAC resin as a 50 % slurry at 4 °C and use it within 1 week.

15. The volume of SCX buffer A to dissolve the sample depends on the HPLC systems. We dissolve the sample in 110 µL of SCX buffer A according to the maximum injection volume (100 µL) of the autosampler in our HPLC system. 100 µL of the sample is loaded onto the HPLC equipment.

16. We use a flow rate of 200 µL/min and four-step linear gradient for the separation, as follows: 0 % B for 30 min, 0–10 % B in 15 min, 10–25 % B in 10 min, 25–40 % B in 5 min, and 40–100 % B in 5 min, and 100 % B for 10 min.

17. We usually combine 75 fractions into 30 fractions. The flow-through fraction is not combined, because polymer-like contaminants are eluted in it. We combined fractions to make peak area of each combined fraction as equal as possible.

18. For example, when we perform non-labeling or metabolic labeling analysis (e.g., SILAC) using the LTQ-Orbitrap Velos mass spectrometer, the operating parameters are set as follows: full MS scans are performed in the Orbitrap mass analyzer (scan range 350–1500 m/z, with 30 K FWHM resolution at 400 m/z). The eight most intense precursor ions are selected for the MS/MS scans. MS/MS scans are performed using collision-induced dissociation (CID). Collision energy is set to 35 %. A dynamic exclusion option is implemented with a repeat count of 1 and exclusion duration of 60 s. The values of automated gain control (AGC) are set to 5.00e+05 for full MS, 1.00e+04 for CID MS/MS.

When we perform chemical labeling analysis (e.g., iTRAQ) using the LTQ-Orbitrap Velos mass spectrometer, the operating parameters are set non-labeling or metabolic labeling analysis except for the following parameters: The five most intense precursor ions are selected for the MS/MS scans. MS/MS scans are performed using collision-induced dissociation (CID) and higher energy collision-induced dissociation (HCD, 7500 FWHM resolution at 400 m/z) for each precursor ion. Collision energy is set to 35 % for CID and 50 % for HCD. The values of automated gain control (AGC) are set to 5.00e + 04 for HCD MS/MS.

19. Analytical column is self-made by packing C18 particles (L-column2 ODS, 3 μm) into a self-pulled needle (200 mm length × 100 μm for the inner diameter). The mobile phases consist of buffers A (0.1 % formic acid, 2 % acetonitrile) and B (0.1 % formic acid, 90 % acetonitrile). Samples are loaded onto the trap column. The nano LC gradient is delivered at 500 nL/min and consists of a linear gradient of buffer B developed from 5 to 30 % B in 135 min. A spray voltage of 2000 V is applied.

20. To identify the phosphopeptides, the CID and/or HCD raw spectra are extracted and searched separately against the Uniprot or IPI database using Proteome Discoverer 1.3 and Mascot v2.4. The precursor mass tolerance is set to 7 ppm and a fragment ion mass tolerance is set to 0.6 Da for CID and 0.01 Da for HCD. The search parameters allow for one missed cleavage for trypsin, fixed modification (carbamido-methylation at cysteine), and variable modifications (oxidation at methionine, phosphorylation at serine, threonine, and tyrosine). Furthermore, when we employ SILAC quantitation, set the appropriate SILAC plex at the precursor ions quantifier node of Proteome Discoverer. In the case of iTRAQ quantitation, iTRAQ labeling at lysine and the N-terminal residue are added to the fixed modification and iTRAQ labeling at tyrosine are added as variable modifications. The score threshold for peptide identification is set at 1 % false discovery rates (FDR).

21. In the case of crude SI peptides, the purities are very different between the products especially for phosphopeptides. In addition, the ionization efficiency depends on peptide sequences. As a result, the signal intensities of SI peptides can be very different. To maintain the robustness of the experimental system, the signal intensity of each SI peptide should be checked by LC-MS/MS before mixing and then the amount of each SI peptide should be adjusted.

22. When we perform serial dilutions of the SI-peptides for the addition of small amounts of peptides, we perform the dilution using a 1 pmol/μL BSA digest as a matrix to prevent adsorption.

23. In our case, the SI peptide mixture is analyzed by LC-MS/MS using LTQ-Orbitrap XL (CID mode) and obtained data is analyzed by Proteome Discoverer.

24. Msf file generated by Proteome Discoverer is opened with Pinpoint software (version 2.3.0, Thermo Scientific) and the list of MS/MS fragment ions derived from SI-peptides is generated. A total of multiple product ions (four to ten product ions) are selected for the SRM transitions of each target peptide based on the following criteria: y-ion series, strong ion intensity, at least two amino acids in length, and no neutral loss fragment. When phosphopeptides contain more than two amino acids of serine, threonine, or tyrosine, it is important to consider the possibility of sequence isomers which have the same amino acid composition, same number of phosphorylation sites, and, however, different phosphorylation sites. In that case, we select site-specific fragments as transitions. Finally, we create SRM method that consists of SRM transitions, which means pairs of m/z of the precursor/product ions, the collision energies (CEs), and retention time.

25. At first, we optimize collision energy (CE) for every SRM transition around the theoretical value calculated according to the formulas $CE = 0.044 \times m/z + 5.5$ for doubly charged precursor ions, and $CE = 0.051 \times m/z + 0.55$ for triply charged precursor ions. In cases which the theoretical value is over 35 eV, the value is set to 35 eV. After this optimization, four most intense transitions are selected for each target peptide.

26. In addition to the SRM method (SRM transitions, CE, and the retention time for each target peptide), the parameters of the instrument are set as follows: a scan width of $0.002\,m/z$, Q1 and Q3 resolution of 0.7 FWHM, cycle time of 1 s, and gas pressure of 1.8 mTorr. Data are acquired in the time-scheduled SRM mode (retention time window: 8 min).

27. We use the TSQ-Vantage triple-quadruple mass spectrometer equipped with the LC system described above. The nanoLC gradient is delivered at 300 nL/min and consists of a linear gradient of mobile phase B developed from 5 to 23 % B in 45 min. A spray voltage of 1800 V is applied.

28. We use Pinpoint software for the analysis of SRM data. SRM transitions with more than 1×10^3 ion intensity at the peak are used for quantitation. We check that the ratios among the peak areas of individual SRM transitions for each targeted phosphopeptide are comparable to those of the corresponding SI peptide.

References

1. Sharma K, D'Souza RC, Tyanova S et al (2014) Ultradeep human phosphoproteome reveals a distinct regulatory nature of tyr and ser/thr-based signaling. Cell Rep 8(5):1583–1594. doi:10.1016/j.celrep.2014.07.036

2. Narumi R, Murakami T, Kuga T et al (2012) A strategy for large-scale phosphoproteomics and SRM-based validation of human breast cancer tissue samples. J Proteome Res 11(11):5311–5322. doi:10.1021/pr3005474

3. Masuda T, Tomita M, Ishihama Y (2008) Phase transfer surfactant-aided trypsin digestion for membrane proteome analysis. J Proteome Res 7(2):731–740. doi:10.1021/pr700658q

4. Rappsilber J, Mann M, Ishihama Y (2007) Protocol for micro-purification, enrichment, pre-fractionation and storage of peptides for proteomics using StageTips. Nat Protoc 2(8):1896–1906. doi:10.1038/nprot.2007.26 1,nprot.2007.261 [pii]

Chapter 8

Testing Suitability of Cell Cultures for SILAC-Experiments Using SWATH-Mass Spectrometry

Yvonne Reinders, Daniel Völler, Anja-K. Bosserhoff, Peter J. Oefner, and Jörg Reinders

Abstract

Precise quantification is a major issue in contemporary proteomics. Both stable-isotope-labeling and label-free methods have been established for differential protein quantification and both approaches have different advantages and disadvantages. The present protocol uses the superior precision of label-free SWATH-mass spectrometry to test for suitability of cell lines for a SILAC-labeling approach as systematic regulations may be introduced upon incorporation of the "heavy" amino acids. The SILAC-labeled cell cultures can afterwards be used for further analyses where stable-isotope-labeling is mandatory or has substantial advantages over label-free approaches such as pulse-chase-experiments and differential protein interaction analyses based on co-immunoprecipitation. As SWATH-mass spectrometry avoids the missing-value-problem typically caused by undersampling in highly complex samples and shows superior precision for the quantification, it is better suited for the detection of systematic changes caused by the SILAC-labeling and thus, can serve as a useful tool to test cell lines for changes upon SILAC-labeling.

Key words Label-free proteomics, SWATH-MS, SILAC, Melanoma cells

1 Introduction

Proteomic analyses are most frequently performed for the differential quantification of proteins from highly complex mixtures such as total cell lysates. Particularly, stable-isotope labeling methods such as ICAT and iTRAQ have been used with huge success for more than a decade [1]. Incorporation of stable-isotope labels by metabolic labeling is even more suited for quantification primarily in cell cultures, as labeling occurs at the earliest time-point possible. Thereby, all subsequent steps may be conducted with pooled samples minimizing sample-specific losses and eliminating run-to-run-variations. Ong et al. [2] introduced the SILAC-technique (Stable Isotope Labeling by Amino acids in Cell culture) using stable-isotope-coded, essential amino acids instead of native amino

Jörg Reinders (ed.), *Proteomics in Systems Biology: Methods and Protocols*, Methods in Molecular Biology, vol. 1394,
DOI 10.1007/978-1-4939-3341-9_8, © Springer Science+Business Media New York 2016

acids within the cell culture medium. Thereby, these "heavy" amino acids replace all the respective "light" amino acids in the proteins over time. Proteins may be analyzed by different separation methods (using top-down or bottom-up proteomics) coupled to mass spectrometry by pooling "heavy" and "light" samples and measuring them together in a single run. Relative quantification is accomplished by integrating the signals and comparing the respective areas for the "heavy" and "light" species. The major advantages of this method are the compatibility to downstream processing methods, elimination of run-to-run-variations, minimization of sample-specific losses, multiplexing capability and -if label-switching [3] is used- easy identification of contaminating proteins. However, the method is also afflicted with several disadvantages such as high cost for stable-isotopes, potential introduction of artificial differences by the stable-isotope-labeled amino acids, conversion of these amino acids to other amino acids and the increased sample complexity as combining "heavy" and "light" samples doubles the amount of analytes in the sample (in case of higher multiplexing even more). Moreover, the higher complexity further increases the probability of overlapping signals in the Full-MS which are used for quantification. Thereby, potentially false quantification might occur. Systematic errors can also occur if incorporation of the "heavy" amino acids has an impact on the abundance of proteins in a particular cell line. Therefore, such an effect on the cell line at hand has to be tested before conducting subsequent SILAC-experiments.

Mainly the high cost of stable-isotope-labeling experiments (SILE) has given rise to the development of label-free quantification methods [4]. SWATH (*S*equential *W*indow *A*cquisition of all *TH*eoretical fragment-ion spectra) mass spectrometry introduced by Gillet et al. [5] is a method of particular interest as it yields a comprehensive view of the analyzed proteome and eliminates the missing-value-problem that typically afflicts information-dependent-acquisition (IDA) workflows [6]. Furthermore, as quantification is done in a selected reaction monitoring-like fashion [7], the quantitative data is obtained at the MS/MS-level and, thus, less prone to interference of fragment ions derived from different precursors. In contrast to SILE, all samples have to be measured individually (no multiplexing) and the so-called "library" has to be generated by a preferably comprehensive IDA-based analysis. Therefore, the needed measurement period on the mass spectrometer for such analyses is typically much longer than for corresponding SILAC-experiments. Furthermore, the demands for reproducibility of all upstream methods are much higher for the SWATH-MS-based approach.

The researcher has to decide which methods to apply for the task at hand. Here, we do not only provide a procedure for comparison of SILAC- and SWATH-MS-based quantification, but also

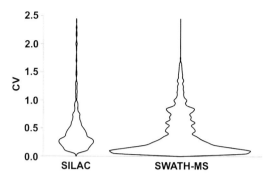

Fig. 1 Distribution of the coefficients of variation (CV) for the SILAC- and the SWATH-based quantification of the same HMB2-cell lysates. Both methods were accomplished in triplicate. Only CVs for proteins identified and quantified in all of the three SILAC-runs were included in this graph. Due to the undersampling in the SILAC-based approach, substantially fewer proteins (719) are contained in the SILAC-violin plot than in the SWATH-violin plot (2202) because the SWATH-methodology eliminates undersampling. Furthermore, the precision of the SWATH-technique (most frequent CV ~8 %) is far superior to the SILAC-method (most frequent CV ~25 %)

a protocol to test for suitability of cells for SILAC-labeling using SWATH-MS, which provides superior quantification results in comparison to typical IDA-based methods.

Our lab usually prefers SWATH-MS over SILAC-labeling for protein quantification from total cell culture lysates. This is not primarily due to the higher cost of SILE, but to the avoidance of missing-values and the superior precision of the quantification (Fig. 1). The protein regulation values from the SWATH- and the SILAC-based approach, however, are very similar (*see* **Note 1**). Therefore, the presented protocol is well suited to test for suitability of cell cultures for SILAC-labeling, e.g., for SILAC-pulse-chase experiments [8] or protein interaction studies based on co-immunoprecipitation [9] that are often poorly reproducible. The superior precision of the SWATH-MS-approach can reveal systematic regulations induced by incorporation of the stable-isotope-label, which might flaw the planned SILAC-based analysis (*see* **Note 2**).

For the testing, the respective cell line is grown on "light" and "heavy" SILAC medium at least in triplicate and analyzed using the SWATH-methodology and pair-wise t-tests for the proteins in the "light" and the "heavy" group are done. The results are corrected for multiple testing using the procedure of Benjamini and Hochberg [11]. The resulting proteins with a significant p-value are hence regulated by the incorporation of the stable-isotope-labeled amino acids.

2 Materials

2.1 Cell Culture for SILAC-Labeling

All items are given for the cell line tested in our lab and should be adapted to the respective cell line to be tested (*see* **Note 3**).

1. HMB2 cell line.
2. Dulbecco's Modified Eagle Medium (DMEM) without arginine and lysine.
3. Penicillin and streptomycin.
4. L-glutamine.
5. $^{13}C_6$,$^{15}N_2$-lysine/$^{13}C_6$,$^{15}N_4$-arginine and the respective unlabeled lysine and arginine.
6. Dialyzed fetal calf serum (FCS).
7. Phosphate buffered saline (PBS).
8. Trypsin.
9. Lysis buffer: 20 mM sodium phosphate, pH 7.4, 0.2 % SDS.
10. Sonicator.
11. Kit for determination of total protein amount or material for your method of choice.

2.2 Filter-Aided Sample Preparation [10]

1. Ultrafiltration devices (30 kDa, 500 μL).
2. 100 mM dithiothreitol (DTT).
3. 300 mM iodoacetamide (IAA).
4. Washing buffer: 8 M urea in 100 mM Tris–HCl, pH 8.5.
5. Digestion buffer: 50 mM ammonium bicarbonate, pH 7.8.
6. Trypsin Gold, mass spectrometry grade, Promega.
7. 5 % formic acid.

2.3 Nano-LC-MS/MS-Analysis

1. HPLC-solvents: A: 0.1 % formic acid; B: 0.1 % formic acid in acetonitrile.
2. Nano-HPLC-system with precolumn concentration.
3. Nano-ESI-QTOF-mass spectrometer of the TripleTOF-series.

2.4 Data Analysis

1. Protein Pilot-software (Sciex).
2. PeakView-software with SWATH-plugin (Sciex).
3. MarkerView-software (Sciex).

3 Methods

3.1 Cell Culture for SILAC-Labeling

1. Seed the cells into "light" and "heavy" SILAC DMEM media supplemented with 10 % dialyzed FCS, 400 U/mL penicillin, 50 μg/mL streptomycin, and 300 μg/mL glutamine at the required density (*see* **Note 4**).

2. Cultivate the cells for at least five cell doubling times with fresh medium every other day to ensure almost complete labeling of all proteins.

3. Check for equal growth of the cells in both media regularly.

4. Wash the cells three times with PBS before harvesting the cells.

5. Detach adherent cells using 500 μg/L trypsin.

6. Lyse the cells by sonication in 1 mL lysis buffer.

7. Determine protein concentrations by your method of choice.

3.2 Filter-Aided Sample Preparation [10] (See Note 5)

1. Add 10 μL DTT solution to 100 μg protein lysate (1 μg/μL) and incubate for 30 min.

2. Add 10 μL IAA solution and incubate at room temperature for 20 min in the dark.

3. Add 5 μL DTT solution and incubate at room temperature for 15 min.

4. Add 100 μL washing buffer and transfer the solution to an ultrafiltration device.

5. Spin through and wash three times with 100 μL washing buffer and additional three times with 100 μL digestion buffer.

6. Add 80 μL digestion buffer and 20 μL trypsin solution (0.1 μg/μL) and incubate at 37 °C over night.

7. Spin through and wash with 100 μL ultrapure water.

8. Acidify the eluate with 5 μL formic acid.

3.3 Nano-LC-MS/ MS-Analysis

HPLC settings can be changed according to the depth of the analysis that shall be achieved; in our lab 2–4 h runs are typically accomplished. The HPLC conditions are the same for the SILAC-runs with combined samples (also used for library generation) and the SWATH-runs of the individual samples (*see* **Note 6**).

1. Load 1 μg of protein digest on a 2 cm, 100 μm I.D. C18-trapping column (particle size 5 μm) with a flow rate of 5 μL/min.

2. Separate the samples on a 25 cm long 75 μm I.D. C18 column (3 μm particle size) at a flow rate of 300 nL/min using a 212 min long gradient from 4 to 40 % B.

3. Operate the mass spectrometer at a TOP25 method (0.25 s for TOF-MS from 350 to 1250 m/z, 0.1 s per MS/MS from 100 to 1500 m/z) for the SILAC-runs (*see* **Note 7**).

4. For the SWATH runs a 50 ms TOF-MS scan is followed by 37 windows of 25 m/z (100 ms per window) spanning the mass range of 350–1250 m/z.

3.4 Data Analysis

1. Search the IDA-data using Protein Pilot against the Uniprot-database.

2. Apply a 1 % false discovery rate (FDR) for the identifications.

3. Use the SWATH-plugin in the PeakView-software to generate a library and extract the quantitative SWATH-data using up to six unique peptides per protein, six transitions per peptide and a retention time window of 10 min.

4. Import the data into MarkerView and normalize to "Total Area Sum".

5. Generate two groups consisting of the "heavy" and the "light" samples, respectively.

6. Do a pairwise-t-test and correct for multiple testing, e.g., by the method of Benjamini and Hochberg [11] or Bonferroni [12].

---

4   Notes

1. Contaminating proteins, e.g., from residual fetal calf serum or faulty sample handling, are more easily spotted in the SILAC-approach using label switching which is strongly recommended in all cases. Contaminations are not regulated in opposite direction upon switch of the labels and can thus be easily identified.

2. In principle, a test for changes introduced by the SILAC-labeling can also be accomplished using a comprehensive SILAC-analysis including a label switch. However, elimination of undersampling issues by SWATH-mass spectrometry, thereby avoiding the missing value problem and the increased quantification precision for highly complex mixtures, result in a higher sensitivity of the SWATH-based approach for detection of stable-isotope-labeling-induced alterations.

3. The presented protocol uses a melanoma cell line (HMB2) and stable-isotope-coded lysine and arginine with a mass difference of 8 Da and 10 Da, respectively, for duplex labeling. It is possible to use higher multiplexing by using for example $^{13}C_6$-lysine and $^{13}C_6$-arginine for a further sample. Deuterated labels may allow even higher multiplexing than 4-plex but deuterated compounds show different retention times in reversed-phase HPLC and are therefore not recommended.

   It is also possible to use only one isotope-coded amino acid for SILAC or use other amino acids than lysine and/or arginine, e.g., leucine. However, a high degree of labeling is only obtained with amino acids that are essential or only produced in negligible amounts by the cell. One has also to keep in mind that when using only one amino acid for labeling not all peptides may be used for the quantification, resulting in less precision.

   Several SILAC-media are commercially available for different cell types but assure that they are deficient of the amino acid carrying the stable isotope-label, also assure the use of dialyzed FCS to avoid incomplete labeling. Furthermore, do not assume that proliferation rates are similar in standard medium and SILAC medium as they might be substantially different and should be determined for the SILAC media to assure sufficient labeling efficiency.

4. The cells should be cultivated in at least three replicates per label. Monitor also possible amino acid interconversions (e.g., arginine to proline) that lead to unexpected labeling of other amino acids. Also other (essential) amino acids than lysine and arginine maybe used for labeling but this type of labeling—in combination with subsequent tryptic digestion—yields a peptide mixture where virtually all peptides can be used for the quantification.

5. Alternatively, also other sample preparation protocols can be used. Gel separations for the generation of the library are also possible and can increase the size of the library enormously due to the enhanced separation. However, additional measurement time is needed for this approach.

6. We used the IDA-runs of the combined SILAC-samples for generation of the library because we wanted to compare the regulation factors obtained from the IDA- and the SWATH-measurements. The size of the library will typically increase if non-combined samples are used for library generation as the complexities of the non-combined samples are lower. Furthermore, combination of the results of different IDA-runs may increase library size. Moreover, newer versions of the PeakView-software (> V2.0) are capable of aligning retention times of the library and the SWATH-runs. Therefore, there is no need to use the same gradient for the library generation and the SWATH-measurements, although slightly wider retention time windows may be necessary in the extraction.

7. Reduction of the total mass range of the SWATH-analysis that is spanned by the different windows can be used to prolong the accumulation time per window (thereby increasing signal-to-noise-ratio) and/or reduction of duty cycle length (thereby increasing the number of points across an LC peak). However, distinct peptides will be lost if the mass range is restricted, although the vast majority of signals are in the range of 400–1000 m/z. Therefore, only very few proteins will escape quantification by these means.

Acknowledgements

This work was supported by the "Deutsche Forschungsgemeinschaft DFG" by the joint research programs "Klinische Forschergruppe 262" and "Sonderforschungsbereich 960," the Wilhelm-Sander foundation, the German Cancer Aid, and Bavarian Systems-Biology Network (BioSysNet).

References

1. Putz S, Reinders J, Reinders Y et al (2005) Mass spectrometry-based peptide quantification: applications and limitations. Expert Rev Proteomics 2(3):381–392

2. Ong SE, Blagoev B, Kratchmarova I et al (2002) Stable isotope labeling by amino acids in cell culture, SILAC, as a simple and accurate approach to expression proteomics. Mol Cell Proteomics 1(5):376–386

3. Ong S, Mann M (2006) A practical recipe for stable isotope labeling by amino acids in cell culture (SILAC). Nat Protoc 1(6):2650–2660. doi:10.1038/nprot.2006.427

4. Neilson KA, Ali NA, Muralidharan S et al (2011) Less label, more free: approaches in label-free quantitative mass spectrometry. Proteomics 11(4):535–553. doi:10.1002/pmic.201000553

5. Gillet LC, Navarro P, Tate S et al (2012) Targeted data extraction of the MS/MS spectra generated by data-independent acquisition: a new concept for consistent and accurate proteome analysis. Mol Cell Proteomics 11(6):O111.016717. doi:10.1074/mcp.O111.016717

6. Gomez YR, Gallien S, Huerta V et al (2014) Characterization of protein complexes using targeted proteomics. Curr Top Med Chem 14(3):344–350

7. Kiyonami R, Domon B (2010) Selected reaction monitoring applied to quantitative proteomics. Methods Mol Biol 658:155–166

8. Fierro-Monti I, Racle J, Hernandez C et al (2013) A novel pulse-chase SILAC strategy measures changes in protein decay and synthesis rates induced by perturbation of proteostasis with an Hsp90 inhibitor. PLoS One 8(11):e80423. doi:10.1371/journal.pone.0080423

9. Takahashi Y (2015) Co-immunoprecipitation from transfected cells. Methods Mol Biol 1278:381–389. doi:10.1007/978-1-4939-2425-7_25

10. Wisniewski JR, Zougman A, Nagaraj N et al (2009) Universal sample preparation method for proteome analysis. Nat Methods 6(5):359–362. doi:10.1038/nmeth.1322

11. Benjamini Y, Hochberg Y (1995) Controlling the false discovery rate: a practical and powerful approach to multiple testing. J Roy Statist Soc Ser B 57(1):289–300. doi:10.2307/2346101

12. Dunn OJ (1961) Multiple comparisons among means. J Am Stat Assoc 56(293):52–64. doi:10.1080/01621459.1961.10482090

Chapter 9

Combining Amine-Reactive Cross-Linkers and Photo-Reactive Amino Acids for 3D-Structure Analysis of Proteins and Protein Complexes

Philip Lössl and Andrea Sinz

Abstract

During the last 15 years, the combination of chemical cross-linking and high-resolution mass spectrometry (MS) has matured into an alternative approach for analyzing 3D-structures of proteins and protein complexes. Using the distance constraints imposed by the cross-links, models of the protein or protein complex under investigation can be created. The majority of cross-linking studies are currently conducted with homobifunctional amine-reactive cross-linkers. We extend this "traditional" cross-linking/MS strategy by adding complementary photo-cross-linking data. For this, the diazirine-containing unnatural amino acids photo-leucine and photo-methionine are incorporated into the proteins and cross-link formation is induced by UV-A irradiation. The advantage of the photo-cross-linking strategy is that it is not restricted to lysine residues and that hydrophobic regions in proteins can be targeted, which is advantageous for investigating membrane proteins. We consider the strategy of combining cross-linkers with orthogonal reactivities and distances to be ideally suited for maximizing the amount of structural information that can be gained from a cross-linking experiment.

Key words Chemical cross-linking, Photo-cross-linking, Photo-affinity labeling, Mass spectrometry, Protein 3D-structure, Protein interactions, Unnatural amino acids

1 Introduction

In the era of systems biology one of the most important tasks becomes the elucidation of protein networks in living cells. Protein–protein interactions in living cells are usually identified using affinity-based methods, such as co-immunoprecipitation or tandem-affinity purification experiments [1]. Yet, as these methods use cell lysates as starting material, the detection of false-positives resulting from a disruption of protein complexes during the lysis procedure presents a serious problem. Moreover, weakly binding substances might get lost during the washing procedures. As an alternative, chemical cross-linking can be employed to covalently fix the interaction partners. Chemical cross-linking relies on the

Jörg Reinders (ed.), *Proteomics in Systems Biology: Methods and Protocols*, Methods in Molecular Biology, vol. 1394, DOI 10.1007/978-1-4939-3341-9_9, © Springer Science+Business Media New York 2016

introduction of a covalent bond between functional groups of amino acids within a protein or between different interaction partners by a chemical reagent [2–5]. After the cross-linking reaction, the proteins of interest are usually enzymatically digested and the resulting peptide mixtures are analyzed by high-resolution mass spectrometry (MS) in a so-called "*bottom-up*" approach [6–8]. Analysis of cross-linked peptides makes use of several advantages associated with MS analysis. First, the mass of the protein or the protein complex under investigation is theoretically unlimited because proteolytic peptide mixtures are analyzed. Second, analysis is generally fast and, third, requires low amounts of protein. The greatest challenge for using the cross-linking method is posed by the high complexity of the created reaction mixtures requiring high-resolution MS techniques for analyzing the cross-linked products.

The functional groups of cross-linking reagents that are commonly used for this technique are amine-reactive *N*-hydroxysuccinimide (NHS) esters (Fig. 1). Also, photo-reactive

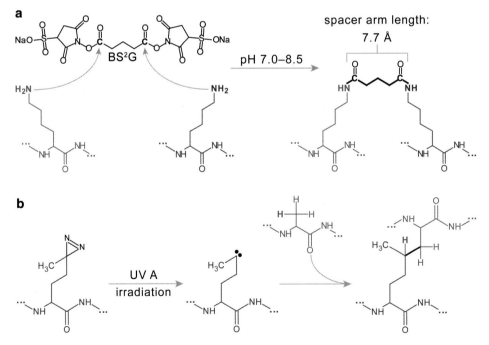

Fig. 1 Amine- and photo-reactive cross-linking of proteins. (**a**) Amine-reactive NHS esters typically connect lysine residues (shown in *blue* and *red*). NHS esters are usually employed as equimolar mixture of non-deuterated and deuterated derivatives to facilitate the identification of cross-linked products in the mass spectra based on their characteristic isotope patterns. The cross-linker BS²G (*black*) possesses a spacer length of 7.7 Å, but Cα–Cα distances of up to 29 Å can be bridged [14]. (**b**) Photo-Leu (*red-black*) as an example for photo-reactive diazirines. Upon irradiation with UV-A light (~365 nm) a reactive carbene (free electron pair shown in *black*) is created that can insert into NH or CH groups (*blue*). The carbene can also create an alkene or react with water, resulting in the formation of an alcohol [30] (*see* Fig. 3)

cross-linkers are increasingly used for photo-cross-linking (or photo-affinity labeling) experiments [9]. The reliable identification of cross-linked products is in most cases achieved by using isotope-labeled, i.e., deuterated cross-linkers, such as BS^3-*D0/D4*, BS^{2}G-*D0/D4*, or DSS-*D0/D12* [10–13], that can bridge lysines with Cα–Cα distances up to ca. 29 Å [14]. Photo-activatable cross-linkers are particularly attractive because they allow cross-linking to proceed in time- and location-specific ways. Here, a covalent linkage is created between a protein and its binding partner upon irradiation by UV-A light. The requirements for the ideal photoaffinity label include its chemical stability prior to photoactivation, its photolysis at wavelengths, which do not cause photochemical damage to the protein, as well as a high reactivity with C-H groups and with nucleophilic X-H bonds. Reproducible UV-induced labeling of target proteins is achieved by phenyl azides, diazirines, and benzophenone photophores [15, 16].

One has to discriminate between three defined goals of cross-linking studies: The first aim is to identify the complex-constituting components by peptide mass fingerprint analysis, which is employed for deriving information of protein interaction networks in cells. The second goal goes beyond a mere identification of interaction partners by pinpointing the interface regions within the protein complexes based on the detected cross-linked amino acids. The third goal is to employ the obtained distance constraint information for creating structural models of the protein complexes with computational methods, such as Rosetta [17, 18].

In this work, we extend the cross-linking tool box from the exclusive use of amine-reactive cross-linkers towards the incorporation of the photo-reactive diazirine amino acids photo-methionine (photo-Met) and photo-leucine (photo-Leu) that can deliver valuable short-distance information [19]. Combined with a mass spectrometric analysis of the created cross-links and computational modeling, detailed insights can be obtained revealing details of the interaction mechanisms between proteins. We exemplify this approach of using cross-linkers with orthogonal reactivities and distances for investigating the interactions between the basement membrane proteins nidogen-1 and laminin γ1 [20, 21].

2 Materials

Prepare all solutions using ultrapure water, i.e., Milli-Q water, and use reagents at the highest available purity. Amine-reactive chemical cross-linkers (NHS esters) should be stored dry under inert gas (helium or nitrogen) and stock solutions have to be prepared in dimethylsulfoxide (DMSO) *directly* before use. The photo-reactive diazirine-containing amino acids L-photo-methionine and L-photo-leucine can be commercially obtained from Thermo Fisher Scientific and have to be stored under light exclusion.

2.1 Chemical Cross-Linking

1. Cross-linking reagent: BS²G-*D0*, and BS²G-*D4* (Thermo Fisher Scientific; *see* **Note 1**).

2. Neat DMSO (*see* **Note 2**).

3. Cross-linking buffer: 20 mM HEPES, pH 7.4 (*see* **Notes 3** and **4**).

4. Quenching solution: 1 M ammonium bicarbonate.

5. Recombinant proteins are prepared in cross-linking buffer, the optimum protein concentration should be between 1 and 10 μM (*see* **Note 4**).

2.2 Incorporation of Photo-reactive Amino Acids into Proteins in HEK 293 Cells

1. Cell culture dishes and Triple Flasks (Thermo Fisher Scientific).

2. Cell culture medium: Dulbecco's Modified Eagle Medium DMEM/F12 (Gibco/Life Technologies), 10 % (v/v) FCS, 1 % (v/v) penicillin/streptomycin solution and 1 μg/μL puromycin.

3. Dulbecco's Modified Eagle Limiting Medium DMEM-LM, leucine- and methionine-depleted (Gibco/Life Technologies).

4. Fetal calf serum (FCS, Biochrom).

5. Penicillin/streptomycin solution.

6. Puromycin.

7. Phosphate-buffered saline (PBS): 140 mM NaCl, 10 mM KCl, 8 mM Na_2HPO_4, 2 mM $KHPO_4$, pH 7.4.

8. 0.05 % (w/v) trypsin–EDTA solution (T/E).

9. 2 M Tris–HCl, pH 8.0.

10. L-photo-leucine and L-photo-methionine (Thermo Fisher Scientific).

11. HEK293 EBNA cells stably transfected with the protein-coding DNA sequence (*see* **Note 9**).

2.3 Photochemical Cross-Linking

After incorporation of the photo-reactive amino acids photo-Met and photo-Leu, we conduct the photo-cross-linking reactions in a home-built UV irradiation chamber [22]. The chamber contains fluorescence lamps (40 W; emission spectrum between 305 and 420 nm, emission maximum at 360 nm (CLEO Performance R UV-A-fluorescence lamp, Phillips)). The UV-A sensor for quantifying the irradiation energy was purchased from Kühnast (*see* **Note 7**).

2.4 SDS–Polyacrylamide Gel Electrophoresis (SDS-PAGE)

1. Milli-Q H_2O.

2. Mini-PROTEAN vertical gel electrophoresis system (Bio-Rad).

3. Precise™ Protein Gel 4–20 % acrylamide (Thermo Fisher Scientific).

4. Running buffer: 100 mM Tris–HCl, 100 mM HEPES, 0.1 % SDS, pH 8.0.

5. Laemmli buffer and 10 % (w/v) SDS solution (store at room temperature, both Bio-Rad).

6. Ammonium persulfate (Sigma): Prepare a 10 % (w/v) solution in Milli-Q H_2O and immediately freeze in single-use (100 μL) aliquots at –20 °C.

7. Page Ruler Protein Ladder (Fermentas).

8. Colloidal Coomassie staining solution: Solution A: 5 % (w/v) Coomassie Brilliant Blue G250 suspended in Milli-Q H_2O. Solution B: 2 % (w/v) ortho-phosphoric acid, 10 % (w/v) ammonium sulfate, dissolved in Milli-Q H_2O.

9. Fixing solution: 40 % (v/v) methanol, 10 % (v/v) acetic acid.

2.5 Proteolytic In-Gel Digestion

1. Porcinetrypsin sequencing grade (Promega). Storage: dissolve 25 μg lyophilized trypsin in 50 μL 1 mM HCl, prepare aliquots containing trypsin (1 μg/2 μL), quick-freeze in liquid nitrogen and store at –80 °C.

2. GluC sequencing grade (Promega, *see* **Note 14**).

3. Chymotrypsin sequencing grade (Promega, *see* **Note 14**).

4. Acetonitrile (ACN) Hypersolv gradient grade (VWR, *see* **Note 15**).

5. Destaining solutions: ACN, 50 % (v/v) ACN in 100 mM ammonium bicarbonate (Sigma) (*see* **Note 15**).

6. Reducing buffer: freshly prepared 10 mM dithiothreitol (DTT) in 100 mM ammonium bicarbonate.

7. Alkylating buffer: freshly prepared 55 mM iodoacetamide (Sigma) in 100 mM ammonium bicarbonate.

8. Digestion buffer, freshly prepared: Dissolve one trypsin or GluC aliquot in 80 μL 50 mM ammonium bicarbonate to obtain a solution of 13 ng/μL, store on ice.

9. Extraction solution: 5 % (v/v) trifluoroacetic acid (TFA, Sigma)–ACN 1:2 (v/v).

2.6 Proteolytic In-Solution Digestion

1. Porcine trypsin sequencing grade (Promega). Storage: dissolve 25 μg lyophilized trypsin in 50 μL 1 mM HCl, prepare aliquots containing 1 μg/2 μL trypsin, quick-freeze in liquid nitrogen and store at –80 °C.

2. GluC sequencing grade (Promega, *see* **Note 14**).

3. Chymotrypsin sequencing grade (Promega, *see* **Note 14**).

4. Denaturing solution: 1.6 M urea in 320 mM ammonium bicarbonate.

5. Reducing buffer: freshly prepared 45 mM DTT in 400 mM ammonium bicarbonate.

6. Alkylating buffer: freshly prepared 100 mM iodoacetamide (Sigma) in 400 mM ammonium bicarbonate.

7. TFA (Sigma).

8. Digestion buffer, freshly prepared: dissolve one trypsin, GluC or chymotrypsin aliquot in 80 μL of 50 mM ammonium bicarbonate to obtain a solution of 13 ng/μL, store on ice.

2.7 Liquid Chromatography/Mass Spectrometry (LC/MS)

In our lab, we use the following LC/MS setup, consisting of an Ultimate 3000 RSLC nano-HPLC system (Dionex) that is directly coupled to the nano-ESI source (EASY Spray source, Thermo Fisher Scientific) of an Orbitrap Fusion Tribrid mass spectrometer (Thermo Fisher Scientific) [23]. Alternatively, an LTQ-Orbitrap XL mass spectrometer (Thermo Fisher Scientific) equipped with nano-ESI source (Proxeon) is employed. Other nano-HPLC systems or mass spectrometers, such as FTICR or Q-TOF instruments equipped with a nano-ESI source, can be used as well.

2.8 Software for Data Acquisition and Analysis

1. Perform MS and MS/MS data acquisition and data analysis with XCalibur 3.0 in combination with DCMS link 2.14.

2. Analyze cross-linked products with StavroX 3.4.5 (www. StavroX.com) [24] that allows an automatic analysis and annotation of MS and MS/MS data.

3. For obtaining optimum results, the following settings should be used: Maximum mass deviation of 3 ppm between calculated and experimental precursor masses and signal-to-noise ratio of ≥2.

4. Consider primary amino groups (Lys side chains and N-termini) as well as hydroxyl groups (Thr, Ser, Tyr) as potential cross-linking sites for BS²G [25, 26].

5. Consider all 20 amino acid residues as cross-linking sites for UV-A-induced cross-linking of photo-Met and photo-Leu.

6. Oxidation of Met should be set as variable modification for all cross-linked proteins.

7. Include carbamidomethylation as fixed modification for Cys.

8. Consider two missed cleavage sites for each amino acid (cleavage sites: Lys and Arg for trypsin; Tyr, Trp, and Phe for chymotrypsin; Glu and Asp for GluC).

3   Methods

3.1 Chemical Cross-Linking

1. Prepare a mixture of your proteins of interest at a specific molar ratio (final concentrations ranging from 1 to 10 μM each) in the cross-linking buffer (e.g., 20 mM HEPES). For equilibration of the protein complex, the mixture is incubated for ca. 15 min at room temperature using an incubation shaker. For each cross-linking condition, a fresh Eppendorf tube should be used. The minimum final volume should be 100 μL per reaction condition.

2. Freshly prepare cross-linker stock solutions (200 mM) in neat DMSO (*see* **Note 2**).

3. Add 1:1 mixtures of non-deuterated (*D0*) and four times deuterated (*D4*) BS²G in 20- to 200-fold molar excess of cross-linker over the protein. Allow the reaction to proceed between 5 and 60 min (*see* **Notes 5** and **6**).

4. For reaction quenching, add 1 M ammonium bicarbonate solution (20 mM final concentration).

5. For storage, quick-freeze the samples in liquid nitrogen and keep at –20 ° C.

3.2 Incorporation of Photo-reactive Amino Acids into Proteins

All steps that require manual handling of cultured cells must be conducted under a laminar flow box. Solutions and materials have to be sterile-packaged or autoclaved prior to use.

1. Add HEK293 EBNA cells to 10 mL of cell culture medium.

2. Culture the cells on a 57 cm² cell culture dish in an incubator with a water-saturated 5 % CO_2 atmosphere at 37 °C.

3. Exchange the medium when the contained pH indicator turns yellow indicating acidification.

4. Monitor the cell growth every other day by checking the cell confluence under a microscope. Once a confluence of 80 % is reached, passage the cells to new dishes:

   (a) Wash the cells with 5 mL PBS.

   (b) Add 1.5 mL T/E and incubate for 5 min at 37 °C to cleave off the cells from the plate.

   (c) Stop proteolysis by adding 8.5 mL of cell culture medium.

   (d) Pellet the cells by centrifuging them for 5 min at $500 \times g$ in a 15-mL Falcon tube.

   (e) Resuspend the cell pellet in 5 mL fresh cell culture medium.

   (f) Distribute the suspension on fresh cell culture dishes.

5. When the cells reach 80 % confluence again, replace the cell culture medium with FCS-free medium (containing only DMEM/F12, penicillin/streptomycin, and puromycin) and incubate them for another 24 h (*see* **Note 10**).

6. Remove the FCS-free medium and add 10 mL of DMEM-LM supplemented with 4 mM L-photo-Leu and 2 mM L-photo-Met (*see* **Notes 8** and **10**).

7. Replace the medium after 2 days, storing the first batch of supernatant at –20 °C.

8. Collect the second batch after 2 days of incubation.

9. Equilibrate the collected medium by adding Tris–HCl buffer (pH 8) to a final concentration of 12 mM.

10. Remove cells and other insoluble components by centrifuging the equilibrated medium for 20 min at $14,000 \times g$ and passing the supernatant through a folded filter. Both steps should be performed at 4 °C.

11. Add NaN_3 to a final concentration of 0.02 % (w/v) to prevent microbial growth in the supernatant.

12. Proceed with the chromatographic purification of the overexpressed protein (*see* **Note 11**).

13. Perform *in-gel* or *in-solution* digestion of the purified protein (according to Subheading 3.4 or 3.5 or 3.6) and analyze the peptide mixtures by LC/MS and LC/MS/MS. Estimate the incorporation rate of the photo-reactive amino acids by comparing the MS intensities of the respective peptide signals containing Leu/photo-Leu and Met/photo-Met (Fig. 2).

3.3 Photochemical Cross-Linking

1. Prepare a mixture of your proteins of interest as described in Subheading 3.1, **step 1**. Samples should be shielded from ambient light during incubation (*see* **Note 8**).

2. Transfer the reaction mixture into a transparent tube and expose it to 8000 mJ/cm^2 UV-A irradiation (*see* **Notes 7** and **12**).

3. For storage, quick-freeze the samples in liquid nitrogen and keep at –20 °C.

3.4 SDS-PAGE

1. Load your samples on a Precise™ Protein Gel 4–20 % acrylamide gradient gel.

2. Perform the gel electrophoresis at a constant current of 25 mA.

3. After the electrophoresis is finished, keep the gel in fixing solution for 1 h.

4. Rinse the gel two times with Milli-Q H_2O.

5. Stain the gel overnight in colloidal Coomassie staining solution.

6. Destain the gel by shaking it in Milli-Q H_2O.

7. Inspect the gel for signals that might represent the cross-linked protein complex (Fig. 3).

3.5 In-Gel Digestion

Working under a laminar flow box wearing gloves and sleeve protectors is recommended in order to avoid keratin and dust contamination of samples. The laminar flow box should include an incubation shaker suitable for Eppendorf tubes.

3.5.1 Excise Protein Bands from the SDS Gel

1. Wash the Coomassie-stained gel twice for 10 min with Milli-Q H_2O on a shaker.

2. Excise gel bands of interest using a sterile scalpel and transfer gel bands to a clean glass plate. Cut the gel band into cubes of ca. 1 mm^3 and transfer them into an Eppendorf tube.

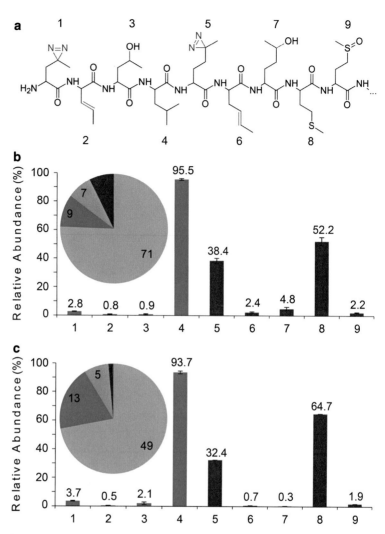

Fig. 2 Incorporation of photo-reactive amino acids into nidogen-1 and laminin γ1. (**a**) Met and Leu variants that were considered during MS analysis, including the reaction products of the diazirine amino acids photo-Met and photo-Leu (1: photo-Leu, 5: photo-Met, 2 and 6: alkene; 3 and 7: alcohol; 4: unmodified Leu; 8 and 9: unmodified and oxidized Met); (**b** and **c**) MS-based quantification of photo-reactive amino acid incorporation. The pie charts show the number of leucines (*blue*) and methionines (*red*) within nidogen-1 (**b**) and laminin γ1 short arm (**c**) that remained unmodified (*light shades*) or were partially replaced by their photo-reactive counterparts (*dark shades*). The bars represent the relative abundance of isoforms of the partially modified peptides, containing the Leu (*dark blue*) and Met (*dark red*) variants listed in (**a**)

3.5.2   Reduce and Alkylate Proteins

1. Add 100 μL Milli-Q H$_2$O to the gel pieces, shake for 10 min.

2. Add 500 μL ACN. Shake for 5 min and discard the liquid.

3. Add 50 μL reduction buffer. Incubate at 56 °C for 30 min, then let the solution cool down to room temperature.

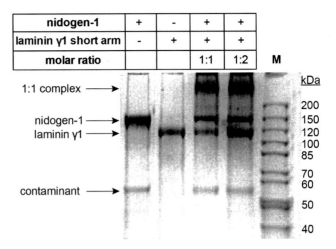

nidogen-1	+	-	+	+	
laminin γ1 short arm	-	+	+	+	
molar ratio			1:1	1:2	M

Fig. 3 SDS-PAGE analysis of nidogen-1 and laminin γ1 short arm after cross-linking with BS²G. Nidogen-1 and laminin γ1 short arm were cross-linked both separately and after co-incubation at molar ratios of 1:1 and 1:2. *M* molecular weight marker

4. Add 500 μL ACN. Shake for 5 min and discard the liquid.

5. Add 50 μL alkylation buffer. Incubate at room temperature in the dark for 30 min.

6. Add 500 μL ACN. Shake for 5 min and discard the liquid.

3.5.3 Wash and Destain Gel Pieces

1. Add 100 μL ACN/100 mM ammonium bicarbonate (1:1), shake for 10–30 min to destain the gel pieces.

2. Add 500 μL ACN. Shake for 5 min and discard the liquid.

3. Repeat **steps 1** and **2** if necessary.

3.5.4 In-Gel Digestion

1. Bring the gel pieces to complete dryness before adding the digestion buffer (protease in ammonium bicarbonate).

2. Per gel band, add 25 μL trypsin digestion buffer and incubate gel pieces for 1 h on ice. In case a double digestion is performed, add GluC (or another protease) first and incubate the gel pieces for 1 h on ice, then add trypsin and incubate for another 1 h on ice.

3. Add 50 mM ammonium bicarbonate. The gel pieces have to be covered completely.

4. Incubate overnight at 37 °C.

3.5.5 Extract Peptides

1. Stop the digestion by adding the double volume of extraction solution.

2. Shake for 15 min at 37 °C, remove the supernatant, and transfer it to a new Eppendorf tube.

3. Repeat **steps 1** and **2** if the gel band is expected to contain a low amount of protein.

4. Concentrate supernatants in a vacuum concentrator to a volume between 60 and 120 μL. Do not concentrate to complete dryness in order to avoid sample loss.

5. Peptide mixtures are ready to be analyzed by nano-HPLC/ nano-ESI-Orbitrap mass spectrometry. Before MS analysis, samples may be stored at –20 °C.

3.6 In-Solution Digestion

1. Incubate proteins for 1 h with a 20-fold excess (v/v) of pre-cooled acetone to precipitate them from solution.

3.6.1 Denature Proteins

2. Centrifuge the sample for 30 min at $15,000 \times g$ (4 °C).

3. Let the pellet dry at room temperature.

4. Solubilize the pellet in 25 μL denaturing solution.

3.6.2 Reduce and Alkylate Proteins

1. Add 7.5 μL of reduction buffer. Incubate at 50 °C for 15 min, then allow the solution to cool down to room temperature.

2. Add 5 μL alkylation buffer. Incubate at room temperature in the dark for 15 min.

3.6.3 In-Solution Digestion

1. Dilute the sample with 60 μL Milli-Q H_2O to avoid denaturation of the proteases by urea.

2. Add trypsin digestion buffer to an enzyme/protein ratio of 1:50 (v/v) and incubate the sample at 37 °C for 2 h. In case a double digestion is performed, add GluC (or another protease) first and incubate for 1 h at room temperature, then add trypsin and incubate at 37 °C for 2 h.

3. Acidify the sample to a pH value below 4 (check with pH paper) and concentrate it in a vacuum concentrator to a volume between 60 and 120 μL.

4. Peptide mixtures are ready to be analyzed by nano-HPLC/ nano-ESI-Orbitrap mass spectrometry. Before MS analysis, samples may be stored at –20 °C.

3.7 Liquid Chromatography/Mass Spectrometry (LC/MS)

1. For LC/MS analysis of the cross-linked peptide mixtures, we use the nano-HPLC/nano-ESI-Orbitrap-MS setup described in Subheading 2.7.

2. Separate the enzymatically digested cross-linked peptide mixture by reversed-phase LC. For this, inject the peptide mixture via an autosampler and load them onto a precolumn (Acclaim PepMap, RP C18, 5 mm × 300 μm, 5 μm, 100 Å).

3. Concentrate and desalt samples during a 15-min washing step with 0.1 % TFA.

4. Separate the peptides by gradient elution (1 % to 40 % B; A: 0.1 % formic acid (FA), B: ACN, 0.08 % FA) over 90 min using a PepMap RSLC C18 column (250 mm × 75 μm, 2 μm, 100 Å) at a flow rate of 300 nL/min.

5. Introduce the peptides into the nano-ESI source (EASY Spray source) of the mass spectrometer. Use a fused-silica emitter at a source voltage of 1.9 kV and set the temperature of the transfer capillary to 275 °C.

6. Record the mass spectra automatically in data-dependent MS/MS mode during gradient elution (Orbitrap Fusion instrument): For each 5-s cycle, one mass spectrum ($R = 120,000$ at m/z 200) is recorded in the m/z range 350–1500 in the Orbitrap analyzer. The most abundant species are isolated in the quadrupole (isolation window 2 Th or larger in case deuterated amine-reactive cross-linkers, such as BS²G-*D0/D4* are used) and fragmented with collision-induced dissociation (CID) at normalized collision energies of 25 % and 35 %. Detection of the fragment ions is performed in the Orbitrap ($R = 15,000$ at m/z 200) using dynamic exclusion for 90 s.

3.8 Data Analysis

Data analysis of cross-linked peptides is performed with the StavroX software allowing lysines and protein N-termini as reaction sites of NHS-ester based cross-linkers (Fig. 1). As NHS-esters have also been found to react with serines, threonines, and tyrosines, these reaction sites should also be taken into account [25, 26]. For photo-cross-links induced by diazirines, all 20 amino acids were considered as potential reaction sites (*see* **Note 13**, Fig. 1).

Please note that trypsin will cleave at modified lysines with low frequency, therefore the number of missed cleavage sites has to be set to a higher value for cross-link identification. Filtering of putative cross-linked candidates is facilitated by the application of isotope-labeled *D0/D4* cross-linkers, which generate a characteristic isotope pattern for cross-linked peptides. To identify this isotope pattern, experimental mass lists obtained by high-resolution mass spectrometry are filtered for monoisotopic deconvoluted masses (using the Proteome Discoverer software, Thermo Fisher Scientific) that show a mass difference of 4 amu. This reduced experimental mass list is then compared to a theoretical mass list of all possibly cross-linked peptides, allowing a maximum mass deviation of 3 ppm. This allows automated identification of potentially cross-linked peptides, for which MS/MS data have to be evaluated. Both the *D0* and the *D4* cross-linked species have to be isolated and fragmented, so the MS/MS data can also be checked for the characteristic isotope patterns (*see* Subheading 3.7, step 6). To further automate the MS/MS-based identification of cross-linked species, the StavroX software calculates the corresponding theoretical fragment ions and compares them to the experimental mass spectra (Fig. 4). Subsequently, StavroX assigns

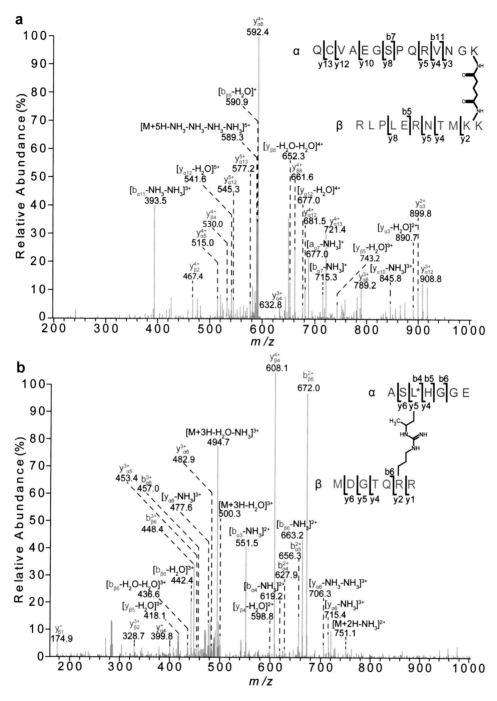

Fig. 4 Reaction products of amine- and photo-reactive cross-linking analyzed by nano-HPLC/nano-ESI-LTQ-Orbitrap MS/MS. (**a**) Two nidogen-1 peptides cross-linked by BS^{2}G. The reaction product comprises the amino acids 407–420 (α-peptide, *red*) and 939–949 (β-peptide, *blue*), in which Lys-407 is connected to Lys-948/949. (**b**) A photo-Leu (L*)-containing laminin γ1 peptide (amino acids 988–994, *red*) cross-linked to a nidogen-1 peptide (amino acids 1033–1039, *blue*). Based on the detected b- and y-type ions that are created by CID, the exact site of cross-linking can be mapped to Arg-1038 in nidogen-1

a score for cross-linked products based on the quality of MS/MS spectra and calculates false discovery rates (FDR). Nevertheless, it is recommended to validate all MS and MS/MS data manually.

3.9 Use Cross-Links as Distance Constraints for Rosetta-Based Computational Modeling

1. Make sure that the numbering of the identified cross-links complies with the atom numbering in the PDB files and/or amino acid sequences used as input for the modeling process (*see* **Note 16**).

2. List all identified cross-links in a tab-separated constraint file, as exemplified in Table 1 (*see* **Note 17**). Constraints obtained from different cross-linking experiments may be listed in the same file.

3. Enforce the specified distance constraints by the command line flag "*-constraints:cst_fa_file</path/to/constraint/file>*", if working with high-resolution models where all amino acid side chain atoms are modeled, or "*-constraints:cst_file</path/to/constraint/file>*", if working with low-resolution models where amino acid side chains are represented by a centroid sphere.

Table 1
Overview of cross-links in the structurally solved region of the nidogen-1/laminin γ1 short arm complex. Cα–Cα distances of cross-linked residues indicated in the crystal structure (PDB entry 1NPE) and in the best-scoring models created by Rosetta

Cross-linked lysines	Cα–Cα distances (Å)		
	1NPE	Model (best atom-pair constraint score)	Model (best total score)
K-948 × K-953	10.4	10.9	11.1
K-1128 × K-1165	13.3	12.3	16.0
K-1072 × K-1128	16.7	19.1	16.2
K-948 × K-1144	17.9	16.4	17.6
K-850 (laminin) × K-1072 (nidogen-1)	20.9	17.5	16.9
K-948 × K-1152	22.2	21.2	22.2
K-1032 × K-1072	27.1	27.1	27.0
K-961 × K-1072	28.7	28.0	28.2
K-864 (laminin) × K-1152 (nidogen-1)	32.2	22.4	27.1
K-850 (laminin) × K-953 (nidogen-1)	33.0	29.5	29.4
K-1032 × K-1152	35.8	35.4	35.4
Photo-L-990 × R-1038	24.7	23.4	23.5
Photo-L-844 (laminin) × K-1072 (nidogen-1)	33.8	19.4	20.8

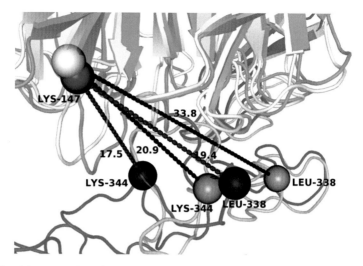

Fig. 5 Incorporation of cross-linking distance constraints into X-ray structures using the Rosetta Relax application. Zoom-in on the high-affinity binding region of nidogen-1 and laminin γ1 short arm (PDB entry 1NPE). Two cross-links (*black dashed lines*), one detected with BS²G (Cα–Cα distance 20.9 Å) and one detected after photochemical cross-linking (Cα–Cα distance 33.8 Å), were mapped in the crystal structure (*gray cartoon representation*). The distance constraint imposed by the photochemical cross-link is clearly violated. In the Rosetta model (*blue cartoon representation*), which was built based on the crystal structure and the experimentally identified cross-links, the Cα–Cα distances of the cross-linked residues are substantially reduced (*red dashed lines*), while secondary structure elements and disulfide bridges are retained as indicated in the crystal structure. Thus, the model reflects a plausible *in solution* conformation of the protein complex (*see* also Table 1)

4. Set the impact of constraint violation on the total score of the created models by means of the command line flags "*-constraints:cst_fa_weight< number>*" or "*-constraints:cst_ weight<number>*", respectively (*see* **Note 18**).

5. Analyze the created models to check for steric clashes and constraint violations. If necessary, adjust the constraint weight (*see* **Note 18**, Fig. 5).

4 Notes

1. NHS ester cross-linkers are highly sensitive to hydrolysis. Therefore, aliquots of the cross-linkers should be prepared and stored in a desiccator under inert gas, e.g., helium or nitrogen.

2. Neat DMSO is required to prepare the NHS cross-linker stock solutions *immediately* before use in order to prevent hydrolysis of the reagents.

3. The pH value of the protein solution should range between ca. 7.0 and 8.5. Please note that an increased pH value will increase the reactivity of amine groups towards NHS esters, but pH values higher than 8.5 should be avoided due to potential stability problems of certain proteins.

4. Tris buffers interfere with amine-reactive cross-linking reagents and should be avoided. Please note that amine-reactive NHS esters also react with hydroxyl groups of serines, threonines, and tyrosines, albeit with a lower frequency compared to amine groups [25, 26]. Considering these amino acids as potential cross-linking sites will increase the amount of information you get from your cross-linking experiment.

5. Using equimolar mixtures of non-deuterated (*D0*) and deuterated (*D4*) cross-linking reagents facilitates the identification of cross-linking products in mass spectra by characteristic mass shifts of 4 amu. Please note that deuterated and non-deuterated species exhibit different retention times in reversed phase-LC, so they may not coelute.

6. For chemical cross-linking with NHS esters, the following parameters are recommended to be optimized for obtaining maximum yields of cross-linked products:

   (a) Molar excess of cross-linker (20- to 200-fold).

   (b) Cross-linking time: Allow the cross-linking reaction to proceed between 5 and 60 min at room temperature. If the proteins of interest are not sufficiently stable at room temperature, conduct the cross-linking reaction at 4 °C for 2 h.

7. For photochemical cross-linking, a UV lamp should be used with a filter blocking wavelengths lower than ca. 300 nm in order to avoid photolytic damage of the protein. During UV-A irradiation, the protein solution should be kept on ice.

8. Avoid the exposure of the samples to ambient light when conducting the photo-cross-linking reactions, i.e., by using light-protected reaction tubes or by wrapping aluminum foil around the reaction tubes. Also, protein production and purification should be conducted under low-light conditions.

9. Protein constructs must comprise an export sequence to facilitate their secretion into the cell culture medium.

10. Incorporation of photo-reactive amino acids is straightforward since it is sufficient to add them to the leucine- and methionine-depleted cell culture media [27]. While the incorporation efficiency is acceptable for photo-Met, yields are comparably low for photo-Leu (Fig. 2). To increase the incorporation efficiency, cells can be temporarily cultured in FCS-free medium before transferring them to photo-reactive amino acid-containing medium, thereby ensuring full depletion of Met

and Leu. However, a prolonged exposure to FCS-free medium will negatively affect the cell survival rate.

11. To the best of our knowledge, all types of affinity purification are compatible with this protocol. Details of the protein purification are not described herein as the purification procedure applied varies for the respective protein construct.

12. The optimum UV irradiation energy might be different for other protein systems. We recommend to optimize the procedure by testing different exposure times, especially when the irradiation device is not equipped with a UV-A sensor.

13. Analysis of the cross-linking datasets with the StavroX software [24] can become quite time-consuming—especially when using photo-reactive amino acids—as a high number of amino acids have to be considered as potential reaction sites.

14. Trypsin cleaves with lower frequency at lysines that are modified by cross-linkers. Therefore, a second protease is required to achieve high proteolytic digestion yields, e.g., GluC (cleaves C-terminally of glutamic and also aspartic acid). As an alternative to trypsin, chymotrypsin (cleaves C-terminally of large and hydrophobic amino acids) might be used, which is especially beneficial for membrane proteins.

15. Organic solvents must not be kept in plastics due to a potential contamination by polymers.

16. Constraint files are compatible with multiple Rosetta applications allowing, for example, protein-protein docking, comparative modeling, and structural optimization based on cross-linking distance constraints (Fig. 5, Table 1). It is beyond the scope of this protocol to describe these procedures in detail. A protocol on Rosetta-based comparative modeling has been published recently [28]. The incorporation of cross-linking constraints into Rosetta workflows has been described elsewhere, including detailed command line flags [18, 19, 29]. For further information on the installation and usage of Rosetta, please consult https://www.rosettacommons.org/docs/latest/.

17. Atom-pair constraints allow specifying experimentally found distances between two atoms, e.g., $C\alpha$–$C\alpha$ distances derived from cross-linking data. Models violating these constraints will be penalized according to the applied function, i.e., their total energy score will increase. The "flat_harmonic" function renders an energy penalty if the modeled $C\alpha$–$C\alpha$ distance exceeds the sum of x_0 and the granted tolerance (Table 2). The respective penalty calculates to

$$\text{atom-pair constraint score} = \frac{C_\alpha - C_\alpha \text{distance} - x_0 - \text{tolerance}}{\text{standard deviation}}$$

Table 2
Information to be included in a Rosetta constraint file. The file should be tab-separated and must not include any column headings. All listed information has to be in accordance with the input PDB file

Constraint type	Atom name 1	Res no 1/ chain ID1	Atom name 2	Res no. 2/ chain ID2	Function type	x_0	SD	Tolerance
Atom pair	CA	1A	CA	1B	FLAT_HARMONIC	20.3	1.0	5.7

Res no residue number, *chain ID* chain identifier, *x0 + tolerance* maximum Cα–Cα distance, *SD* standard deviation

18. Increasing the constraint weight and decreasing the standard deviation will lead to a higher energy penalty for constraint violation, favoring models that fulfill all experimental constraints. Careful optimization of both settings is recommended since too harsh penalties entail the danger that realistic models, which marginally exceed the distance limits, are overseen during analysis. Of note, forcing Rosetta to comply with cross-links that turn out to be sterically impossible may lead to a distortion of the model and a loss in secondary structure [19].

Acknowledgements

This work is funded by the BMBF (Pronet-T3). We would like to thank Dr. Christian Ihling for LC/MS analyses, Dr. Knut Kölbel for fruitful discussions, and Dirk Tänzler for excellent technical support. Prof. Jens Meiler and Dr. David Nanneman, Vanderbilt University, are acknowledged for introducing P.L. into Rosetta, Prof. Gunter Fischer and Dr. Cordelia Schiene-Fischer, Max-Planck Forschungsstelle Halle, are acknowledged for generously providing their cell culture facilities. We are indebted to Prof. Mats Paulsson and Dr. Frank Zaucke; University of Cologne, for their help with laminin and nidogen expression and purification.

References

1. Gavin A, Bösche M, Krause R et al (2002) Functional organization of the yeast proteome by systematic analysis of protein complexes. Nature 415(6868):141–147. doi:10.1038/415141a

2. Petrotchenko EV, Borchers CH (2010) Crosslinking combined with mass spectrometry for structural proteomics. Mass Spectrom Rev 29(6):862–876. doi:10.1002/mas.20293

3. Sinz A (2014) The advancement of chemical cross-linking and mass spectrometry: from single proteins to protein interaction networks. Expert Rev Proteomics 11:733–743

4. Tang X, Bruce JE (2010) A new cross-linking strategy: protein interaction reporter (PIR) technology for protein-protein interaction studies. Mol Biosyst 6(6):939–947. doi:10.1039/b920876c

5. Calabrese AN, Pukala TL (2013) Chemical cross-linking and mass spectrometry for the structural analysis of protein assemblies. Aust J Chem 66(7):749. doi:10.1071/CH13164

6. Kalisman N, Adams CM, Levitt M (2012) Subunit order of eukaryotic TRiC/CCT chaperonin by cross-linking, mass spectrometry, and combinatorial homology modeling. Proc

Natl Acad Sci U S A 109(8):2884–2889. doi:10.1073/pnas.1119472109

7. Chen ZA, Jawhari A, Fischer L et al (2010) Architecture of the RNA polymerase II-TFIIF complex revealed by cross-linking and mass spectrometry. EMBO J 29(4):717–726. doi:10.1038/emboj.2009.401

8. Leitner A, Joachimiak LA, Bracher A et al (2012) The molecular architecture of the eukaryotic chaperonin TRiC/CCT. Structure 20(5):814–825. doi:10.1016/j.str.2012.03.007

9. Tanaka Y, Bond MR, Kohler JJ (2008) Photocrosslinkers illuminate interactions in living cells. Mol Biosyst 4(6):473–480. doi:10.1039/b803218a

10. Müller DR, Schindler P, Towbin H et al (2001) Isotope-tagged cross-linking reagents. A new tool in mass spectrometric protein interaction analysis. Anal Chem 73(9):1927–1934

11. Rinner O, Seebacher J, Walzthoeni T et al (2008) Identification of cross-linked peptides from large sequence databases. Nat Methods 5(4):315–318. doi:10.1038/nmeth.1192

12. Schmidt A, Kalkhof S, Ihling C et al (2005) Mapping protein interfaces by chemical cross-linking and Fourier transform ion cyclotron resonance mass spectrometry: application to a calmodulin/adenylyl cyclase 8 peptide complex. Eur J Mass Spectrom (Chichester, Eng) 11(5):525–534. doi:10.1255/ejms.748

13. Schulz DM, Kalkhof S, Schmidt A et al (2007) Annexin A2/P11 interaction: new insights into annexin A2 tetramer structure by chemical crosslinking, high-resolution mass spectrometry, and computational modeling. Proteins 69(2):254–269. doi:10.1002/prot.21445

14. Merkley ED, Rysavy S, Kahraman A et al (2014) Distance restraints from crosslinking mass spectrometry: mining a molecular dynamics simulation database to evaluate lysine-lysine distances. Protein Sci 23(6):747–759. doi:10.1002/pro.2458

15. Brunner J (1993) New photolabeling and cross-linking methods. Annu Rev Biochem 62:483–514. doi:10.1146/annurev.bi.62.070193.002411

16. Weber PJ, Beck-Sickinger AG (1997) Comparison of the photochemical behavior of four different photoactivatable probes. J Pept Res 49(5):375–383

17. Doberenz C, Zorn M, Falke D et al (2014) Pyruvate formate-lyase interacts directly with the formate channel FocA to regulate formate translocation. J Mol Biol 426(15):2827–2839. doi:10.1016/j.jmb.2014.05.023

18. Herzog F, Kahraman A, Boehringer D et al (2012) Structural probing of a protein phosphatase 2A network by chemical cross-linking and mass spectrometry. Science 337(6100):1348–1352. doi:10.1126/science.1221483

19. Lössl P, Kölbel K, Tänzler D et al (2014) Analysis of nidogen-1/laminin γ1 interaction by cross-linking, mass spectrometry, and computational modeling reveals multiple binding modes. PLoS One 9:e112886

20. Paulsson M, Aumailley M, Deutzmann R et al (1987) Laminin-nidogen complex. Extraction with chelating agents and structural characterization. Eur J Biochem 166(1):11–19

21. Takagi J, Yang Y, Liu J et al (2003) Complex between nidogen and laminin fragments reveals a paradigmatic beta-propeller interface. Nature 424(6951):969–974. doi:10.1038/nature01873

22. Schaks S, Maucher D, Ihling CH et al (2012) Investigation of a calmodulin/peptide complex by chemical cross-linking and high-resolution mass spectrometry. In: Koenig S (ed) Biomacromolecular mass spectrometry: tips from the bench. Nova Science, Hauppauge, NY, pp 1–18

23. Senko MW, Remes PM, Canterbury JD et al (2013) Novel parallelized quadrupole/linear ion trap/Orbitrap tribrid mass spectrometer improving proteome coverage and peptide identification rates. Anal Chem 85(24):11710–11714. doi:10.1021/ac403115c

24. Götze M, Pettelkau J, Schaks S et al (2012) StavroX—a software for analyzing crosslinked products in protein interaction studies. J Am Soc Mass Spectrom 23(1):76–87. doi:10.1007/s13361-011-0261-2

25. Mädler S, Bich C, Touboul D et al (2009) Chemical cross-linking with NHS esters: a systematic study on amino acid reactivities. J Mass Spectrom 44(5):694–706. doi:10.1002/jms.1544

26. Kalkhof S, Sinz A (2008) Chances and pitfalls of chemical cross-linking with amine-reactive N-hydroxysuccinimide esters. Anal Bioanal Chem 392(1–2):305–312

27. Suchanek M, Radzikowska A, Thiele C (2005) Photo-leucine and photo-methionine allow identification of protein-protein interactions in living cells. Nat Methods 2(4):261–267. doi:10.1038/nmeth752

28. Combs SA, Deluca SL, Deluca SH et al (2013) Small-molecule ligand docking into comparative models with Rosetta. Nat Protoc 8(7):1277–1298. doi:10.1038/nprot.2013.074

29. Kahraman A, Herzog F, Leitner A et al (2013) Cross-link guided molecular modeling with ROSETTA. PLoS One 8(9):e73411. doi:10.1371/journal.pone.0073411

30. Kölbel K, Ihling CH, Sinz A (2012) Analysis of peptide secondary structures by photoactivatable amino acid analogues. Angew Chem Int Ed Engl 51(50):12602–12605. doi:10.1002/anie.201205308

Chapter 10

Tissue MALDI Mass Spectrometry Imaging (MALDI MSI) of Peptides

Birte Beine*, Hanna C. Diehl*, Helmut E. Meyer, and Corinna Henkel

Abstract

Matrix assisted laser desorption/ionization mass spectrometry imaging (MALDI MSI) is a technique to visualize molecular features of tissues based on mass detection. This chapter focuses on MALDI MSI of peptides and provides detailed operational instructions for sample preparation of cryoconserved and formalin-fixed paraffin-embedded (FFPE) tissue. Besides sample preparation we provide protocols for the MALDI measurement, tissue staining, and data analysis. On-tissue digestion and matrix application are described for two different commercially available and commonly used spraying devices: the SunCollect (SunChrom) and the ImagePrep (Bruker Daltonik GmbH).

Key words MALDI imaging, MSI, Peptide, Trypsin, Matrix application, ImagePrep, SunCollect, Formalin-fixed paraffin-embedded (FFPE), Cryoconserved

1 Introduction

Mass Spectroscopy Imaging (MSI) started off in 1962 with the usage of secondary ion mass spectrometry (SIMS) as ion source [1]. The term MSI generally encompasses SIMS, desorption electrospray ionization (DESI) and matrix assisted laser desorption/ionization (MALDI) mass spectrometry (MS) whereas all three ionization techniques provide insights into the molecular content of an intact piece of tissue while preserving the spatial resolution [2]. In case of MALDI MSI a laser with a micron dimension between 10 and 150 μm rasters over a predefined area of a tissue covered with matrix. An individual mass spectrum consisting of several mass-to-charge (m/z) ratios and intensities is generated at each position. A color scale is used to represent the signal intensity at each position. Basically three different mass analyzers (time of flight (TOF), quadrupole ion trap and Fourier transform ion cyclotron resonance (FT-ICR)) are used in combination with the

---

* The authors contributed equally to the manuscript.

Jörg Reinders (ed.), *Proteomics in Systems Biology: Methods and Protocols*, Methods in Molecular Biology, vol. 1394, DOI 10.1007/978-1-4939-3341-9_10, © Springer Science+Business Media New York 2016

mentioned ion sources. Different classes of molecules have been analyzed by MALDI MSI so far as for example proteins [3], peptides [4], lipids [5], and metabolites [6]. The possibility to acquire spatially resolved mass spectra directly is a big advantage in comparison to tedious laser microdissection of defined tissue areas of interest. MALDI MSI is a complementary method to liquid chromatography (LC)-MS approaches and for example protein arrays and can efficiently close the gap between histology and mass spectrometry.

Several different workflows for MALDI MSI were established in the last years and methods get more and more precise regarding preparation techniques and laser focus [6, 7]. A perfect tissue preparation for peptide MALDI MSI would consist of small trypsin droplets and small matrix crystals to result in optimal spatial resolution. Several papers in the beginning of the 2000th century describe experiments with a raster width of 200 or even 300 μm [8]. Over the years methods have changed and other lasers as for example the Smartbeam laser were generated to achieve a better performance with regard to spatial resolution. Additionally the way of trypsin and matrix application significantly influences the limitation of spatially resolved images, e.g., due to large matrix crystals which lower the overall resolution or trypsin droplets bigger than 100 μm [9, 10] which in return limit the spatial resolution too. Those problems can be avoided by using automatic sample preparation devices as for example the ImagePrep or SunCollect, which both work with a defined nebulized spray of trypsin and matrix, resulting in small trypsin droplets and matrix crystals. An alternative are spotting instruments with a raster width of about 100 μm or more [9], whereas the resolution is mainly limited by the spotter itself, because the distance from one spot to the next is quiet large. Nevertheless the advantage is to deposit a large amount of trypsin at one defined position and thus obtain better digestion efficiency. Another option to apply the matrix is sublimation by means of dry matrix application [11] but then trypsin still has to be sprayed or spotted so that only trypsin deposition determines the spatial resolution.

Most people in the community tend to work with automatic spraying devices self-made or provided by a company. For peptide MALDI MSI experiments the well-known matrices HCCA (α-cyano-4-hydroxycinnamic acid) or DHB (2,5-dihydroxybenzoic acid) are used. HCCA is slightly preferred because of its small crystal size [7, 12].

There are numerous different peptide MALDI MSI protocols published which makes it hard to define a standard procedure. The workflow generally consists of different washing and drying steps (or deparaffinization and antigen retrieval for FFPE tissue), trypsin application and incubation under humid and warm conditions for digestion, matrix application, measurement and data analysis. Ideally one would optimize all these different steps for every used tissue type. Diehl and coworkers performed an extensive experimental setup with 69 rat brains, to test different digestion times, different matrices, and quality of matrix and enzyme [7]. According to their results and other published data washing steps, type of enzyme, digestion time, matrix application, and type of matrix are major factors, which can essentially influence the outcome of your experiment [13].

This chapter focuses on the use of MALDI MSI for the analysis of peptides from fresh frozen and formalin-fixed paraffin-embedded (FFPE) tissue samples and provides operation procedures from sample preparation to measurement and data analysis. Moreover a list of necessary material and reagents will be given and notes on possible pitfalls to be avoided.

2 Materials

Prepare all solutions using ultrapure water (prepare by purifying deionized water to obtain a sensitivity of 18 MΩ cm at 21 °C) and analytical or mass spectrometry (MS) grade reagents. Prepare and store all reagents at room temperature unless stated otherwise.

2.1 Instruments

1. Ultrasonic water bath.

2. Shandon Pathcentre Tissue Processor (Thermo Fisher Scientific, Germany).

3. Microtome for sectioning FFPE tissue (e.g., RM 2155, Leica Instruments).

4. Cryostat for sectioning cryoconserved tissue (e.g., HM550, Thermo Fisher Scientific).

5. Water bath and flattening table for FFPE tissue sectioning.

6. Two small paint brushes.

7. Incubator (using at 37 °C and 60 °C).

8. Digital pH meter.

9. Microwave oven (e.g., LG MS-202VUT, Seoul, South Korea).

10. Glass staining jar for histology.

11. Flatbed scanner (e.g., Epson Scan Photo V600, Suwa, Japan).

12. Spraying devices ImagePrep (Bruker Daltonik GmbH, Germany) and SunCollect (SunChrom Wissenschaftliche Geräte GmbH, Germany).

13. MTP Slide Adapter II (#235380, Bruker Daltonik GmbH).

14. MALDI-MS, e.g., ultrafleXtreme (Bruker Daltonik GmbH).

2.2 Material for Collection and Sectioning of Cryoconserved Tissue

1. Liquid nitrogen or isopentane.
2. Phosphate buffered saline (PBS).
3. Acetone, 100 %, room temperature.
4. Optimal cutting temperature (OCT) compound.
5. Conductive indium tin oxide (ITO) coated glass slides (Bruker Daltonik GmbH).

2.3 Material for Tissue Fixation and Embedding

1. Phosphate buffered saline (PBS).
2. Formalin (4 %) for fixation.
3. Ethanol for dehydration (70 %, 96 %, 100 %).
4. Xylene for dehydration (100 %).
5. Paraffin wax for embedding (Richard-Allan Scientific™ Histoplast Paraffin, Thermo Fisher Scientific).
6. Conductive indium tin oxide (ITO) coated glass slides (Bruker Daltonik GmbH).

2.4 Solutions for Deparaffinization and Antigen Retrieval of FFPE Tissue (See Note 1)

1. Xylene, 100 %.
2. Ethanol, 100 %.
3. Ammonium bicarbonate (NH_4HCO_3), 0.01 M.
4. Citric acid monohydrate, 0.01 M, pH 6.0.

2.5 Solutions for Trypsin Application

1. NH_4HCO_3, 0.05 M.
2. Acetonitrile (ACN).
3. Trifluoroacetic acid (TFA).
4. Methanol.

2.6 Materials for Setting Reference Points on the ITO Slide

1. Liquid Tip Ex (water based) to set marks for geometric orientation of the tissue.
2. Solvent-resistant pen to set fine marks for coregistration (e.g., laboratory marker, #6130603, Paul Marienfeld GmbH & Co. KG).

2.7 Matrix Solution (See Note 2)

1. For tissue preparations using α-cyano-4-hydroxycinnamic acid (HCCA): 7 mg/ml HCCA, 50 % ACN, 0.2 % trifluoroacetic acid (TFA).
2. For tissue preparations using 2,5-dihydroxybenzoic acid (DHB): 30 mg/ml DHB, 50 % methanol, 1 % TFA.

2.8 Histological Staining: Hematoxylin–Eosin (HE) Staining

1. Ready to use Meyer's hematoxylin stain (Sigma-Aldrich, St. Louis, MO, USA).
2. Eosin Y.
3. Tap water.
4. Distilled Water.

5. Ethanol, 100 %.

6. Xylene, 100 %.

7. Mounting medium (e.g., Richard-Allan Scientific Cytoseal XYL, Thermo Fisher Scientific).

8. Cover glass.

2.9 External Calibration Standard for the MALDI Instrument

1. Peptide standard II (Bruker Daltonik GmbH).

2. 0.1 % TFA.

2.10 Data Analysis Software

1. FlexImaging, version 4.1 (Bruker Daltonik GmbH).

2. FlexControl, version 3.7 (Bruker Daltonik GmbH).

3. SCiLS Lab, version 2014b (SCiLS GmbH, Bremen, Germany).

3 Methods

3.1 Fresh Frozen Tissue Sample Preparation

The following description is based on the assumption that all steps of tissue preparation need to be performed including the tissue conservation process. If you work for example with clinical samples this process is in most cases already done and you can skip Subheading 3.1.1 and start directly with the sectioning of the tissue (Subheading 3.1.2) (*see* **Note 3**).

3.1.1 Sample Collection and Snap Freezing

Ideal samples are snap or flash frozen in the gas phase of either liquid nitrogen or isopentane directly after dissection in order to avoid autolysis and to preserve the molecular composition of the tissue (*see* **Note 4**).

1. In case of nitrogen do not drop the sample directly into the nitrogen to avoid cracks within the sample. Rather place the sample if small enough in a tube or vial and cap it tightly or wrap in aluminum foil and submerge in liquid nitrogen for flash freezing (*see* **Note 5**).

2. Supervise the temperature with a thermometer when using isopentane to make sure that it does not reach −78 °C (sublimation temperature of CO_2) (*see* **Note 6**).

3. If samples are contaminated with blood it is advisable to briefly wash the tissue in cold phosphate buffered saline (PBS) prior to freezing.

4. Place frozen samples in pre-labeled and prechilled containers. The samples can be stored at −80 °C or used for subsequent cutting of tissue sections.

| *3.1.2 Fresh Frozen Tissue Sectioning* | A cryostat is used to cut the samples into 10 μm thin tissue sections. |

1. Remove the frozen sample from the freezer and allow it to equilibrate in the cryostat chamber to the desired temperature for approximately 30 min, depending on the size of the sample (*see* **Note 7**).

2. Fix the sample on the metal holder using optimal cutting temperature (OCT) compound (*see* **Note 8**). Complete embedding of the sample in OCT should be avoided since it can lead to suppressed ion formation and intensity during the measurement.

3. Cut samples into 10 μm thin tissue slices using the glass anti-roll plate to prevent upward curling of the cut section. Remove the glass plate and directly pick up the section from the stage onto a conductive indium tin oxide (ITO) glass slide (*see* **Note 9**). Static attraction will draw the section to adhere quickly to the warm slide. Folding and curling of the section has to be avoided during this "thaw mounting" process. Assure complete adhesion of the tissue by placing the back of the slide shortly on the back of the hand. Perform this step also inside the cryostat (*see* **Note 10**).

4. Let the sample dry inside the cryostat (*see* **Note 11**). Then place the dried sample into a precooled, labeled air tight container. Samples can either be processed right away or in that condition be stored at ultra-low temperature (–80 °C) until analysis (*see* **Note 12**).

3.1.3 Drying and Washing of the Sample Prior to Enzyme Application

1. Samples taken from the –80 °C freezer are dried in a vacuum desiccator for about 30 min (*see* **Note 13**).

2. The slides are washed for 1 min per step in a series of ethanol steps 70 %/70 %/100 %. Slightly moving of the slides while in solution enhances the washing process. This step is primarily done to remove salts, lipids and other contaminants.

3. After washing dry the tissue once again in the vacuum desiccator for about 30 min.

3.1.4 Setting Reference Points and Scanning

Reference points for orientation and instrument settings need to be placed on the glass slide close to the tissue prior to enzyme and matrix application. Figure 1 shows and example for reference point placement.

1. Draw crosses with Tip Ex (water based) in some distance to the tissue to prevent contamination (*see* **Note 14**).

2. Then write letters using a solvent-resistant pen very close to the tissue (*see* Fig. 1).

3. Scan the tissue with 2400 dpi resolution using a flatbed scanner. This scan is used for teaching in order to transfer the

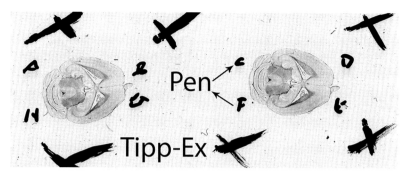

Fig. 1 Sample slide with reference points. The crosses are made with Tipp-Ex. The *letters* are written very close to the tissue using a solvent-resistant pen

geometry of the tissue to the measurement software of the MALDI instrument.

The tissue preparation continuous with the enzyme application (Subheading 3.3).

3.2 FFPE Tissue Sample Preparation

The following description is based on the assumption that all steps of tissue preparation need to be performed including the fixation and embedding process of the tissue. If you work for example with clinical samples in most of the cases these processes are already done in the clinic and you can skip reading Subheading 3.2.1 (*see* **Note 3**).

3.2.1 Tissue Fixation, Dehydration and Paraffin Embedding

1. After dissection of the tissue transfer the sample immediately into 4 % formaldehyde. To remove blood or other fluids perform a brief washing step with ice cold phosphate buffered saline (PBS) prior to fixation.

2. A standardized routine fixation, dehydration, and paraffin embedding in pathology department is performed overnight in an automated manner (e.g., using the Shandon Pathcentre Tissue Processor, Thermo Fisher Scientific). A typical program for tissue fixation is displayed in Table 1 (*see* **Note 15**).

3.2.2 FFPE Tissue Sectioning

It is much easier to generate very thin sections (2–6 μm) from FFPE tissue than from cryoconserved tissue. Instead of using a cryostat at subzero temperatures, FFPE sectioning is performed using a microtome at room temperature.

1. Preheat the water bath and flattening table for tissue sectioning to 39 °C.

2. Place the paraffin block with the tissue into the specimen holder (*see* **Note 16**).

3. Cut the sample into 5 μm thin sections and transfer the sections into the water bath using small paint brushes. Wait until

Table 1

Program for tissue fixation using the Shandon Pathcentre. The protocol was kindly provided by the pathology department of the University Hospital RWTH Aachen (Germany) directed by Prof. Dr. med. R. Knüchel-Clarke

Step	Solution	Concentration (%)	Incubation (min)
1	Formalin	4	80
2	Formalin	4	55
3	Ethanol	70	80
4	Ethanol	96	55
5	Ethanol	100	55
6	Ethanol	100	55
7	Ethanol	100	55
8	Xylene	100	55
9	Xylene	100	55
10	Xylene	100	55
11	Histoplast	100	45
12	Histoplast	100	45
13	Histoplast	100	45
14	Histoplast	100	45

the tissue is smoothed then use a conductive ITO slide to pick up the whole tissue section (*see* **Note 17**).

4. Place the ITO slide onto the preheated flattening table and leave it there until the surplus water is evaporated.

5. For optimal adhesion of the tissue onto the ITO slide incubation at 37 °C for 12 h is advisable.

6. The slides and the tissue block can be stored for several weeks, airtight and dry at 4 °C.

3.2.3 Tissue Deparaffinization and Rehydration

Before performing antigen retrieval it is necessary to remove the paraffin wax and rehydrate the tissue effectively.

1. Incubate the slide for 1 h at 60 °C prior to paraffin wax removal to assure optimal adhesion of the tissue onto the glass slide.

2. Transfer the slide into a glass staining jar filled with 100 % xylene and incubate the sample for 5 min (*see* **Note 18**).

3. Place the slide in a second glass jar with fresh 100 % xylene and incubate another 5 min.

4. Incubate the slide twice in 100 % ethanol for 2 min using another staining glass jar. Use fresh solution for the second step. Afterwards allow the tissue to dry completely.

5. For rehydration of the tissue place the slide in a glass jar filled with 0.01 M NH_4HCO_3 and leave it for 3 min. Repeat this step once again with fresh 0.01 M NH_4HCO_3 solution.

3.2.4 Antigen Retrieval

In this protocol antigen retrieval is achieved using a heat induced citric acid antigen retrieval method [14].

1. To prevent bubble formation during the antigen retrieval procedure prepare a histology glass jar with a rack filled up with dummy glass slides leaving one space free for your sample.

2. Place the slide into 0.01 M citric acid buffer (pH 6) and put the glass jar into the microwave. Start the 600 W program for 3 min. Afterwards reduce the power to 90 W for 10 more minutes (*see* **Note 19**). Avoid strong boiling.

3. To proceed with the antigen retrieval prepare a heat plate to 150 °C and place the glass jar onto the heat plate and incubate for another 30 min directly after the incubation in the microwave. Let the sample cool down for 15 min on the bench top.

4. To optimize the pH for the subsequent enzymatic digestion, incubate the sample twice in 0.01 M NH_4HCO_3 solution once the citric acid buffer has cooled down. Prepare a histology glass jar with the solution. Use fresh solution for the second 1 min incubation.

3.2.5 Setting Reference Points and Scanning

Apply the reference points and scan the tissue as described in Subheading 3.1.4.

The tissue preparation continuous with the enzyme application and digest of the tissue (*see* Subheading "Protease Application and Digest—SunCollect/SunDigest" or "Protease Application and Digest—ImagePrep").

3.3 Enzyme and Matrix Application for Peptide MALDI MSI

This protocol provides the experimental procedures of enzyme and matrix application with two different instruments which are the most common commercially available devices for these applications; the SunCollect (SunChrom) and the ImagePrep (Bruker Daltonik GmbH).

There are two main differences comparing both instruments regarding their spraying mechanism.

The *SunCollect* device uses pneumatic atomization to spray enzyme and matrix solution onto the sample. The spray head is flexible and moves quickly and continuously during the preparation process over the tissue. The spaying process is regulated by software and individual spraying sessions can be programmed. Parameters which can be modified are for example the distance of the spaying head to the sample, the speed of the spay head while moving over the tissue and the liquid flow per minute. Those are only some of the parameters which can be modified for optimizing the spraying result. For detailed information please read the manufactures' manual.

The ImagePrep device is an automatic spraying device, which uses vibrational vaporization for spray production. The spray head is fixed and the spray must be optimized to reach the sample which must be positioned in a defined distance towards the spray head. The ImagePrep software is preset with some default methods for protease and matrix applications and is controlled by a sensor which is positioned near the tissue. Each method consists of one or more phases each with individual spraying, incubating or drying conditions. Parameters of the different phases can be individually modified as well as the number of spraying phases can be changed. As before for detailed information please read the manufactures' manual. Figure 2 provides an overview of the two instruments together with a detailed view of the actual spraying device.

3.3.1 Protease Application and Digest

Enzyme application is a crucial step in tissue preparation for peptide MALDI MSI. There are several parameters which need to be considered: optimal digestion of the proteins requires a humid environment but delocalization due to diffusion needs to be minimized to remain the spatial resolution of the peptides. Therefore the spray for enzyme application needs to produce very small droplets and at the same time provide sufficient humidity for enzyme activity. In the following section we describe procedures that fulfill these requirements either using the SunCollect (Subheading "Protease Application and Digest—SunCollect/SunDigest") or the ImagePrep (Subheading "Protease Application and Digest—ImagePrep") spraying device.

The enzymatic digestion requires an environment with high relative humidity (RH) and temperature. When using trypsin, temperatures of 37 °C are the lowest temperature level for sufficient digestion. In the MALDI MSI community a lot of self-made instrumentations (digestion chambers) for efficient enzyme digestion are used. In this protocol we provide two different types of digestion chambers: the SunDigest, a sensor regulated and programmable automatic device and a self-made digestion chamber using pipette boxes.

Protease Application and Digest—SunCollect/SunDigest

1. Prechill a 250 μl syringe on ice (*see* **Note 20**).

2. Prepare 20 μg of trypsin in 200 μl cold buffer (0.05 M NH_4HCO_3 and 5 % ACN) and fill the syringe without bubbles.

3. Place the syringe into the syringe pump of the SunCollect, connect it with the capillary and place a cold pack on top of the syringe to prevent enzyme activity.

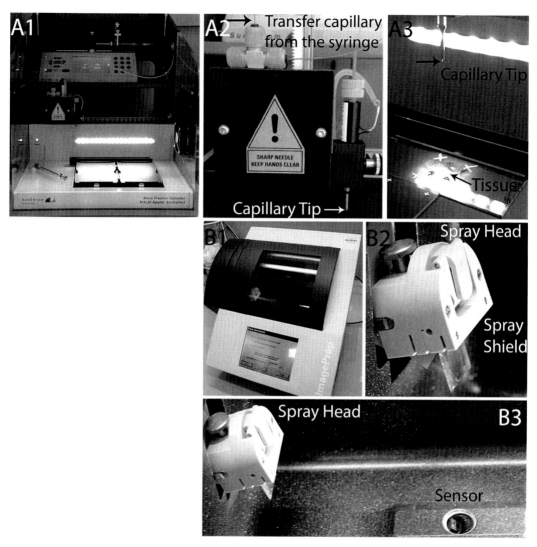

Fig. 2 SunCollect and ImagePrep instrument. In addition to an overview of the two instruments are detailed views of the spraying devices given. A1–A3 show the SunCollect device. B1–B3 illustrates the ImagePrep device

4. After placing the sample into the instrument it is necessary to insert the coordinates of the tissue position into the software. For that follow the instructions of the manual.

5. Use the software to design a spraying protocol for enzyme application of your sample (*see* **Note 21**).

6. Before starting the program place an empty glass slide below the spray and check for a continuous and fine spray (*see* **Note 22**).

7. Remove the glass slide and start the spray. The spray head will move over the entire predefined area in fixed lines thus covering

the whole sample with enzyme solution. It is necessary to apply several layers of enzyme solution (*see* **Note 23**).

8. Prepare the SunDigest before continuing with the transfer of the sample into the digestion chamber. Apply approximately 5 ml water onto the felt pads (*see* **Note 24**).

9. Choose the parameters for digestion (temperature, relative humidity (RH) and time). The instrument regulates the RH and prevents condensation by active heating of the lid. Two sensors regulate the heat inside the instrument; one is placed above the other below the sample inside the heating block (*see* **Note 25**). The measured data are recorded and graphically displayed as quality control for the user.

10. Check the graphical summary of all parameters and wait until the instrument executed the chosen settings (Chamber temperature, base temperature, RH and lid temperature) (*see* **Note 26**).

11. After digestion (*see* **Note 27**) remove the slide from the SunDigest and start the matrix application (*see* Subheading 3.4.1).

Protease Application and Digest— ImagePrep

1. Test the spray performance by loading 200 μl 0.05 M NH_4HCO_3 buffer directly into the spray head. Hold the head in a horizontal position while pipetting. Afterwards seal the head with the empty glass bottle in the metal surrounding and set the power to 38 % and the modulation to 0 % in the adjustment menu of the ImagePrep. Stop the time while spraying. The spray should reach the sensor at the round glass plate in the middle of the spraying chamber thereby evenly distributing the solution. The entire volume should be finished within a time frame of 20 s. If not adjust the spray power and repeat the test.

2. Wipe the inside of the chamber to remove the moist and clean the area of the sensor in the middle of the chamber carefully.

3. Put the ITO slide on the placement area for the slide above the light-scattering sensor without covering the sensor itself with the tissue.

4. Load the prepared trypsin solution (200 μl) in the same manner as the test solution and start the Bruker standard program for digest: *trypsin_deposition_nsh01*.

5. Start the NIVI-Logger (part of the ImagePrep Software that runs on a computer) to monitor the spray cycles. The program should reach about 20 spray cycles.

6. Upon completion transfer the slide to a humidity chamber and let the digestion take place at 37 °C for as long as desired (*see* **Note 28**).

3.4 Matrix Application

Matrix application is another critical step of the sample preparation for MALDI MSI. During matrix application it is necessary to use solvents and water to extract the peptides from the tissue and to induce the cocrystallization of the matrix and the analyte. Crystal size formation and extraction efficiency have a large impact on the MSI result. Thus it is the overall goal during matrix application to achieve efficient peptide extraction while maintaining good spatial resolution.

3.4.1 Matrix Application—SunCollect

1. Place the tissue in the SunCollect instrument. When teaching was applied for enzyme application no further teaching is necessary. Otherwise teach the instrument (see Subheading "Protease Application and Digest—SunCollect/SunDigest").

2. Use a 2500 µl syringe for matrix application and fill it with matrix solution. Place the syringe into the pump and connect it with the capillary (*see* **Note 29**).

3. Use the software to design a spraying program for matrix application of your sample (*see* **Note 30**). The program for matrix application differs considerably to the one for enzyme application. To initiate crystal formation flow rates of 10–20 µl/min should be applied. The flow rate per minute increases with the number of layers. The maximal flow rate to prevent diffusion should be 40 µl/min.

4. Hold an empty glass slide below the spray and check for a continuous and fine spray.

5. Remove the glass slide and start the matrix application program.

6. After matrix application fix the slide into the MTP Slide Adapter II. Apply a peptide standard in the middle of the steel frame of the adapter for external calibration: according to the dried droplet method, mix equal volumes (1 µl) of calibration standard sample with matrix solution and dry at room temperature.

7. Clean the capillary by filling the syringe with 70 % ACN. Use a flow rate of 40 µl/min and let it run for at least 15 min. While flushing the capillary with ACN clean the spray tip with the same solvent using a pipette. Let several drops run down the capillary tip without touching the end of the capillary with the pipette tip.

Continue with the MALDI MSI measurement (*see* Subheading 4).

3.4.2 Matrix Application—ImagePrep (See Note 31)

1. Start off with testing the sprayer. Fill the glass vial 2/3 with methanol and choose in the display of the ImagePrep *Spray Head → Adjust → Test Spray*. First test 100 % and then 10 %. With the first setting the spray should reach the round plate on the opposite site of the chamber whereas the 10 % should at least

reach the end of the placement area (with the sensor in the middle) for the slide. If not adjust the offset from for example 4–5 %.

2. Remove the methanol residues by wiping the inside of the chamber with a paper towel.

3. Adjust the power and modulation for the matrix. Load the prepared matrix into the glass vial and choose *adjustment* in the display of the instrument. By default is the power set to 25 and the modulation to 30 for DHB and 15 and 40 for HCCA. The matrix spray should reach the end of the placement area of the slide. If necessary adjust the power or modulation (*see* **Note 32**).

4. Once again clean the inside of the chamber especially the area above the sensor. Afterwards place the sample slide on top without covering the sensor with the tissue (*see* **Note 33**). If more than one tissue is on the slide, position the sensor in between the two samples. Lay a cover glass on top of the slide, above the sensor.

5. Start the NIVI-Logger (part of the ImagePrep software that runs on a computer) to monitor the spray cycles.

6. Start the run using the Bruker standard program *DHB_for_digest_nsh01* or *HCCA_nsh04*.

7. Wait at least until the first 3 cycles of the second phase are finished and control the profile of the logger. An increase of the measured curve should be visible if not adjust the spray power boost (*see* **Note 34**).

8. Once the program has finished, stop the logger, take out the slide and remove the matrix from the short edges of the slides with 100 % ethanol before fixing the slide into the MTP Slide Adapter II (*see* **Note 35**).

9. Apply a peptide standard in the middle of the steel frame of the adapter for external calibration: according to the dried droplet method, mix equal volumes (1 µl) of calibration standard sample with matrix solution and dry at room temperature.

10. Clean the inside of the ImagePrep with methanol by running the corresponding *clean* program. Afterwards dry the inside of the chamber with a paper towel.
    Continue with the MALDI MSI measurement.

3.5 MALDI MSI Measurement

The following section describes the peptide MALDI MSI measurement using a MALDI-TOF mass spectrometer (ultrafleXtreme, Bruker Daltonik GmbH) controlled by the Bruker software FlexControl and FlexImaging.

1. Open FlexControl.

2. Load the MTP Slide Adapter II with the fixed ITO slide into the instrument.

3. Open FlexImaging and create a file for your sample, chose the desired *autoXecute* method and raster width. Use the Tip Ex crosses on the slide to coordinate the sample and correlate it to the previously made scan of the slide.

4. Define the measurement region on the tissue.

5. In FlexControl: choose reflector mode with a set mass range of m/z 600–4000. Save those settings in a FlexControl method. The method should also contain a statistical recalibration of the spectra. For this purpose integrate the flexAnalysis method *ImageID_StatRecal.FAMSMethod* [15].

6. Load the generated FlexControl method and measure the external calibrants on the target frame. The error of each peptide should not exceed 20 ppm. Apply the calibration and save the method.

7. Finally determine the necessary laser power for the tissue. Save the method and start the run using FlexImaging.

3.6 HE-Staining of the Analyzed Tissue

The unique characteristic of MALDI MSI compared to other mass spectrometry based analyses is the ability to maintain the spatial resolution of the analyzed molecules. It is thus possible to compare the MS data directly to the histological features, which are visible after staining the tissue. Standard histological staining protocols can be used to stain the tissue after the measurement. Check which kind of staining is the best to answer your questions. In the following we provide a fast hematoxylin–eosin (HE) staining which is suitable for many tissue types. Hematoxylin stains nuclei blue, while eosin results in a pink staining of the connective tissue.

1. Remove the matrix by washing the tissue in 70 % ethanol. For this purpose move the slide up and down for several times.

2. Incubate the tissue in Meyers' hematoxylin solution for 1 min (*see* **Note 36**).

3. Incubate the tissue in tab water for 4 min. This step causes a pH shift and intensifies the staining of hematoxylin.

4. Dip the slide twice into 0.25 % eosin solution.

5. Dip the slide twice into distilled water.

6. Repeat dipping the slide in 70 % ethanol.

7. Incubate the slide for 20 s each in 96 %, 100 %, and again 100 % ethanol.

8. For complete dehydration incubate the slide 5 min in 100 % xylene. Then dry the slide completely.

9. Put one drop of mounting medium on the tissue and place a cover slip on top. Try to avoid air bubbles since they will dete-

riorate the quality of subsequent microscopy. To remove bubbles press slightly onto the glass. Surplus mounting media can be removed after complete drying (2 h at room temperature) by scraping it off the surface using another glass slide.

3.7 Data Analysis Using SCiLS Lab

For the analysis of the MALDI MSI data we use SCiLS Lab. This software allows you to analyze several tissues simultaneously, which is important when comparing different samples in one study or comparing different methods.

1. Import the *mis-files* from FlexImaging into SCiLS Lab. You can arrange your samples in the way you want them to be displayed (rotation is possible as well). SCiLS Lab converts the data into a so-called *h5-file*. The generated *h5-file* contains all the data you need for further analysis (*see* **Note 37**).

2. In SCiLS Lab you have several options for data analysis. To gain a first impression it is a good starting point to run the "*Segmentation pipeline*" and let the software perform all necessary steps automatically. This pipeline contains preprocessing steps such as baseline removal, normalization and the generation of an overview spectrum for all the samples in your data set. Furthermore peak picking, alignment, spatial denoising and segmentation are performed.

3. To evaluate different data sets statistical and structure analysis are suitable tools. The software provides a large amount of analysis tools so that only some selected examples can be provided. For statistical analysis two different component analyses can be performed in SCiLS Lab: PCA (*principal component analysis*) and pLSA (*probabilistic latent semantic analysis*) whereby the latter one can also be seen as a form of structure analysis (*see* **Note 38**).

4. When you use a model tissue like rat brain it is possible to judge the quality of your different methods using an additional strategy: counting the number of m/z values, which show a structured image for a well-known and highly abundant protein, e.g., myelin basic protein (MBP) in rat brain (*see* **Note 39**). For this strategy you have to first perform an in-silico digest of MBP with trypsin using for example MS-Digest from the *ProteinProspector* website (http://prospector.ucsf.edu/prospector/cgi-bin/msform.cgi?form=msdigest) to obtain a list of MBP specific peptides (*see* **Note 40**). For the in-silico digest defined settings must be used: maximal number of missed cleavages 1, oxidation of methionine as variable modification, peptide mass range 800–4000 Da and MALDI-TOF/TOF as instrument type. In case of MBP this results in a list of 45 peptides respectively 45 monoisotopic m/z values.

This list can be imported into the h5 data set (*see* **Note 41**). It is then possible to visualize the images of each individual *m/z* value allowing for example a peak shift of ±0.3 Da. It is now possible to count the number of visible MPB structures and compare them between you different experiments.

4   Notes

1. All solutions used for deparaffinization besides ammonium bicarbonate can be used several times. When you change the tissue type you should also change the solutions. The citric acid buffer can be stored for 1 week, check pH prior to use.

2. Matrix solution should be sonicated for approximately 15 min prior to use. There should be no visible matrix particles left. The matrix solution can be used up to 7 days but should be kept in the dark in an air tight glass bottle at 4 °C to avoid volatilization of the solvent.

3. When you contribute in clinical studies you usually get samples that are already sectioned. If that is the case it is important to note that the people preparing you samples stick to the rules which need to be fulfilled when working with MALDI MSI samples like section thickness and awareness of temperatures during the preparation process.

4. For beginners in the field of MALDI MSI we recommend to start with commercially available rat brain samples which are the most common model sample in MALDI MSI. In this way errors during sample collection can be avoided. The sample is fairly easy to cut and the user can focus on MALDI MSI preparation parameter optimization.

5. For a rat brain a 100 ml beaker will be sufficient for freezing.

6. Independent of the freezing method be careful and avoid morphological distortion or damage of the sample. Freezing artifacts in the middle of a tissue occur more frequently when large objects are flash frozen. Guideline: at least one dimension of the sample must be below 10 mm. Whole rat brains can be successfully frozen.

7. The optimal temperature for sectioning depends mainly on the type and nature of the tissue. Reference charts for temperature settings are generally provided in the manual of the producer of the cryostat.

8. Tissue Tek® is a useful OCT material.

9. Avoid placing more than one tissue sample on the same slide when using the ImagePrep device for subsequent spraying.

10. It has been reported that ITO slides modified with poly-L-lysine improved the adherence of the cells to the substrate [16].

11. Do not place the slide on the metal ground of the cryostat while letting the cut tissue section dry inside the instrument. The direct contact with the metal will lead to freeze artifacts. We therefore recommend placing a vacuum formed slide holder inside the cryostat to overcome this problem.

12. In general it is advisable to use flash frozen samples right away for imaging rather than storing them long term. The same applies for the cut tissue sections on ITO glass slides. In case of long-term storage we recommend to put each slide in an individual container even though there might be space for more but by doing so we avoid temperature changes when opening the box to remove single slides. We found LockMailer™ with screw caps to be the best storage device for that purpose.

13. Drying of the samples is also possible in a regular desiccator.

14. Use only little amount of Tip Ex to maintain an even surface otherwise you will have problems to cover the subsequently HE stained sample with a cover glass after MALDI MSI measurement.

15. Fixation time can be reduced to ~30 min in case of small biopsy samples whereas some bloody or fatty tissues as well as some fetal tissues require longer fixation times. Make sure to check the optimal conditions for your tissue sample prior to fixation.

16. We recommend removing surplus paraffin wax at the edges of the block to minimize the sectioning area.

17. If the flattening of the tissue is not sufficient after 5–10 min, temperature needs to be optimized.

18. We recommend using a histology glass jar with molded glass cover in which the slides are in an upright position.

19. During this process it is necessary to constantly check for bubble formation and to remove bubbles by simply pulling the slide holder upwards.

20. Put the syringe into a plastic bag to avoid contamination.

21. To find the perfect enzyme spray settings for your tissue you should test one to three slides before you prepare a real sample. Test enzyme spray settings and digestion settings and compare the MALDI MSI results in regard to intensities, spatial resolution and digest efficiency.

22. We spray for at least 1 min before starting the program.

23. Clean the capillary flow by inserting a syringe with 70 % ACN after every use. Use a flow rate of 40 μl/min for 5 min.

24. Estimate the amount of water on the felt pads of the SunDigest. Your glove should be wet after touching the pads.

25. Fill both racks with the same number of slides to ensure homogenous distribution of heat and humidity. We use the program at 50 °C and 95 % RH, 2 h for cryoconserved tissue, 1 h for FFPE tissue.

26. For quality control of the digest it is reasonable to apply for example *bovine serum albumin* standard onto a small part of the tissue. You can use the same settings for the application of standard as for trypsin digest (we advise to use only one layer of standard and therefore use a suitable amount of protein in the solution).

27. Clean the capillary flow by inserting a syringe with 70 % ACN after every use. Use a flow rate of 40 μl/min for 5 min.

28. An empty pipette box can be used as a humidity chamber, by filling the lower part of the box with water (fill height ~ 1 cm), placing the slide on top of the perforated top and closing the lid before placing the box into an incubator for digestion.

29. Before spraying the matrix we run 70 % ACN for at least 5 min.

30. Before starting with your sample of interest you need to prepare some test slides. Use these slides to verify different parameters like height of the spray head (Z position), flow rate, speed and number of layers. Check the crystal shape and size under a microscope and chose the one which fits best for your purpose (examples given in Fig. 3). In general the crystals should be very small and show a homogenous distribution over the whole sample. In Fig. 3 we provide examples of matrix crystals. Be aware that you should check the resulting MALDI MSI image as well. Based on the results shown in Fig. 3 we decided to use the z position 24. We think using z position 24 in this case results in the best density, distribution and crystal shape of HCCA. Comparing the matrix distribution in different magnitudes is imported to gain a sufficient overview. It is for example easier to identify clogging when using a lower magnitude. We advise to regularly check matrix distribution patterns, crystal size and shape under the microscope as quality control.

31. Based on our observations the adjustment of the spray settings strongly varies between different persons in the sense of what is considered to be a good spray. Consequently we strongly advise that all spraying operations within a project should be carried out by the same person.

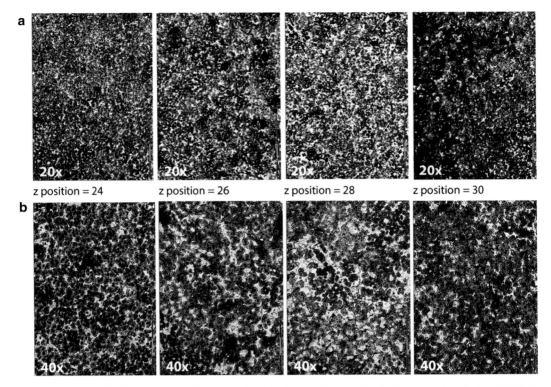

Fig. 3 Overview of different crystal shapes and distributions using the SunCollect spraying device. HCCA (7 mg/ml, 50 % ACN, 0.2 % TFA) matrix was applied onto rat brain tissue sections only the parameter of the z position during matrix spray was changed (*see* **Note 29**). (**a**) and (**b**) display images with different magnitude (×20 and ×40)

32. When adjusting the settings (power and modulation) it can be helpful to place an empty glass slide on the intended position to control the crystal size and density of the sprayed matrix.

33. Best results were obtained when placing the tissue directly in front of the sensor.

34. It is advisable to observe from time to time the actual spraying process inside the chamber since the logger alone is sometimes not sufficient for judgment.

35. Add an additional fifth phase if the matrix appears still uneven and not dense enough.

36. You can intensify the staining of hematoxylin in two ways: increase the incubation time for hematoxylin and/or increase the time for the following incubation step in tap water.

37. Please check the system requirements for running SCiLS Lab before you start.

38. Principle component analysis (PCA) is a widely used component analysis method to assess the overall structure in a data set. The components resulting from a PCA are statistically

uncorrelated and arranged so that the first principal component has the largest variance within the data set. Each subsequent component has the largest variance within the remaining orthogonal dimension of the data set. Thus a PCA is able to uncover and visualize variability within the data. Scaling the data before calculating the PCA is often useful; variances can be scaled before calculating the PCA for spatial and spectral features. Use the aligned peaks of the preprocessed datasets for computation of PCA for individual and mean spectra [7]. Using the probabilistic latent semantic analysis (pLSA) both, score images and loadings can be interpreted in terms of mass spectra intensities [17]. They are non-negative for pLSA in comparison to PCA. The results of the pLSA can therefore be interpreted as spatial tissue components and their corresponding mass distribution in the tissue. Before starting the pLSA estimate the optimal number of pLSA components. Therefore it is helpful to know the anatomical features of your used sample (for rat brain the Allen mouse brain atlas is a good source (http://mouse.brain-map.org/)). When you try to establish a protocol in your lab keep in mind, that you should produce technical replicates (e.g., triplicates) of each tested condition and measure the samples in an alternating order.

39. MBP is a suitable protein for such analysis because it is one of the most abundant proteins of the myelin membrane of the central nervous system (CNS). Thus it is guaranteed that the protein is present in all the samples for comparison and secondly that the strong expression would facilitate detection of MBP during MALDI MSI measurement. Information about the anatomical distribution of the protein in the brain can be obtained from for example the Allen mouse brain atlas.

40. The necessary protein sequence information (FASTA format) can be obtained from the *UniProt/Swiss-Prot* data base in this case one would use *Rattus norvegicus* as taxonomy.

41. All you need for the generation of the peak list into your h5 file is an excel csv.-file with the *m/z* of interest. Copy the column with the *m/z* values and paste them during the import process into the import window. To do so open in SCiLS: File → Import MZ Range from CSV and paste the copied column.

Acknowledgement

We would like to thank Julian Elm for his skilful technical assistance during the method development process. We thank Dennis Trede for excellent support dealing with SCiLS Lab. Furthermore we like to thank Prof. Dr. med. Ruth Knüchel-Clarke for providing the protocol for the fixation process of FFPE tissue. H.D. was financially supported by PURE (Protein research Unit Ruhr within Europe).

References

1. Castaing R, Slodzian G (1962) Microanalyse par emission ionique secondaire. J Microscopie 1:395–410

2. Trim PJ, Djidja MC, Muharib T et al (2012) Instrumentation and software for mass spectrometry imaging—making the most of what you've got. J Proteomics 75(16):4931–4940. doi:10.1016/j.jprot.2012.07.016

3. Groseclose MR, Andersson M, Hardesty WM et al (2007) Identification of proteins directly from tissue: in situ tryptic digestions coupled with imaging mass spectrometry. J Mass Spectrom 42(2):254–262. doi:10.1002/jms.1177

4. Monroe EB, Annangudi SP, Hatcher NG et al (2008) SIMS and MALDI MS imaging of the spinal cord. Proteomics 8(18):3746–3754. doi:10.1002/pmic.200800127

5. Murphy RC, Hankin JA, Barkley RM (2009) Imaging of lipid species by MALDI mass spectrometry. J Lipid Res 50 Suppl:S317–S322. doi:10.1194/jlr.R800051-JLR200

6. Khatib-Shahidi S, Andersson M, Herman JL et al (2006) Direct molecular analysis of whole-body animal tissue sections by imaging MALDI mass spectrometry. Anal Chem 78(18):6448–6456. doi:10.1021/ac060788p

7. Diehl HC, Beine B, Elm J et al (2015) The challenge of on-tissue digestion for MALDI-MSI—a comparison of different protocols to improve imaging experiments. Anal Bioanal Chem. doi:10.1007/s00216-014-8345-z

8. Oezdemir RF, Gaisa NT, Lindemann-Docter K et al (2012) Proteomic tissue profiling for the improvement of grading of noninvasive papillary urothelial neoplasia. Clin Biochem 45(1–2):7–11. doi:10.1016/j.clinbiochem.2011.09.013

9. Casadonte R, Caprioli RM (2011) Proteomic analysis of formalin-fixed paraffin-embedded tissue by MALDI imaging mass spectrometry. Nat Protoc 6(11):1695–1709. doi:10.1038/nprot.2011.388

10. Delvolve AM, Woods AS (2011) Optimization of automated matrix deposition for biomolecular mapping using a spotter. J Mass Spectrom 46(10):1046–1050. doi:10.1002/jms.1986

11. Yang J, Caprioli RM (2011) Matrix sublimation/recrystallization for imaging proteins by mass spectrometry at high spatial resolution. Anal Chem 83(14):5728–5734. doi:10.1021/ac200998a

12. Jaskolla TW, Karas M, Roth U et al (2009) Comparison between vacuum sublimed matrices and conventional dried droplet preparation in MALDI-TOF mass spectrometry. J Am Soc Mass Spectrom 20(6):1104–1114. doi:10.1016/j.jasms.2009.02.010

13. Ait-Belkacem R, Sellami L, Villard C et al (2012) Mass spectrometry imaging is moving toward drug protein co-localization. Trends Biotechnol 30(9):466–474. doi:10.1016/j.tibtech.2012.05.006

14. Gustafsson OJ, Eddes JS, Meding S et al (2013) Matrix-assisted laser desorption/ionization imaging protocol for in situ characterization of tryptic peptide identity and distribution in formalin-fixed tissue. Rapid Commun Mass Spectrom 27(6):655–670. doi:10.1002/rcm.6488

15. Mann M, Hojrup P, Roepstorff P (1993) Use of mass spectrometric molecular weight information to identify proteins in sequence databases. Biol Mass Spectrom 22(6):338–345. doi:10.1002/bms.1200220605

16. Zavalin A, Todd EM, Rawhouser PD et al (2012) Direct imaging of single cells and tissue at sub-cellular spatial resolution using transmission geometry MALDI MS. J Mass Spectrom 47(11):1473–1481. doi:10.1002/jms.3108

17. Hanselmann M, Kirchner M, Renard BY et al (2008) Concise representation of mass spectrometry images by probabilistic latent semantic analysis. Anal Chem 80(24):9649–9658. doi:10.1021/ac801303x

Chapter 11

Ethyl Esterification for MALDI-MS Analysis of Protein Glycosylation

Karli R. Reiding, Emanuela Lonardi, Agnes L. Hipgrave Ederveen, and Manfred Wuhrer

Abstract

Ethyl esterification is a technique for the chemical modification of sialylated glycans, leading to enhanced stability when performing matrix-assisted laser desorption/ionization (MALDI)-mass spectrometry (MS), as well as allowing the efficient detection of both sialylated and non-sialylated glycans in positive ion mode. In addition, the method shows specific reaction products for $\alpha 2,3$- and $\alpha 2,6$-linked sialic acids, leading to an MS distinguishable mass difference. Here, we describe the ethyl esterification protocol for 96 glycan samples, including enzymatic N-glycan release, the aforementioned ethyl esterification, glycan enrichment, MALDI target preparation, and the MS(/MS) measurement.

Key words Sialic acid (N-acetylneuraminic acid), Stabilization, Linkage-specific, Peptide-N-glycosidase F (PNGase F), N-glycan release, Ethyl esterification (EE), Solid-phase extraction (SPE), Cotton, Sepharose, Hydrophilic interaction liquid chromatography (HILIC), Matrix-assisted laser desorption/ionization (MALDI), Mass spectrometry (MS)

1 Introduction

Protein glycosylation stands for a group of important and highly complex post-translational modifications, which can have a profound effect on many characteristics of the conjugate, including folding, solubility, plasma half-life, and receptor binding activity [1–3]. Changes in glycosylation depend on genetic, physiological and environmental factors, and are known to occur under influence of biological processes like aging, adolescence, and pregnancy, as well as in neoplastic or inflammatory diseases [4–8]. As such, the study of glycosylation has the potential of pinpointing relevant biomarkers (for a particular disease or physiological state), but also to assist personalized medicine by separating patients according to disease etiology [9, 10]. Adding to this the importance of biopharmaceutical glycosylation, the need for high-throughput methodologies for glycan analysis becomes apparent [11].

Jörg Reinders (ed.), *Proteomics in Systems Biology: Methods and Protocols*, Methods in Molecular Biology, vol. 1394, DOI 10.1007/978-1-4939-3341-9_11, © Springer Science+Business Media New York 2016

Of the methods to study glycosylation, mass spectrometry (MS) has proven to be extremely fast and efficient for achieving compositional assignment [12, 13]. It has, however, the downside of not providing information on monosaccharide linkage positions, and is hampered by the innate instability and charge of glycans carrying a sialic acid [13, 14]. This is evident in matrix-assisted laser desorption/ionization (MALDI)-MS, where sialylated glycan species are affected by ionization bias, neutral salt adduction and loss of sialic acid residues. Several strategies can be employed to overcome these problems, among which the chemical modification of the causative sialic acid carboxyl groups into esters or amides [15–17]. In some cases these reactions can make use of the specific chemical characteristics of α2,3- and α2,6-linked sialic acids, as α2,3-linked variants may undergo an intramolecular lactonization, while the α2,6-linked variants are susceptible to intermolecular reactions with alcohols or amines [18, 19]. The respective water loss and ester/amide formation introduce a mass difference to the isomers, which can then be used for MS assignment.

Recently we published a rapid and robust method for linkage-specific stabilization of sialylated glycans, using the combination of 1-ethyl-3-(3-dimethylaminopropyl)carbodiimide (EDC) and 1-hydroxybenzotriazole (HOBt) in an environment of ethanol [20]. Full reaction specificity was demonstrated between α2,3- and α2,6-linked sialylation isomers, and the products could be stably analyzed by reflectron mode MALDI-time-of-flight (TOF)-MS. Notably, the protocol was shown resistant to impurities, making it suitable for the analysis of free glycans from a wide range of sources, and has proven informative for the study of glycans obtained from plasma as well as from specific proteins like immunoglobulin G [20, 21].

Here the steps required in the protocol will be thoroughly discussed, including N-glycan release by peptide-N-glycosidase F (PNGase F), ethyl esterification, purification by hydrophilic interaction liquid chromatography (HILIC) solid phase extraction (SPE), MALDI target spotting, and concluding with a section on MS(/MS) measurement and quality control. The focus will be on a high-throughput 96-well variant of the protocol, describing Sepharose beads for the HILIC enrichment. However, when performing the protocol for a low number of samples, the use of cotton HILIC microtips (using the same solutions) is easier and faster, while producing highly comparable results [22].

2   Materials

Ultrapure deionized water is required for the preparation of all solutions ($\geq$18 MΩ at 25 °C). The reagent volumes here described are sufficient for 100 samples, and can be up- or downscaled unless otherwise specified. Take note when performing the protocol on very low sample quantities, as pipetting imprecision and ethanol evaporation start playing a larger role. Importantly, make sure that all glassware is clean before use or risk polymer contamination in the resulting mass spectra.

2.1 PNGase F N-Glycan Release

1. 5× phosphate buffered saline (PBS): Order directly, or prepare as 50 mM Na_2HPO_4 9 mM KH_2PO_4 685 mM NaCl 13.5 mM KCl, pH 7.4. 5× PBS can be stored at room temperature.

2. 2 % (*w/v*) sodium dodecyl sulfate (SDS) in water. Weigh 200 mg SDS on an analytical balance and transfer to a 15 mL tube (*see* **Note 1**). Add 9.8 mL water. Store at room temperature.

3. 4 % (*w/v*) Nonidet P-40 (NP-40) substitute (Sigma-Aldrich, Steinheim, Germany) in water. Pipette 400 mg NP-40 substitute to a 15 mL tube on an analytical balance (*see* **Note 2**). Add 9.6 mL water. Store at room temperature.

4. Release solution: In a 1.5 mL tube, thoroughly mix 400 μL 4 % NP-40 substitute, 400 μL 5× PBS, and 80 μL PNGase F (Roche Diagnostics, Mannheim, Germany) (*see* **Note 3**).

5. Polypropylene (PP) 96-well plate.

6. Adhesive tape for plate sealing.

7. Ovens at 37 °C and 60 °C.

2.2 Ethyl Esterification

1. Esterification reagent: 250 mM EDC hydrochloride (Fluorochem, Hadfield, UK) 250 mM HOBt monohydrate (Sigma-Aldrich) in ethanol. Weigh 67.56 mg HOBt and transfer to a 2 mL tube. Weigh 95.85 mg EDC and transfer to the same tube (*see* **Note 4**). Add 2 mL ethanol and mix thoroughly (*see* **Note 5**). The reagent can be stored at –20 °C (*see* **Note 6**).

2. 96-well plate (PP).

3. Adhesive tape for plate sealing.

4. An oven at 37 °C.

2.3 Sepharose HILIC Solid Phase Extraction

1. Sepharose bead slurry: 25 % Cl-4B Sepharose beads (45–165 μm, GE Healthcare, Uppsala, Sweden) in 20 % ethanol. Transfer 500 μL beads to a 2 mL tube. Spin down the beads and remove the supernatant. Add 1.5 mL 20 % ethanol. Make fresh before using.

2. 85 % acetonitrile (ACN). Measure 150 mL water in a 50 mL graduated cylinder and add to a 1 L screw-cap glass bottle. Using a 1 L graduated cylinder, measure 850 mL ACN and add to the glass bottle. Store at 4 °C, but bring to room temperature before usage.

3. 85 % ACN 0.1 % trifluoroacetic acid (TFA). Prepare as *2*, add 1 mL TFA (use a chemical fume hood to contain and exhaust volatile dangerous chemicals). Store at 4 °C, but bring to room temperature before usage.

4. 96-well filter plate (0.7 mL/well, PE frit, Orochem, Naperville, IL).

5. 2× 96-well plate (PP).

6. Adhesive tape for plate sealing.

7. Lint-free paper.

8. Vacuum manifold equipped with liquid collection box.

9. Shaking platform.

10. Centrifuge equipped with plate-holder inserts.

2.4 Target Plate Spotting

1. Matrix solution: 5 mg/mL 2,5-dihydroxybenzoic acid (2,5-DHB) (Bruker Daltonics, Bremen, Germany) 1 mM NaOH in 50 % ACN. Weigh 5 mg 2,5-DHB on an analytical balance, and dissolve in 1 mL 50 % ACN. Prepare a 100 mM NaOH solution by adding 5 µL 50 % NaOH to 947 µL water and mixing thoroughly. Of this, add 10 µL to the matrix solution (*see* **Note 7**). Can be stored at –20 °C.

2. Ethanol.

3. MTP AnchorChip 800/384 TF MALDI target (Bruker Daltonics).

3   Methods

All steps can be performed at room temperature unless otherwise noted.

3.1 PNGase F N-Glycan Release

1. Add 4 µL 2 % SDS to each well of a 96-well plate (*see* **Note 8**).

2. Add 2 µL glycoprotein sample to each well and mix briefly by pipetting up and down (*see* **Notes 9** and **10**).

3. Cover the plate with adhesive tape to prevent evaporation (*see* **Note 11**).

4. Incubate the plate for 10 min in a 60 °C oven to denature the proteins (*see* **Note 12**).

5. Remove the plate from the oven, and allow it to return to room temperature (±5 min).

6. Carefully remove adhesive tape, making sure any condensation on the tape does not cause cross-contamination of the samples.

7. Add 4 µL release solution to each well (*see* **Note 8**).

8. Cover the plate with new adhesive tape to prevent evaporation.

9. Incubate the plate for 16 h (overnight) in a 37 °C oven. After this, the PNGase F-released N-glycan mixture may be stored at –20 °C until further treatment.

3.2 Ethyl Esterification

1. Add 20 μL esterification reagent to each well of a 96-well plate (*see* **Note 8**).

2. Add 1 μL of PNGase F-released N-glycan sample to each well, and mix briefly by pipetting (*see* **Notes 9** and **13**).

3. Cover the plate with adhesive tape to prevent evaporation.

4. Incubate the plate for 1 h in a 37 °C oven (*see* **Note 14**).

5. Proceed with HILIC SPE on the now ethyl esterified samples. Do not store the samples under esterification reaction conditions.

3.3 Sepharose HILIC Solid Phase Extraction

1. Carefully remove the adhesive tape and discard it (*see* **Note 10**).

2. Add 20 μL pure ACN to each sample and mix briefly by pipetting up and down (*see* **Notes 9** and **15**).

3. Place a 96-well filter plate on a vacuum manifold equipped with a liquid collection box.

4. Transfer 20 μL Sepharose bead slurry to each well of the 96-well filter plate. The Sepharose will sediment rapidly, keep it thoroughly suspended by mixing the solution before each pipetting step (*see* **Note 9**).

5. Apply low vacuum until the liquid has flown through (*see* **Note 16**). However, make sure the filters and beads remain moist (*see* **Note 17**). Any flow-through building up in the collection box can best be discarded after **steps 8** and **15** when the filter plate is removed from the vacuum manifold.

6. Add 100 μL water to each well, then apply low vacuum until the liquid has flown through (*see* **Note 9**). Repeat two additional times.

7. Add 100 μL of 85 % ACN to each well, then apply low vacuum until the liquid has flown through. Repeat two additional times (*see* **Note 9**).

8. Press the filter plate briefly on lint-free paper to remove any solution adhering to the bottom, then place the filter plate on an empty 96-well plate (*see* **Note 18**).

9. Briefly mix the ethyl esterified samples, then transfer them to the filter plate (±38 μL). Apply the samples directly to the beads (*see* **Note 9**).

10. Place the filter plate (on the 96-well plate) on a shaking platform, and incubate for 5 min at maximum velocity (*see* **Note 19**).

11. Transfer the filter plate back to the vacuum manifold and apply low vacuum until the liquid has flown through.

12. Apply 100 μL 85 % ACN 0.1 % TFA to each well, then apply low vacuum until the liquid has flown through. Repeat two additional times (*see* **Notes 9** and **20**).

13. Apply 100 μL 85 % ACN to each well, then apply low vacuum until the liquid has flown through. Repeat two additional times (*see* **Note 9**).

14. Apply additional vacuum to remove residual fluid from the filter plate.

15. Press the filter plate briefly on lint-free paper to remove solution adhering to the bottom, then place it on a clean (new) 96-well plate (*see* **Note 21**).

16. Add 30 μL water to each well for elution (*see* **Note 9**).

17. Place the stacked plates (filter plate on the 96-well plate) on a shaking platform, and incubate for 5 min at maximum velocity (*see* **Note 22**).

18. Transfer the stacked plates to a centrifuge equipped with plate-holder inserts. Use a counter-weight if only one plate will be centrifuged (*see* **Note 23**).

19. Centrifuge for 1 min at $200 \times g$ (*see* **Note 24**).

20. The enriched ethyl esterified N-glycans are now in the flow-through. Do not store the samples before measurement, as this may over time lead to degradation of the lactonized reaction products.

3.4 MALDI Target Spotting

1. Spot 1 μL matrix solution on an AnchorChip MALDI target (*see* **Note 9**).

2. Immediately add 1 μL purified ethyl esterified N-glycan sample on the same spot (*see* **Notes 9** and **25**).

3. Allow the spots to dry by air.

4. Recrystallize the spots by briefly tapping them with a pipette tip containing 0.2 μL ethanol. No pressure is required to transfer the ethanol, and a new tip has to be used for each spot (*see* **Notes 9**, **26** and **27**).

3.5 MALDI-MS Measurement

1. Measure the samples in positive ion mode (*see* **Note 28**). When preforming TOF-MS the resolution can, due to the derivatization step, be increased by reflectron mode measurement (as compared to linear mode measurement) without leading to visible metastable decay of the sialylated glycan species.

2. Before sample measurement, use a standard for calibration (*see* **Note 29**).

3. Assess laser power requirements for the specific matrix/sample combination (*see* **Note 30**).

4. Record the mass spectra using a random walking algorithm in order to sample a large part of the spot and limit measurement variation.

5. When analyzing the spectra, take into account the changes in sialic acid mass due to the ethyl esterification. The sialic acid residues have undergone linkage-specific modification, with the carboxyl groups of α2,6-linked sialic acids having formed ethyl esters, while α2,3-linked sialic acids are now lactonized. Consequentially, while unmodified *N*-acetylneuraminic acid residues have a mass increment of 291.10 Da, after the reaction the α2,3-linked and α2,6-linked variants will have increment masses of 273.08 Da and 319.13 Da respectively (Δ46.04 Da).

6. Be sure to perform quality control on the spectra. While the protocol should give only minor side reactions, a number of potential signals have proven to be informative for troubleshooting purposes and are listed in Table 1.

7. Fragmentation spectra can be obtained by laser- or collision-induced dissociation MS/MS. A list of the most common mass differences suitable for structural assignment is provided in Table 2.

4 Notes

1. Care should be taken when weighing the SDS powder. Avoid inhalation, as SDS may cause irritation of the respiratory tract.

2. NP-40 (substitute) is a very viscous solution, and is not easy to pipette. The tip of a pipette tip can be cut off to widen it and improve the flow.

3. PNGase F is best removed from the freezer just prior to adding it to the release solution. It should be returned to –20 °C immediately afterwards to protect its integrity.

4. Optionally use a flow cabinet for EDC handling. While only of mild toxicity, the chemical produces a strong odor.

5. When preparing EDC/HOBt in larger volumes, store it in a glass bottle. Be sure to rinse the bottle beforehand with ethanol to prevent contamination of the reagent.

6. The stability of the reagent was tested at various conditions, and while reactivity will remain secured for over three months when stored at –20 °C, repeated exposure to room temperature will remove reactivity within a day. When preparing a large quantity of reagent that needs storage, be sure to aliquot prior to storage.

Table 1
Potential satellite signals relative to expected glycan peaks. Whereas amidation and incomplete esterification affect only the sialic acids, and therefore change the relative distribution of the analyzed glycans, the reducing end losses and salt variation have an impact on all species and preserve the relative distribution. In addition to the displayed masses, the presence of peaks in the spectrum with significantly lower resolution (metastable peaks) may also indicate incomplete esterification or too high laser power settings

Δmass (Da)	Explanation	Preventive measure
−367.15	Loss of fucosylated reducing end *N*-acetylglucosamine	Lower MALDI laser power
−221.09	Loss of reducing end *N*-acetylglucosamine when not fucosylated	Lower MALDI laser power
−101.05	^{0,2}A crossring fragment of reducing end *N*-acetylglucosamine	Lower MALDI laser power
−29.02	Amidation of α2,6-linked sialic acid rather than ethyl esterification	Limit ammonium-based buffering during sample preparation or perform desalting before ethyl esterification
−6.05	Incomplete ethyl esterification of α2,6-linked sialic acids with neutral proton to sodium exchange at the carboxylic acid	Increase amount of reagent relative to sample
15.97	$[M+K]^+$ ionization rather than $[M+Na]^+$ ionization	Increase NaOH concentration in matrix or dilute sample before spotting
17.03	Amidation of α2,3-linked sialic acid rather than lactonization	Limit ammonium-based buffering during sample preparation or perform desalting before ethyl esterification
39.99	Incomplete lactonization of α2,3-linked sialic acids with neutral proton to sodium exchange at the carboxylic acid	Increase amount of reagent relative to sample
132.19	Unidentified ionization variant rather than $[M+Na]^+$ ionization	Increase NaOH concentration in matrix or dilute sample before spotting
357.18	Unidentified reducing end modification	Perform glycan enrichment before ethyl esterification

7. NaOH is added to the matrix to promote $[M+Na]^+$ ionization, particularly limiting the formation of $[M+K]^+$ ions.

8. This step can rapidly be performed with a repeating pipette. Do not touch the liquid at the bottom of the wells, but rather pipette to the side of wells taking a new side for each new solution added to the mix.

9. This step can rapidly be performed with a multichannel pipette (8 or 12-channel). Samples can directly be transferred from one 96-well plate to the next, while stock solutions can be put into reservoirs before pipetting.

Table 2
The most commonly observed structurally informative mass differences from a given precursor ion. Unless multiple fucoses are present on a structure, assigning a fucose to an antenna requires a −221.09 signal to prove the lack of core fucosylation. Observed losses can be verified at the lower *m/z* range of the spectrum as [M + Na]$^+$ ions

Δmass (Da)	Explanation
−221.09	Reducing end *N*-acetylglucosamine when not fucosylated
−319.13	α2,6-linked *N*-acetylneuraminic acid
−365.13	*N*-acetyllactosamine (LacNAc) (unsialylated antenna)
−367.15	Reducing end *N*-acetylglucosamine when fucosylated
−511.19	Fucosylated LacNAc (Lewis A or X)
−638.22	Antenna carrying an α2,3-linked *N*-acetylneuraminic acid
−684.26	Antenna carrying an α2,6-linked *N*-acetylneuraminic acid
−784.28	Antenna carrying both an α2,3-linked *N*-acetylneuraminic acid and a fucose (sialyl Lewis A or X)

10. Upscaling the volume of the PNGase F release is not a problem, but downscaling leads to issues with evaporation—keep 10 μL as a minimum total volume unless using smaller wells (e.g., a 384-well plate).

11. Adhesive tape is a source of contamination when not handled correctly. Condensation and droplets on the tape can bring polymers into a sample, while imprecision in replacement can lead to cross-contamination. Rather use new tape for every step in the protocol.

12. Use a heating source that has homogeneous temperature distribution (oven), or a source featuring top-heating (PCR machine). Bottom-heating-only will cause significant condensation on the plate seal, increasing the chance of contamination.

13. 1:10 sample/reagent still yields complete ethyl esterification under all tested conditions, and 1:20 has been chosen to have a margin of error. Lowering the relative amount of reagent may be attractive when starting with high sample volumes, but may result in incomplete reactions (most likely due to increased water content).

14. The temperature and time requirements of the ethyl esterification are lenient, as half an hour at room temperature has already shown to yield reaction completeness for the standards tested. 1 h and 37 °C were chosen to be controllable conditions suitable for a wide range of samples.

15. The ethyl esterified glycans are retained on Sepharose (or cotton) directly from a 50:50 ACN/ethanol solution.

16. The vacuum can be just enough to cause flow, which will lead to maximum interaction time. The vacuum can even be so low as to not even register on the pressure gauge.

17. Take care not to let the beads dry, as it will decrease the efficiency of all interactions in the purification protocol, including the sample binding and elution, as well as cause clogging.

18. This empty 96-well plate can be reused across experiments.

19. Sample binding is a critical step in the purification process, so be sure to allow for at least 5 min of incubation.

20. While the ion pairing agent TFA does increase the overall purity of the recovered glycan mixture, the acidic conditions it causes can potentially lead to hydrolysis of the esters formed by the ethyl esterification. While we did not observe this breakdown, be sure to limit the exposure time of the samples to the acidic conditions, and rapidly proceed towards the washing steps without TFA.

21. Droplets of organic solution remaining at the bottom of the plate may cause cross-contamination, or mix with the eluent, having a deleterious effect on the glycan profiles obtained.

22. Elution is a critical step in the purification process, be sure to allow for at least 5 min of incubation.

23. A counterweight plate can be prepared by stacking a filter plate and elution plate, and adding 30 μL of water as with the actual samples. This counterweight can be reused for future experiments.

24. Should water remain in the filter plate (which may happen if those wells had fallen dry before) try spinning for an additional 1 min or at higher g-force value.

25. The organics in the matrix will start evaporating rapidly when on the plate, changing the eventual crystallization conditions. It is recommended to always spot the sample straight after spotting the matrix, and only then moving on to the next sample.

26. Recrystallization increases the homogeneity of the sample, and facilitates automatic measurement. Recrystallization with ethanol will not work on a MALDI target without having a hydrophilic patch within a hydrophobic layer (like an AnchorChip target, but unlike a polished steel target) as the ethanol will not be contained on the spot.

27. Next to recrystallization, alternative matrices can be used to create a homogeneous matrix layer. One example would be super-DHB (a 9:1 mixture of 2,5-DHB and 2-hydroxy-5-methoxybenzoic acid)

(Sigma-Aldrich) which leads to a similar increase in shot-to-shot repeatability. As downside, super-DHB is less stable in solution than regular 2,5-DHB, and therefore has to be prepared freshly more often.

28. While native sialylated glycans preferentially ionize as $[M-H]^-$ and require measurement in negative ion mode, after ethyl esterification positive ion mode can be used for both sialylated and unsialylated species. In addition, the increased stability after modification allows the use of a reflectron for enhanced resolution.

29. One example of a standard would be peptide calibration standard (Bruker Daltonics), but any sample can be used that is amenable to ionization and contains known masses. Preparing an ethyl esterified glycan standard would increase the similarity to the actual sample in both the crystallization and ionization conditions, and may be preferential.

30. For ethyl esterified samples the laser power generally needs to be higher than for underivatized or reducing end-labeled samples.

Acknowledgements

This work was supported by the European Union (Seventh Framework Programme HighGlycan project, grant number 278535).

References

1. Wormald MR, Dwek RA (1999) Glycoproteins: glycan presentation and protein-fold stability. Structure 7(7):R155–R160

2. Crocker PR, Paulson JC, Varki A (2007) Siglecs and their roles in the immune system. Nat Rev Immunol 7(4):255–266. doi:10.1038/nri2056

3. Alessandri L, Ouellette D, Acquah A et al (2012) Increased serum clearance of oligomannose species present on a human IgG1 molecule. MAbs 4(4):509–520. doi:10.4161/mabs.20450

4. Moremen KW, Tiemeyer M, Nairn AV (2012) Vertebrate protein glycosylation: diversity, synthesis and function. Nat Rev Mol Cell Biol 13(7):448–462. doi:10.1038/nrm3383

5. Lauc G, Essafi A, Huffman JE et al (2010) Genomics meets glycomics—the first GWAS study of human N-Glycome identifies HNF1alpha as a master regulator of plasma protein fucosylation. PLoS Genet 6(12), e1001256. doi:10.1371/journal.pgen.1001256

6. Vanhooren V, Desmyter L, Liu XE et al (2007) N-glycomic changes in serum proteins during human aging. Rejuvenation Res 10(4):521–531a. doi:10.1089/rej.2007.0556

7. Reis CA, Osorio H, Silva L et al (2010) Alterations in glycosylation as biomarkers for cancer detection. J Clin Pathol 63(4):322–329. doi:10.1136/jcp.2009.071035

8. Knezevic A, Gornik O, Polasek O et al (2010) Effects of aging, body mass index, plasma lipid profiles, and smoking on human plasma N-glycans. Glycobiology 20(8):959–969. doi:10.1093/glycob/cwq051

9. Adamczyk B, Tharmalingam T, Rudd PM (2012) Glycans as cancer biomarkers. Biochim Biophys Acta 1820(9):1347–1353. doi:10.1016/j.bbagen.2011.12.001

10. Albrecht S, Unwin L, Muniyappa M et al (2014) Glycosylation as a marker for inflammatory arthritis. Cancer Biomark 14(1):17–28. doi:10.3233/CBM-130373

11. Jefferis R (2009) Glycosylation as a strategy to improve antibody-based therapeutics. Nat Rev Drug Discov 8(3):226–234. doi:10.1038/nrd2804

12. Huffman JE, Pucic-Bakovic M, Klaric L et al (2014) Comparative performance of four methods for high-throughput glycosylation analysis of immunoglobulin G in genetic and epidemiological research. Mol Cell Proteomics 13(6):1598–1610. doi:10.1074/mcp.M113.037465

13. Harvey DJ (1999) Matrix-assisted laser desorption/ionization mass spectrometry of carbohydrates. Mass Spectrom Rev 18(6):349–450. doi:10.1002/(SICI)1098-2787(1999)18:6<349::AID-MAS1>3.0.CO;2-H

14. Powell AK, Harvey DJ (1996) Stabilization of sialic acids in N-linked oligosaccharides and gangliosides for analysis by positive ion matrix-assisted laser desorption/ionization mass spectrometry. Rapid Commun Mass Spectrom 10(9):1027–1032. doi:10.1002/(SICI)1097-0231(19960715)10:9<1027::AID-RCM634>3.0.CO;2-Y

15. Miura Y, Shinohara Y, Furukawa J et al (2007) Rapid and simple solid-phase esterification of sialic acid residues for quantitative glycomics by mass spectrometry. Chemistry 13(17):4797–4804. doi:10.1002/chem.200601872

16. Liu X, Li X, Chan K et al (2007) "One-pot" methylation in glycomics application: esterification of sialic acids and permanent charge construction. Anal Chem 79(10):3894–3900. doi:10.1021/ac070091j

17. Liu X, Qiu H, Lee RK et al (2010) Methylamidation for sialoglycomics by MALDI-MS: a facile derivatization strategy for both alpha2,3- and alpha2,6-linked sialic acids. Anal Chem 82(19):8300–8306. doi:10.1021/ac101831t

18. Wheeler SF, Domann P, Harvey DJ (2009) Derivatization of sialic acids for stabilization in matrix-assisted laser desorption/ionization mass spectrometry and concomitant differentiation of alpha(2 → 3)- and alpha(2 → 6)-isomers. Rapid Commun Mass Spectrom 23(2):303–312. doi:10.1002/rcm.3867

19. Alley WR Jr, Novotny MV (2010) Glycomic analysis of sialic acid linkages in glycans derived from blood serum glycoproteins. J Proteome Res 9(6):3062–3072. doi:10.1021/pr901210r

20. Reiding KR, Blank D, Kuijper DM et al (2014) High-throughput profiling of protein N-glycosylation by MALDI-TOF-MS employing linkage-specific sialic acid esterification. Anal Chem 86(12):5784–5793. doi:10.1021/ac500335t

21. Bondt A, Rombouts Y, Selman MH et al (2014) IgG Fab glycosylation analysis using a new mass spectrometric high-throughput profiling method reveals pregnancy-associated changes. Mol Cell Proteomics. doi:10.1074/mcp.M114.039537

22. Selman MH, Hemayatkar M, Deelder AM et al (2011) Cotton HILIC SPE microtips for microscale purification and enrichment of glycans and glycopeptides. Anal Chem 83(7):2492–2499. doi:10.1021/ac1027116

Chapter 12

Characterization of Protein *N*-Glycosylation by Analysis of ZIC-HILIC-Enriched Intact Proteolytic Glycopeptides

Gottfried Pohlentz*, Kristina Marx*, and Michael Mormann

Abstract

Zwitterionic hydrophilic interaction chromatography (ZIC-HILIC) solid-phase extraction (SPE) combined with direct-infusion nanoESI mass spectrometry (MS) and tandem MS/MS is a well-suited method for the analysis of protein *N*-glycosylation. A site-specific characterization of *N*-glycopeptides is achieved by the combination of proteolytic digestions employing unspecific proteases, glycopeptide enrichment by use of ZIC-HILIC SPE, and subsequent mass spectrometric analysis. The use of thermolysin or a mixture of trypsin and chymotrypsin leads per se to a mass-based separation, that is, small nonglycosylated peptides and almost exclusively glycopeptides at higher *m/z* values. As a result of their higher hydrophilicity *N*-glycopeptides comprising short peptide backbones are preferably accumulated by the ZIC-HILIC-based separation procedure. By employing this approach complications associated with low ionization efficiencies of *N*-glycopeptides resulting from signal suppression in the presence of highly abundant nonglycosylated peptides can be largely reduced. Here, we describe a simple protocol aimed at the enrichment of *N*-glycopeptides derived from in-solution and in-gel digestions of SDS-PAGE-separated glycoproteins preceding mass spectrometric analysis.

Key words *N*-glycosylation, Glycopeptides, In-solution digestion, In-gel digestion, NanoESI MS, CID

1 Introduction

Glycosylation is found in over 50 % of all eucaryotic proteins and is described as the most complex form of posttranslational modification leading to a heterogeneous expression of glycoproteins as mixtures of glycoforms. The biological role of glycans is highly variable since glycoproteins have a ubiquitous and complex nature. They occur inside cells and in extracellular fluids and are embedded in cell membranes [1]. The glycan moieties are involved in determination of the physicochemical properties of glycoproteins, e.g., charge, size, accessibility, structure, and solubility. Therefore, they

---

*The first two authors (Pohlentz and Marx) contributed equally and share the first authorship.

Jörg Reinders (ed.), *Proteomics in Systems Biology: Methods and Protocols*, Methods in Molecular Biology, vol. 1394,
DOI 10.1007/978-1-4939-3341-9_12, © Springer Science+Business Media New York 2016

can influence structural and modulatory functions which are important for enzymes, structural proteins, or hormones or are directly involved in specific recognition of glycans serving as receptors [2, 3]. Thus, glycoproteins are crucial for the development, growth, function, or survival of an organism. Hence, their relationship of structure, location, and function is still an important feature in life sciences [4].

The main types of glycosylation are *N*-glycosylation—where the glycan is covalently attached to asparagine residues within the consensus-sequence NXS/T with X as any amino acid except proline—and *O*-glycosylation where the glycan is linked covalently to serine and/or threonine residues of eucaryotic glycoproteins [3].

N-glycosylation is subdivided into three types, viz. complex type, high mannose type, and hybrid type, with all sharing a common trimannose-chitobiose core whereas *O*-glycans show a wide variety of core structures [2].

Techniques and methodologies aimed at glycosylation analysis require high sensitivity to provide the detection and separation of large molecules containing very small structural differences. Various analytical techniques have been described to determine protein *N*-glycosylation [5–9]. Many protocols involve a deglycosylation step using specific glycosidases (e.g., PNGase F) prior to proteolytic digestion. However, this approach limits the structural information since evidence on the glycan-protein linkage with respect to site specificity is lost [10, 11]. This problem can be compassed by direct inspection of intact glycopeptide ions derived from proteolytic digestions by use of nanoESI mass spectrometry (MS) which can provide information on glycan structure and specific glycan-protein linkage site [12]. However, direct mass spectrometric analysis may still be hampered by low ionization efficiencies of glycopeptides, signal suppression as a result of highly abundant nonglycosylated peptides, and a lower overall abundance of individual glycosylated peptides due to their high structural heterogeneity [13]. These problems can be overcome either by use of high-performance liquid chromatography-mass spectrometry (HPLC-MS) [14–16] or by enrichment of glycoproteins and glycopeptides employing solid-phase extraction (SPE) [17, 18]. Recently, zwitterionic hydrophilic interaction liquid chromatography (ZIC-HILIC) has been introduced for the selective enrichment of hydrophilic analytes such as glycans or glycopeptides based on the strong interaction with zwitterionic sulfobetaine moieties present at the surface of the stationary phase. The first reports on SPE of tryptic glycopeptides by use of ZIC-HILIC described their selective enrichment followed by enzymatic deglycosylation and subsequent separate mass spectrometric analysis of glycans and deglycosylated peptides. The data obtained allowed for an identification of *N*-glycosylation sites and glycan characterization though a site-specific assignment of glycosylation was not possible [19–21].

Recently, we have shown that the combination of less specific proteases like thermolysin, elastase, or the use of a mixture of trypsin and chymotrypsin and ZIC-HILIC SPE leads to specific enrichment of glycopeptides obtained by in-solution digestions [13, 22–24]. Shorter peptide backbones lead to higher hydrophilicity of the analytes and therefore stronger interactions with the stationary phase. However, this protocol cannot solve the problem of analyzing complex glycoprotein mixtures. Separation of glycoproteins in mixtures prior to their enzymatic degradation, followed by enrichment of glycopeptides and characterization by mass spectrometry, is a prerequisite to obtain an entire and unambiguous structural characterization. Sodium dodecyl sulfate-polyacrylamide gel electrophoresis (SDS-PAGE) is a commonly used method for separation of proteins and also applied to glycoproteins. Separated (glyco)proteins can be submitted to in-gel digestion and glycopeptides can be selectively enriched by use of ZIC-HILIC [25, 26]. A significant improvement of the method can be achieved if HILIC SPE is combined with a desalting step using C18-reversed-phase (RP) SPE (vide infra).

Here, we present a simple protocol for in-solution and in-gel digestion of glycoproteins followed by selective glycopeptide enrichment employing ZIC-HILIC SPE. An overview of the workflow is depicted in Fig. 1. We have chosen bovine IgG as a representative example. The nanoESI mass spectra obtained from in-solution and in-gel tryptic/chymotryptic digestions of bovine IgG (in-gel digestion of the heavy chain) obtained after different steps of SPE using ZIC-HILIC and/or C18-RP as stationary phase are depicted in Fig. 2. All detected glycopeptides are listed in Table 1. Table 2 gives a synopsis of the different glycan structures at specific glycosylation sites independent of the length of the peptide backbone. For a simplified assignment of different glycopeptide species the following notation has been used: The first number of the code represents the glycan structure listed in Table 2 (column 1). The following letter refers to the glycosylation site (cf. Table 1, columns 1 and 2) and the indices correspond to the different peptide backbones (cf. Table 1, columns 1 and 3).

The three most abundant glycopeptides (No. 9, 10, 11; cf. Table 1 and Fig. 2a) can be detected in their ionized form immediately after in-solution digest or in-gel digest with C18-RP SPE (ZipTip C18) purification without any glycopeptide enrichment. Figure 2b shows the advantages of ZIC-HILIC separation after in-solution digest (B1) compared to in-solution digest without further preparation (B2). Glycopeptides give rise to intense signals and the number of detected glycoforms increases significantly after ZIC-HILIC SPE separation (cf. Table 2, column 4 and 5). The best results for in-gel digests of IgG heavy chains could be observed after combined purification, i.e., desalting with C18-RP SPE

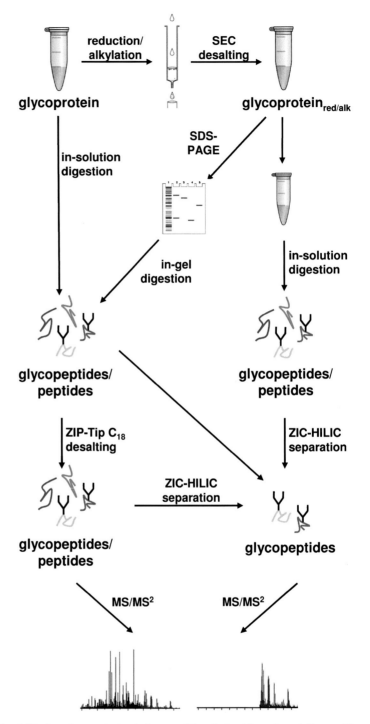

Fig. 1 Strategy for analysis of glycopeptides derived from in-solution and in-gel digestions of glycoproteins

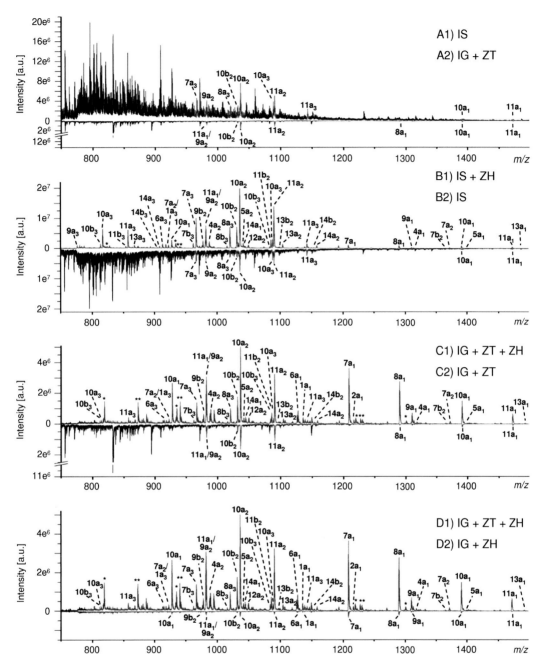

Fig. 2 Direct-infusion nanoESI mass spectra obtained from tryptic/chymotryptic digestions of bovine IgG preparations analyzed by different methods: IS = In-solution digest, IG = In-gel digest, ZT = ZipTip desalting, ZH = ZIC-HILIC enrichment. The code used for peak annotation corresponds to the glycan structures listed in Table 2 (column 1)—first number, the glycosylation site (*see* Table 1, columns 1 and 2)—letter, and the peptide backbone (*see* Table 1, columns 1 and 3)—indices

Table 1

Synopsis of detected glycopeptides derived from in-solution and in-gel tryptic/chymotryptic digestions of bovine IgG (in-gel digestion of heavy chain) after enrichment (for details see Table 2)

Peak assignment[a]	Glycosylation site	Peptide[b]	HexNAc	Hex	Fuc	m/z (calc+2)	m/z (exp+2)	m/z (calc+3)	m/z (exp+3)	m/z (calc+4)	m/z (exp+4)
Bovine IgG1											
1a$_1$	N$_{183}$	EEQFN#STYR	3	3	0	1134.96	1135.02				
2a$_1$	N$_{183}$	EEQFN#STYR	3	4	0	1215.99	1216.06				
4a$_1$	N$_{183}$	EEQFN#STYR	4	4	0	1317.53	1317.46				
5a$_1$	N$_{183}$	EEQFN#STYR	4	5	0	1398.55	1398.48				
6a$_1$	N$_{183}$	EEQFN#STYR	3	2	1	1126.96	1127.15				
7a$_1$	N$_{183}$	EEQFN#STYR	3	3	1	1207.99	1207.90				
8a$_1$	N$_{183}$	EEQFN#STYR	3	4	1	1289.02	1288.96				
9a$_1$	N$_{183}$	EEQFN#STYR	4	3	1	1309.53	1309.45				
10a$_1$	N$_{183}$	EEQFN#STYR	4	4	1	1390.56	1390.49	927.37	927.36		
11a$_1$	N$_{183}$	EEQFN#STYR	4	5	1	1471.58	1471.55	981.39	981.37		
13a$_1$	N$_{183}$	EEQFN#STYR	5	4	1	1492.10	1491.98				
14a$_1$	N$_{183}$	EEQFN#STYR	5	5	1	1573.12	1573.05	1049.08	1049.08		
3a$_2$	N$_{183}$	TRPKEEQFN#STY	4	3	0			933.40	933.38		
4a$_2$	N$_{183}$	TRPKEEQFN#STY	4	4	0			987.42	987.41		
5a$_2$	N$_{183}$	TRPKEEQFN#STY	4	5	0			1041.44	1041.43		
7a$_2$	N$_{183}$	TRPKEEQFN#STY	3	3	1	1371.09	1371.03	914.39	914.38		
9a$_2$	N$_{183}$	TRPKEEQFN#STY	4	3	1			982.09	982.07		
10a$_2$	N$_{183}$	TRPKEEQFN#STY	4	4	1			1036.10	1036.10		
11a$_2$	N$_{183}$	TRPKEEQFN#STY	4	5	1			1090.12	1090.13		
12a$_2$	N$_{183}$	TRPKEEQFN#STY	5	3	1			1049.78	1049.76		
13a$_2$	N$_{183}$	TRPKEEQFN#STY	5	4	1			1103.80	1103.79		
14a$_2$	N$_{183}$	TRPKEEQFN#STY	5	5	1			1157.82	1157.81		
1a$_3$	N$_{183}$	TRPKEEQFN#STYR	3	3	0			917.74	917.71		
2a$_3$	N$_{183}$	TRPKEEQFN#STYR	3	4	0			971.76	971.73		
6a$_3$	N$_{183}$	TRPKEEQFN#STYR	3	2	1			912.41	912.38		
7a$_3$	N$_{183}$	TRPKEEQFN#STYR	3	3	1			966.43	966.43		
8a$_3$	N$_{183}$	TRPKEEQFN#STYR	3	4	1			1020.45	1020.44		
9a$_3$	N$_{183}$	TRPKEEQFN#STYR	4	3	1			1034.12	1034.12	775.84	775.83

Code	Site	Peptide											
10a3	N_{183}	TRPKEEQFN#STYR	4	4	1					1088.14	1088.14	816.36	816.34
11a3	N_{183}	TRPKEEQFN#STYR	4	5	1					1142.16	1142.17	856.87	856.86
13a3	N_{183}	TRPKEEQFN#STYR	5	4	1							867.13	867.09
14a3	N_{183}	TRPKEEQFN#STYR	5	5	1							907.64	907.66
Bovine IgG2													
7b2	N_{183}	TRPNEEQFN#STY	3	3	1	1364.06	1363.95						
8b2	N_{183}	TRPNEEQFN#STY	3	4	1	1445.09	1445.01						
9b2	N_{183}	TRPNEEQFN#STY	4	3	1	1465.60	1465.53	977.40	977.41				
10b2	N_{183}	TRPNEEQFN#STY	4	4	1	1546.63	1546.62	1031.42	1031.42				
11b2	N_{183}	TRPNEEQFN#STY	4	5	1	1627.65	1627.57	1085.44	1085.45				
12b2	N_{183}	TRPNEEQFN#STY	5	3	1			1045.10	1045.10				
13b2	N_{183}	TRPNEEQFN#STY	5	4	1			1099.11	1099.12				
14b2	N_{183}	TRPNEEQFN#STY	5	5	1			1153.13	1153.15				
7b3	N_{183}	TRPNEEQFN#STYR	3	3	1			961.74	961.73				
8b3	N_{183}	TRPNEEQFN#STYR	4	4	1			1015.76	1015.77				
9b3	N_{183}	TRPNEEQFN#STYR	4	3	1			1029.44	1029.45				
10b3	N_{183}	TRPNEEQFN#STYR	4	4	1			1083.45	1083.47			812.84	812.84
11b3	N_{183}	TRPNEEQFN#STYR	4	5	1							853.36	853.37
14b3	N_{183}	TRPNEEQFN#STYR	5	5	1							904.13	904.13

[a] Annotation code of glycopeptides: First number indicates glycan structure (*see* Table 2), *letter* denotes glycosylation site, *suffix* indicates peptide backbone

[b] # indicates occupied glycosylation site

Table 2
Detected glycan structures in glycopeptides derived from in-solution and in-gel tryptic/chymotryptic digestions of bovine IgG (in-gel digestion of heavy chain)

No./glycosylation site[a]	Glycan	IS	IS+ZH	IG+ZT	IG+ZH	IG+ZT+ZH
Bovine IgG						
1 $^1N_{183}$		−	+	−	+	+
$^2N_{183}$		−	−	−	−	−
2 $^1N_{183}$		−	+	−	−	+
$^2N_{183}$		−	−	−	−	−
3 $^1N_{183}$		−	+	−	−	−
$^2N_{183}$		−	−	−	−	−
4 $^1N_{183}$		−	+	−	−	+
$^2N_{183}$		−	−	−	−	−
5 $^1N_{183}$		−	+	−	−	+
$^2N_{183}$		−	−	−	−	−
6 $^1N_{183}$		−	+	−	+	+
$^2N_{183}$		−	−	−	−	−
7 $^1N_{183}$		+	+	−	+	+
$^2N_{183}$		−	+	−	−	+
8 $^1N_{183}$		+	+	+	+	+
$^2N_{183}$		−	+	−	−	+
9 $^1N_{183}$		+	+	+	+	+
$^2N_{183}$		−	+	−	+	+
10 $^1N_{183}$		+	+	+	+	+
$^2N_{183}$		+	+	+	+	+
11 $^1N_{183}$		+	+	+	+	+
$^2N_{183}$		−	+	−	−	+
12 $^1N_{183}$		−	+	−	−	+
$^2N_{183}$		−	+	−	−	−
13 $^1N_{183}$		−	+	−	−	+
$^2N_{183}$		−	+	−	−	−
14 $^1N_{183}$		−	+	−	−	+
$^2N_{183}$		−	+	−	−	+

Sample preparation techniques: IS=In-solution, IG=In-gel, ZT=ZipTip, ZH=ZIC-HILIC. Glycan structures depicted using the recommendation of the Consortium for Functional Glycomics [30]. *Blue square*: N-acetylglucosamine, *green circle*: mannose, *yellow circle*: galactose, *red triangle*: fucose

[a] *Prefix* indicates IgG 1 and IgG 2, respectively; *suffix* denotes glycosylation site

preceding ZIC-HILIC SPE separation (cf. Fig. 2c/d; Table 2, column 6–8). The direct separation with ZIC-HILIC after digests seems to be hampered caused by electrostatic (ionic) interactions which occur in the presence of high concentrations of salt ions in the sample after gel separation.

Eventually, the application of the presented methods in combination with direct-infusion nanoESI MS enables multiple CID experiments and therefore facilitates the analysis of *N*-glycans of individual glycoproteins separated by SDS-PAGE. Henceforward, these methods may also support the analysis of unknown glycoproteins in complex glycoprotein mixtures.

2   Materials

All aqueous solutions should be prepared using ultrapure water (18.2 MΩ×cm at 25 °C, Merck Millipore Synergy Ultrapure Water Systems, Billerica, MA, USA) and analytical grade solvents and reagents should be employed. All solutions should be prepared freshly prior to use at ambient temperature.

2.1   Reduction and Alkylation

1. Reduction and alkylation buffer: 6.0 M guanidinium hydrochloride, 250 mM trizma base/HCl, and 65 mM dithiothreitol, pH 8.6.

2. Iodoacetamide.

2.2   Size-Exclusion Chromatography

1. Sephadex G-25 size-exclusion chromatography columns (illustra NAP™ 5 columns, volume 0.5 ml, GE Healthcare Europe GmbH, Freiburg, Germany). Alternatively, polyacrylamide gel columns (Micro Bio-Spin Biogel P-6 chromatography columns, Bio-Rad Laboratories GmbH, München, Germany) can be used (*see* **Note 1**).

2. Column equilibration and elution buffer: 50 mM Ammonium bicarbonate.

2.3   In-Solution Digestion

1. Digestion buffer: 10 mM ammonium bicarbonate.

2. Proteases: Trypsin (0.1 µg/µl in 1 mM HCl), sequencing grade (Roche Diagnostics, Mannheim, Germany);
    chymotrypsin (0.1 µg/µl in 1 mM HCl), sequencing grade (Roche Diagnostics, Mannheim, Germany);
    thermolysin (0.5 µg/µl in H$_2$O) (Sigma-Aldrich Chemie GmbH, Taufkirchen, Germany) (*see* **Note 2**).

2.4   In-Gel Digestion

1. Destaining solutions: 100 mM Ammonium bicarbonate/acetonitrile (50/50) and neat acetonitrile.

2. Digestion buffer: 10 mM Ammonium bicarbonate.

3. Proteases: Trypsin (0.1 µg/µl in 1 mM HCl), sequencing grade (Roche Diagnostics, Mannheim, Germany); chymotrypsin (0.1 µg/µl in 1 mM HCl), sequencing grade (Roche Diagnostics, Mannheim, Germany); thermolysin (0.5 µg/µl in H_2O) (Sigma-Aldrich Chemie GmbH, Taufkirchen, Germany) (*see* **Note 2**).

4. Extraction buffers: Acetonitrile/water + formic acid (50/50 + 5), acetonitrile/water + formic acid (80/20 + 5), acetonitrile.

2.5 Desalting of Extracted Proteolytic Peptides

1. Solid-phase extraction C18-RP tips: ZipTip C18, Tip Size P10, (Merck Chemicals GmbH, Schwalbach, Germany).

2. Wetting solution: Acetonitrile.

3. Binding solution: Trifluoroacetic acid (0.1 %).

4. Equilibration solution: Trifluoroacetic acid (0.1 %).

5. Wash solution: Trifluoroacetic acid (0.1 %).

6. Elution solution 1: Trifluoroacetic acid (0.1 %)/acetonitrile (50/50).

7. Elution solution 2: Trifluoroacetic acid (0.1 %)/acetonitrile (20/80).

8. Elution solution 3: Acetonitrile.

2.6 ZIC-HILIC Enrichment of Proteolytic N-glycopeptides

1. Solid-phase extraction ZIC-HILIC tips: ZIC-HILIC ProteaTip, 10–200 µl (dichrom GmbH, Marl, Germany).

2. Binding solution: Acetonitrile/water + formic acid (80/20 + 2).

3. Equilibration solution: Acetonitrile/water + formic acid (80/20 + 2).

4. Wash solution: Acetonitrile/water + formic acid (80/20 + 2).

5. Elution solution: Water + formic acid (98/2).

2.7 Nano Electrospray Mass Spectrometry (NanoESI MS)

1. Sample buffer: Acetonitrile/water + formic acid (50/50 + 2).

2. Mass spectrometer: Quadrupole time-of-flight (Q-TOF) mass spectrometer (Micromass, Manchester, UK) equipped with a Z-spray source in the positive ion mode. Typical source parameters: source temperature: 80 °C, desolvation gas (N_2) flow rate: 75 l/h, capillary voltage: 1.1 kV, cone voltage: 30 V. Low energy CID parameters: collision gas (Ar) pressure: 3.0×10^{-3} Pa, collision energies: 20–40 eV (E_{lab}).

3 Methods

All procedures should be performed at ambient temperature unless otherwise specified.

3.1 Reduction and Alkylation of Glycoproteins

1. For in-solution reduction and alkylation 1 nmol of glycoprotein is dissolved in 200 µl of a mixture of 6.0 M guanidinium hydrochlorid, 250 mM trizma base/HCl, and 65 mM dithiothreitol.

2. The resulting mixture is incubated for 1 h at 56 °C.

3. 2.5 mg Iodoacetamide is added and the resulting mixture is incubated for 45 min under exclusion of light.

3.2 Desalting by Size-Exclusion Chromatography (Sephadex G-25)

1. Allow excess packing buffer to drain from the NAP™ 5 column by gravity to the top of the gel bed.

2. Apply 2.5 ml of 50 mM ammonium bicarbonate and allow the buffer to drain out by gravity to the top of the gel bed. Repeat this step four times.

3. Carefully add the sample obtained under Subheading 3.1 to the column and allow the sample to penetrate the gel bed completely.

4. Apply 500 µl of 50 mM ammonium bicarbonate and allow the buffer to drain out by gravity to the top of the gel bed. The flow through is dicarded.

5. Place an appropriate collection tube under the column, apply 500 µl of 50 mM ammonium bicarbonate, and allow the buffer to drain out by gravity to the top of the gel bed.

6. The sample is dried by removal of the solvent in vacuo by use of a centrifugal evaporator.

3.3 Desalting by Size-Exclusion Chromatography (Biogel P-6)

The experimental procedure follows the manufacturer's instructions.

1. Resuspend settled gel by sharply inverting the Micro Bio-Spin Biogel P-6 chromatography column several times and remove residual air bubbles by tapping the column.

2. Snap off the tip of the column and place in a 2 ml collection tube.

3. Remove the cap and allow excess packing buffer to drain from column by gravity to the top of the gel bed. If the buffer flow does not start immediately push back the cap on the column and remove again to initiate the buffer flow.

4. The packing buffer is discarded and the column is placed back in the collection tube. The column is centrifuged in a microcentrifuge for 2 min at $1000 \times g$. Subsequently, the flow through is discarded and the column is placed back in the collection tube.

5. Apply 500 µl of 50 mM ammonium bicarbonate, drain buffer by centrifugation for 1 min at $1000 \times g$, discard buffer, and place back column in the collection tube. Repeat this step four times.

6. Place column in an appropriate collection tube (1.5 ml), carefully add the sample obtained under Subheading 3.1 (20–

75 μl) to the column, and allow the sample to penetrate the gel bed completely. Centrifuge the assembly for 4 min at $1000 \times g$.

7. The sample is dried by removal of the solvent in vacuo by use of a centrifugal evaporator.

3.4 In-Solution Digestion

1. For in-solution digestion 100 pmol of either the reduced and alkylated or the untreated glycoprotein is dissolved in 20 μl 10 mM ammonium bicarbonate (*see* **Note 3**).

2. Shake the mixture for 7 min at 95 °C (*see* **Note 4**) and chill tubes to room temperature.

3. Add 1 μl of the protease solution and incubate the reaction mixture overnight at 37 °C in a shaker (750 rpm) (*see* **Note 5**). If thermolysin is employed as protease incubate at 65 °C in a shaker (750 rpm).

4. The sample is dried by removal of the solvent in vacuo by use of a centrifugal evaporator.

5. Add 50 μl water and dry in vacuo by use of a centrifugal evaporator. Repeat this step at least once.

3.5 In-Gel Digestion

If reduced and alkylated (glyco-)proteins have been separated by use of one-dimensional or two-dimensional sodium dodecyl sulfate-polyacrylamide gel electrophoresis (SDS-PAGE) and visualized by either silver or Coomassie® brilliant blue staining glycopeptides are formed by in-gel digestion (*see* **Note 3**). The in-gel digestion procedure follows the method described by Shevchenko et al. [27].

1. Excise protein bands or spots by use of a clean scalpel.

2. Cut bands into small pieces (*see* **Note 6**) and transfer gel cubes into a reaction tube of appropriate volume (0.5 or 1.5 ml).

3. Add 100 μl of 100 mM ammonium bicarbonate/acetonitrile (50/50) and shake for 30 min. Subsequently, remove destaining solution.

4. Add 500 μl of acetonitrile and shake for 10 min until gel pieces shrink and become opaque. Remove supernatant and dry gel pieces in vacuo by use of a centrifugal evaporator.

5. Add 30 μl of 10 mM ammonium bicarbonate and 10 μl of protease solution and incubate gel pieces for 30 min on an ice bath.

6. If digestion buffer solution is absorbed entirely add an appropriate amount of 10 mM ammonium bicarbonate and protease solution (3/1, v/v) to cover gel pieces completely. Incubate reaction mixture for 30 min on an ice bath (*see* **Note 7**).

7. Add an appropriate amount of 10 mM ammonium bicarbonate to cover gel pieces completely and incubate the reaction mixture overnight at 37 °C in a shaker (750 rpm). If thermolysin is employed as protease incubate at 65 °C in a shaker (750 rpm).

8. Allow reaction mixture to chill to ambient room temperature and spin down gel pieces and solution.

9. Withdraw supernatant and collect the solution in an appropriate collection tube.

10. Add 100 μl acetonitrile/water (50/50) containing 5 % formic acid and shake for 15 min at 37 °C. Withdraw supernatant and collect the extract (*see* **Note 8**).

11. Add 100 μl acetonitrile/water (80/20) containing 5 % formic acid and shake for 15 min at 37 °C. Withdraw supernatant and collect the extract (*see* **Note 8**).

12. Add 100 μl of neat acetonitrile and shake for 15 min at 37 °C. Withdraw supernatant and collect the extract (*see* **Note 8**).

13. Combine extracts and supernatant obtained under Subheading 3.5, **step 9**, and remove solvents in vacuo by use of a centrifugal evaporator.

3.6 Desalting of Extracted Proteolytic Peptides

Proteolytic peptides obtained from in-gel-digestions are desalted prior to glycopeptide enrichment by reversed-phase solid-phase extraction employing pipette tips containing an immobilized C18 resin, viz. C18 ZipTip pipette tips. The experimental procedure follows the manufacturer's instructions.

1. Samples are dissolved in 10 μl of trifluoroacetic acid (0.1 %).

2. Attach C18 ZipTip pipette tip to a compatible 10 μl pipettor, aspirate 10 μl wetting solution, and dispense to waste. Repeat this step three times.

3. Aspirate three times 10 μl equilibration solution and dispense to waste.

4. Load sample by aspirating and dispensing the sample solution at least ten times.

5. Aspirate 10 μl wash solution and dispense into a fresh reaction tube of appropriate volume. Repeat this step three times (*see* **Note 9**).

6. Peptides are released by consecutively aspirating and dispensing five times 10 μl of elution solution 1, 2, and 3. Eluates are collected and combined in a fresh reaction tube of appropriate volume and finally dried in vacuo by use of a centrifugal evaporator.

3.7 ZIC-HILIC Enrichment of Proteolytic N-glycopeptides

The following procedure can be directly applied to *N*-glycopeptides derived from in-solution digestion of glycoproteins (Subheading 3.4) or to *N*-glycopeptides obtained by in-gel digestions purified by reversed-phase solid-phase extraction.

1. Samples are dissolved in 15 μl of binding solution.

2. Attach ZIC-HILIC ProteaTip pipette tip to a compatible pipettor, aspirate 15 μl equilibration solution, and dispense to waste. Repeat this step five times.

3. Load sample by aspirating and dispensing the sample solution (15 μl) at least 20 times.

4. Aspirate 15 μl wash solution and dispense in a fresh reaction tube of appropriate volume. Repeat this step five times (*see* **Note 9**).

5. Glycopeptides are released by consecutively aspirating and dispensing ten times 15 μl of the elution solution. Eluates are collected in a fresh reaction tube of appropriate volume and finally dried in vacuo by use of a centrifugal evaporator.

3.8 Nano Electrospray Mass Spectrometry (NanoESI MS)

1. Mass spectra of glycopeptides enriched by ZIC-HILIC obtained under electrospray conditions in positive ion mode typically exhibit the analytes in charge states from +2 to +4. Analytes are mainly found in their corresponding protonated form or as ionic species with one or two protons replaced by sodium ions. Subsequent to the application of a mass deconvolution glycosylated species are noticed straightforward by identifying series of ions harboring the same peptide backbone but different glycan isoforms linked to the same glycosylation site. These analyte molecules differ by typical mass increments characteristic for glycan building blocks, i.e., 132.042 Da (pentose), 146.058 (deoxyhexose), 162.053 Da (hexose), 203.079 Da (*N*-acetylhexosamine), 291.095 Da (neuraminic acid), or combinations of these masses. Collisional activation of selected glycopeptide precursor ions mainly gives rise to fragment ions originating from cleavage of glycosidic bonds. Typically, B- and Y-type fragment ions lead to intense signals in the CID spectra of glycopeptide ions (nomenclature according to Domon and Costello [28]). While the latter comprise the reducing end and thus the peptide moiety B-type oxonium ions harbor the non-reducing end of the glycan and exhibit characteristic m/z-values: m/z([HexNAc$-H_2O+H$]$^+$): 204.087, m/z([NeuAc$-H_2O+H$]$^+$): 292.103, m/z([HexNAc-Hex$-H_2O+H$]$^+$): 366.139, m/z([HexNAc-Hex$_2-H_2O+H$]$^+$): 528.192, etc. *N*-glycan biosynthesis leads to a common core sugar sequence attached to a highly conserved sequon, viz. Manα1–6(Man α1–3)Manβ1–4GlcNAcβ1–4GlcNAcβ1–Asn–X–Ser/Thr

(X = any amino acid except proline) which is typically extended to finally yield either high-mannose-, complex-, or hybrid-type glycans. Combining this blueprint with the information deduced from the appearance of specific B- and Y-type fragment ions finally leads to a structural assignment of the *N*-glycan structure. Loss of the entire oligosaccharide chain liberates the deglycosylated peptide residue that might be identified by its exact mass if the amino acid sequence of the protein under inspection is known, thus giving rise to the glycosylation site (*see* **Note 10**).

4   Notes

1. The use of micro Bio-Spin Chromatography columns packed with polyacrylamide gel (Bio-Gel P-6) for removal of salts and other low-molecular-weight compounds by size-exclusion chromatography is recommended for the purification of lectins and other carbohydrate-binding proteins. These analytes are often prone to strong binding to the Sephadex G-25 gel matrix and (glyco-) protein recovery is typically very low.

2. Protease solutions may be stored at −20 °C for 1 month without significant loss of catalytic activity.

3. If the current study also aims at characterization and identification of disulfide bridges in glycoproteins one aliquot of protein should be submitted to proteolytic digestion without a preceding reduction and alkylation step. Under nanoESI-CID conditions fragmentation of intra- and inter-peptide disulfide bonds of proteolytic peptides provides sufficient information for their determination. Collisional activation of proteolytic peptides comprising a disulfide bridge gives rise to a set b- and of y-type fragment ions which typically allow the determination of the sequence of the amino acids located outside the disulfide loop. Additionally, fragment ion spectra reveal the presence of low-abundance fragment ions formed by the cleavage of peptide bonds within the disulfide loop. These fragmentations are preceded by asymmetric cleavage of the disulfide bridge, giving rise to a modified cysteine containing a disulfohydryl substituent and a dehydroalanine residue on the remote cleavage site [22–24, 29].

4. Thermal (glyco-)protein denaturation typically increases accessibility of cleavage sites and thus improves efficiency of proteolytic digestion.

5. The selection of an appropriate protease strongly depends on the primary structure of the (glyco-)protein under inspection. Since the use of trypsin as protease usually gives rise to very large glycopeptide species digestions using unspecific proteases such as thermolysin or a 1:1 mixture of trypsin and chymotrypsin furnish rather short nonglycosylated peptides and glycopeptides of significantly higher molecular mass. Owing to the high hydrophilicity of *N*-glycopeptides harboring a large glycan moiety and a short peptide stretch an efficient enrichment by ZIC-HILIC can be achieved [13].

6. If by accident gel bands/spots were divided into very small pieces leading to clogging of regular pipette tips microloader tips (Eppendorf AG, Hamburg, Germany) can be used instead.

7. If – after 30 min – protease solution is not absorbed entirely supernatant should be removed by use of either a regular pipette tip or a microloader tip (*see* **Note 6**), discard supernatant. According to our experience this step largely reduces the occurrence of autoproteolytic peptide ions in the resulting mass spectra.

8. Notwithstanding the protocol described by Shevchenko et al. [27] proteolytic peptides are extracted by use of solvents with decreasing polarity.

9. Wash solutions should be retained until the actual sample has been analyzed to avoid unintended sample loss.

10. Most low-energy CID spectra of *N*-glycopeptide ions also contain fragment ions formed by cleavage of the peptide backbone. These ionic species can be used to deduce amino acid sequence stretches that lead to an unambiguous identification of the glycosylation site (for an example refer to ref. 22).

References

1. Rudd PM, Dwek RA (1997) Glycosylation: heterogeneity and the 3D structure of proteins. Crit Rev Biochem Mol Biol 32:1–100

2. Dwek RA (1996) Glycobiology: toward understanding the function of sugars. Chem Rev 96:683–720

3. Varki A, Cummings RD, Esko JD et al (2009) Essentials of glycobiology, 2nd edn. Cold Spring Harbor Laboratory Press, Cold Spring Harbor, NY

4. Hart GW, Copeland RJ (2010) Glycomics hits the big time. Cell 143:672–676

5. Geyer H, Geyer R (2006) Strategies for analysis of glycoprotein glycosylation. Biochim Biophys Acta 1764:1853–1869

6. Marino K, Bones J, Kattla JJ et al (2010) A systematic approach to protein glycosylation analysis: a path through the maze. Nat Chem Biol 6:713–723

7. Zauner G, Selman MHJ, Bondt A et al (2013) Glycoproteomic analysis of antibodies. Mol Cell Proteomics 12:856–865

8. Desaire H (2013) Glycopeptide analysis, recent developments and applications. Mol Cell Proteomics 12:893–901

9. Alley WR Jr, Mann BF, Novotny MV (2013) High-sensitivity analytical approaches for the structural characterization of glycoproteins. Chem Rev 113:2668–2732

10. Haslam SM, North SJ, Dell A (2006) Mass spectrometric analysis of *N*- and *O*-glycosylation of tissues and cells. Curr Opin Struct Biol 16:584–591

11. Künneke K, Pohlentz G, Schmidt-Hederich A et al (2004) Recombinant human laminin-5 domains. Effects of heterotrimerization, proteolytic processing, and *N*-glycosylation on alpha3beta1 integrin binding. J Biol Chem 279:5184–5193

12. Henning S, Peter-Katalinić J, Pohlentz G (2007) Structure elucidation of glycoproteins by direct nanoESI MS and MS/MS analysis of proteolytic glycopeptides. J Mass Spectrom 42:1415–1421

13. Neue K, Mormann M, Peter-Katalinić J et al (2011) Elucidation of glycoprotein structures by unspecific proteolysis and direct nanoESI mass spectrometric analysis of ZIC-HILIC-enriched glycopeptides. J Proteome Res 10:2248–2260

14. Wuhrer M, Deelder AM, Hokke CH (2005) Protein glycosylation analysis by liquid chromatography-mass spectrometry. J Chromatogr B 825:124–133

15. Zauner G, Koeleman CAM, Deelder AM et al (2013) Nano-HPLC-MS of glycopeptides obtained after non-specific proteolysis. In: Kohler JJ, Patrie SM (eds) Mass spectrometry of glycoproteins, vol 951, Methods in molecular biology. Springer, Heidelberg, pp 113–127

16. Wuhrer M, de Boer AR, Deelder AM (2009) Structural glycomics using hydrophilic interaction chromatography (HILIC) with mass spectrometry. Mass Spectrom Rev 28:192–206

17. Chen CC, Su WC, Huang BY et al (2014) Interaction modes and approaches to glyco-

peptide and glycoprotein enrichment. Analyst 139:688–704

18. Huang BY, Yang CK, Liu CP et al (2014) Stationary phases for the enrichment of glycoproteins and glycopeptides. Electrophoresis 35:2091–2107

19. Hägglund P, Bunkenborg J, Elortza F et al (2004) A new strategy for identification of *N*-glycosylated proteins and unambiguous assignment of their glycosylation sites using HILIC enrichment and partial deglycosylation. J Proteome Res 3:556–566

20. Thaysen-Andersen M, Højrup P (2006) Enrichment and characterization of glycopeptides from gel-separated glycoproteins. Am Biotechnol Lab 24:14–17

21. Picariello G, Ferranti P, Mamone G et al (2008) Identification of *N*-linked glycoproteins in human milk by hydrophilic interaction liquid chromatography and mass spectrometry. Proteomics 8:3833–3847

22. Gnanesh Kumar BS, Pohlentz G, Schulte M et al (2014) Jack bean α-mannosidase: amino acid sequencing and *N*-glycosylation analysis of a valuable glycomics tool. Glycobiology 24:252–261

23. Sharma A, Pohlentz G, Bobbili KB et al (2013) The sequence and structure of snake gourd (*Trichosanthes anguina*) seed lectin, a three-chain nontoxic homologue of type II RIPs. Acta Crystallogr D Biol Crystallogr 69:1493–1503

24. Reissner C, Stahn J, Breuer D et al (2014) Dystroglycan binding to α-neurexin competes with neurexophilin-1 and neuroligin in the brain. J Biol Chem 289:27585–27603

25. Gnanesh Kumar BS, Pohlentz G, Mormann M et al (2013) Characterization of α-mannosidase from *Dolichos lablab* seeds: partial amino acid sequencing and *N*-glycan analysis. Protein Expr Purif 89:7–15

26. Gnanesh Kumar BS, Pohlentz G, Schulte M et al (2014) *N*-glycan analysis of mannose/glucose specific lectin from *Dolichos lablab* seeds. Int J Biol Macromol 69:400–407

27. Shevchenko A, Tomas H, Havlis J et al (2006) In-gel digestion for mass spectrometric characterization of proteins and proteomes. Nat Protoc 1:2856–2860

28. Domon B, Costello CE (1988) A systematic nomenclature for carbohydrate fragmentations in FAB-MS MS spectra of glycoconjugates. Glycoconjugate J 5:397–409

29. Mormann M, Eble J, Schwoeppe C et al (2008) Fragmentation of intra-peptide and inter-peptide disulfide bonds of proteolytic peptides by nanoESI collision-induced dissociation. Anal Bioanal Chem 392:831–838

30. Consortium for Functional Glycomics. CFG functionalglycomicsgateway. http://www.functionalglycomics.org/static/consortium/CFGnomenclature.pdf. Accessed 20 July 2012

Chapter 13

Simple and Effective Affinity Purification Procedures for Mass Spectrometry-Based Identification of Protein-Protein Interactions in Cell Signaling Pathways

Julian H.M. Kwan and Andrew Emili

Abstract

Identification of protein-protein interactions can be a critical step in understanding the function and regulation of a particular protein and for exploring intracellular signaling cascades. Affinity purification coupled to mass spectrometry (APMS) is a powerful method for isolating and characterizing protein complexes. This approach involves the tagging and subsequent enrichment of a protein of interest along with any stably associated proteins that bind to it, followed by the identification of the interacting proteins using mass spectrometry. The protocol described here offers a quick and simple method for routine sample preparation for APMS analysis of suitably tagged human cell lines.

Key words Affinity purification-mass spectrometry, Cell signaling pathway, Liquid chromatography-mass spectrometry

1 Introduction

The identification and mapping of protein-protein interactions, which underpin many fundamental biological processes including intracellular signaling pathways, is an important aspect of understanding protein function and regulation. Affinity purification-mass spectrometry (APMS) is a powerful tool for identifying the components of multi-protein complexes [1–8]. The fundamental basis of APMS is the use of one or more high-affinity recognition reagents (such as an antibody) to selectively enrich a particular target protein, together with physically associated factors, relative to the myriad of functionally unrelated proteins present in a cell extract or tissue lysate.

The endogenous form of the protein of interest may be enriched using a specific antibody. However this approach may not always be the most expedient since the generation of antibodies of sufficient specificity and affinity can be difficult and expensive.

Jörg Reinders (ed.), *Proteomics in Systems Biology: Methods and Protocols*, Methods in Molecular Biology, vol. 1394,
DOI 10.1007/978-1-4939-3341-9_13, © Springer Science+Business Media New York 2016

A popular and effective alternative is to use molecular cloning techniques to introduce an affinity tag (e.g., epitope or small protein like GFP) as a fusion to the open reading frame corresponding to the protein of interest. Using a tag/antibody pair that has been established to exhibit sufficient affinity and specificity avoids the need to generate a specific antibody for each target protein, and can lead to more efficient and consistent APMS results [8]. In this protocol, we suggest the use of FLAG or GFP epitope tags.

There are many tools, options, and caveats for the generation of suitable cell lines or tissue samples expressing an affinity-tagged protein of interest. These might range from endogenous tagging of loci in transgenic animals [9] to transient transfection of plasmids over-expressing a particular target in cell culture [10]. The protocol reported here provides a simple procedure that we have found to be effective for APMS analysis of tagged proteins expressed in the common tissue culture cell line HEK 293T. In principle the protocol can be easily adapted to other cell lines or biological samples with two key considerations. The first requirement is that the protein of interest must be solubilized under gentle lysis conditions, which must not be too harsh so as to disrupt physiologically relevant protein-protein interactions. The second requirement is that the sample remains compatible with subsequent mass spectrometry analysis; this might mean taking steps to reduce the levels of potential contaminants, such as detergents, which are detrimental to protein identification.

When performed properly, affinity purification will enrich proteins (i.e., prey) that are physically bound to a protein of interest (i.e., the bait); these are the interacting proteins one intends to identify. However, many other nonspecific proteins will also bind (i.e., artifacts). The inclusion of control purifications (e.g., no tag/bait, or irrelevant bait) is important for distinguishing genuine protein-protein interactions from nonspecific proteins. In the simplest interpretation of APMS data, proteins identified together with the bait are considered to interact unless they are also detected with the negative control(s). There are two key considerations for why this interpretation might not be sufficient. First, the bait may specifically interact with a prey protein that is also found in the control, but at a significantly lower level. In such a scenario, considering the prey's enrichment relative to the control is important. The second consideration takes into account the inconsistent recovery and identification of some proteins by APMS due to variations in experimental and biological conditions. The same baits processed repeatedly over several different experiments may each identify preys that are unique to the run. This is true of both the experimental and control samples. To address this source of spurious results (i.e., potential false positives), one must consider detection reproducibility across runs. This variability also underlines the importance of performing independent biological replicates, such as by performing two sets of APMS experiments starting from different cell culture batches on different days.

Different algorithms can be used to assign a score to each potential protein interaction, which represents how likely a prey protein is to genuinely interact with the bait. Popular algorithms, such as compPASS [6] and SAINT [11, 12], take into account both the reproducibility and the enrichment relative to controls to systematically assign a score which indicates the likelihood that a particular candidate interactor is specific (i.e., a genuinely interacting prey protein). Such algorithms are particularly useful for analysis of larger APMS datasets, with multiple baits, since they can use data from each other purification as additional controls between baits and thus generate a more robust census of nonspecifically adsorbed proteins versus biologically relevant interactors captured by your experimental procedure.

The steps described in the protocol that follows cover generating soluble protein cell lysate, capturing the protein of interest on a solid support, washing to remove unbound proteins, and processing bound proteins for mass spectrometry identification. Since the methods of data acquisition and final format of results will vary based on the instrumentation, protocols, and software favored by the mass spectrometry (MS) facility, this protocol will not cover the details of MS procedures and data analysis.

2   Materials

1. Sample: 2×150 mm dishes per sample of HEK 293T cells (80–90 % confluent) expressing GFP or FLAG epitope-tagged bait protein of interest.

2. Lysis buffer: TBS (30 mM Tris–HCl, pH 7.5, 150 mM NaCl) with 0.5 % Nonidet P40 and protease and phosphatase inhibitors.

3. Affinity media: Antibody (Life Technologies anti-GFP #G10362, or anti-FLAG #F1804 as appropriate for affinity tag), Dynabeads Protein G (Life Technologies #10003D).

4. Wash buffer: TBS (30 mM Tris–HCl, pH 7.5, 150 mM NaCl) without detergent and protease and phosphatase inhibitors (*see* **Note 1**).

5. Trypsin digestion buffer: 50 mM ammonium bicarbonate (*see* **Note 2**).

6. Proteolytic digestion: Trypsin sequencing grade (Roche) (*see* **Note 3**).

7. Digestion termination: Formic acid (LC/MS grade).

8. Desalting media: C-18 cartridge (10–200 L NuTip; Glygen Corp #NT2C18.96).

9. Solution A: H_2O (HPLC grade) 0.1 % formic acid.

10. Solution B: 70 % Acetonitrile, 0.1 % formic acid.

3   Methods

1. Remove media from cell culture plates (*see* **Note 4**).

2. Lyse cells on dish in 1 ml cold lysis buffer per 150 mm dish, and incubate for 5 min on ice or in cold room/fridge (*see* **Note 5**).

3. Scrape lysate into pre-chilled 1.5 ml microcentrifuge tubes (*see* **Note 6**).

4. Spin down cell debris (20 min, 20,000 × g, 4 °C) (*see* **Note 7**).

5. Transfer supernatant to fresh pre-chilled tubes; the pellet of cell debris may be discarded (*see* **Note 8**).

6. Add 1 μg of appropriate antibody to each sample (*see* **Note 9**).

7. Incubate samples with end-over-end rotation at 4 °C (cold room) for 1–2 h.

8. Wash 40 μl Protein G (Dynabead) bead slurry per sample with lysis buffer and resuspend in same volume of lysis buffer (*see* **Note 10**).

9. Add 40 μl washed Dynabead Protein G slurry to each sample and incubate with rotation at 4 °C for an additional 1 h (*see* **Note 11**).

10. Collect beads and wash with 1.5 ml cold wash buffer (repeat) (*see* **Note 12**).

11. Resuspend beads in 1 ml wash buffer and transfer to a new pre-chilled microfuge tube (*see* **Note 13**).

12. Wash beads with 400 μl trypsin digestion buffer (*see* **Note 14**).

13. Resuspend beads in 20 μl trypsin digestion buffer and add 750 ng trypsin (*see* **Note 15**).

14. Digest with rotation (end over end) at 37 °C for 4 h to overnight (*see* **Note 16**).

15. Magnetize beads, transfer supernatant to a fresh tube, and then add another 750 ng trypsin. Incubate at 37 °C with agitation for 4 h (*see* **Note 17**).

16. Add formic acid (2 % final concentration) to sample to terminate digestion.

17. Condition desalting media using multiple rounds (10×) of aspiration and discharge of 200 μl solution B (repeat) (*see* **Note 18**).

18. Condition desalting media by 10× aspiration and discharge of 200 μl solution A (repeat 2× with new aliquots of solution A) (*see* **Note 19**).

19. Load sample onto desalting media by 20–50× aspiration and discharge of digested sample (*see* **Note 20**).

20. Wash sample by 10× aspiration and discharge of 200 μl solution A (repeat 2×) (*see* **Note 21**).

21. Elute sample by 20× aspiration and discharge of 20 μl solution B.

22. Lyophilize eluate using a vacuum concentrator. After drying, samples may be stored at −20 °C until resuspension in 20 μl of solution A immediately prior to analysis by liquid chromatography-mass spectrometry (*see* **Note 22**).

4   Notes

1. Wash buffer omits detergent to reduce detergent contamination that can hinder mass spectrometry identification. Protease and phosphatase inhibitors may also be omitted because the sample is not expected to be exposed to the buffer for an extended period of time.

2. Make with HPLC-grade water, pH should be about 8.0, and do not adjust.

3. The lyophilized trypsin should be resuspended in trypsin digestion buffer at 0.5 μg/μl (instead of acetic acid as recommended by the manufacturer).

4. You may rinse the cells gently with cold PBS to remove remaining media.

5. This lysis condition is suitable for solubilization of cytoplasmic proteins, but is not well suited for solubilization of nuclear or membrane proteins.

6. Since lysis occurs in a small volume, use a cell lifter or rubber policeman to scrape cell lysate to the bottom of the dish; this insures maximum sample recovery.

7. This spin can be done in a refrigerated tabletop centrifuge.

8. Input amount can be scaled up or down as needed; for 2× 150 mm plates per sample (suggested), the supernatant from two microfuge tubes is combined in a 5 or 15 ml tube.

9. Use an antibody that has been successful for immunoprecipitation experiments. We typically use Life Technologies anti-GFP (#G10362) or anti-FLAG (#F1804) as capture reagents.

10. To wash magnetic beads, use a magnet bar to collect beads to the tube wall for 1 min; while the beads are magnetized remove the supernatant. The tube may be briefly centrifuged to collect supernatant at the bottom of the tube, re-magnetize, and remove the remaining supernatant. Next, remove the tube from the magnet and resuspend the beads in fresh buffer by inversion or agitation.

11. End-over-end rotation is important to keep the beads in solution to maximize target recovery.

12. Use a magnetic bar designed for magnetic bead collection in 1.5 ml tubes. If the initial sample volume is greater than 1.5 ml, magnetize the first 1.5 ml to collect the beads and discard the supernatant. Repeat until all beads have been collected, and then proceed with washes.

13. Changing tubes is an important step to reduce contamination from nonspecifically adsorbed proteins and detergents that adhere to the tube wall.

14. Using trypsin digestion buffer for the final wash reduces the carryover of salt present in the lysis and wash buffers.

15. Ensure that the lyophilized trypsin was resuspended in trypsin digestion buffer and not acetic acid; since the digestion volume is small the addition of acid would inhibit digestion.

16. Briefly centrifuge tube to collect digestion buffer and beads to the bottom of the tube prior to incubation with rotation; surface tension will keep the sample at the bottom of the tube. End-over-end rotation is required to keep beads in solution for efficient digestion.

17. The supernatant now contains the digested sample. This additional incubation step allows more time for digestion to produce peptides for mass spectrometry identification.

18. Desalting media attach to micropipette, aspirate, and discharge a single aliquot of 200 µl solution B in a microfuge tube containing excess solution B to avoid forming bubbles. Repeat with a new aliquot of solution B.

19. Be careful not to contaminate solution A with carryover of solution B that may adhere to outside of tip (use separate aliquots as needed).

20. The digested peptides will bind to the C-18 tip, allowing salt and other contaminants to be washed away. Try to avoid aspirating air and generating bubbles. Longer/more aspiration and discharge will allow more peptides to bind the C-18 media.

21. Do not confuse or contaminate solution A with solution B. Solution B will elute the peptides.

22. 8 µl of resuspended sample is generally sufficient for mass spectrometry identification.

References

1. Butland G, Peregrín-Alvarez JM, Li J et al (2005) Interaction network containing conserved and essential protein complexes in Escherichia coli. Nature 433:531–537. doi:10.1038/nature03239

2. Zhao R, Davey M, Hsu Y-C et al (2005) Navigating the chaperone network: an integrative map of physical and genetic interactions mediated by the hsp90 chaperone. Cell 120:715–727. doi:10.1016/j.cell.2004.12.024

3. Gavin A-C, Aloy P, Grandi P et al (2006) Proteome survey reveals modularity of the yeast cell machinery. Nature 440:631–636. doi:10.1038/nature04532

4. Krogan NJ, Cagney G, Yu H et al (2006) Global landscape of protein complexes in the yeast Saccharomyces cerevisiae. Nature 440: 637–643. doi:10.1038/nature04670

5. Sardiu ME, Cai Y, Jin J et al (2008) Probabilistic assembly of human protein interaction networks from label-free quantitative proteomics. Proc Natl Acad Sci U S A 105:1454–1459. doi:10.1073/pnas.0706983105

6. Sowa ME, Bennett EJ, Gygi SP, Harper JW (2009) Defining the human deubiquitinating enzyme interaction landscape. Cell 138:389–403. doi:10.1016/j.cell.2009.04.042

7. Behrends C, Sowa ME, Gygi SP, Harper JW (2010) Network organization of the human autophagy system. Nature 466:68–76. doi:10.1038/nature09204

8. Varjosalo M, Sacco R, Stukalov A et al (2013) Interlaboratory reproducibility of large-scale human protein-complex analysis by standard-ized AP-MS. Nat Methods 10:307–314. doi:10.1038/nmeth.2400

9. Roncagalli R, Hauri S, Fiore F et al (2014) Quantitative proteomics analysis of signalo-some dynamics in primary T cells identifies the surface receptor CD6 as a Lat adaptor-independent TCR signaling hub. Nat Immunol 15:384–392. doi:10.1038/ni.2843

10. Guruharsha KG, Rual J-F, Zhai B et al (2011) A protein complex network of Drosophila melanogaster. Cell 147:690–703. doi:10.1016/j.cell.2011.08.047

11. Choi H, Larsen B, Lin Z-Y et al (2011) SAINT: probabilistic scoring of affinity purification-mass spectrometry data. Nat Methods 8:70–73. doi:10.1038/nmeth.1541

12. Teo G, Liu G, Zhang J et al (2014) SAINTexpress: improvements and additional features in Significance Analysis of INTeractome software. J Proteomics 100:37–43. doi:10.1016/j.jprot.2013.10.023

Chapter 14

A Systems Approach to Understand Antigen Presentation and the Immune Response

Nadine L. Dudek, Nathan P. Croft, Ralf B. Schittenhelm, Sri H. Ramarathinam, and Anthony W. Purcell

Abstract

The mammalian immune system has evolved to respond to pathogenic, environmental, and cellular changes in order to maintain the health of the host. These responses include the comparatively primitive innate immune response, which represents a rapid and relatively nonspecific reaction to challenge by pathogens and the more complex cellular adaptive immune response. This adaptive response evolves with the pathogenic challenge, involves the cross talk of several cell types, and is highly specific to the pathogen due to the liberation of peptide antigens and their presentation on the surface of affected cells. Together these two forms of immunity provide a surveillance mechanism for the system-wide scrutiny of cellular function, environment, and health. As such the immune system is best understood at a systems biology level, and studies that combine gene expression, protein expression, and liberation of peptides for antigen presentation can be combined to provide a detailed understanding of immunity. This chapter details our experience in identifying peptide antigens and combining this information with more traditional proteomics approaches to understand the generation of immune responses on a holistic level.

Key words Major histocompatibility complex, Human leukocyte antigens, Peptide ligands, Mass spectrometry, Antigen presentation

1 Introduction

The human major histocompatibility complex (MHC) is located on the short arm of chromosome 6 and encompasses around 4 Mbp or 0.1 % of the genome. Around 220 genes have been identified in this region and at least 10 % of these genes have a direct function in the immune response to pathogens or the regulation of immunity. The human MHC can be divided into three regions which encode the class I, class II, and class III human leukocyte antigen (HLA) gene products. HLA molecules demonstrate tremendous polymorphism, which reflects the natural evolution of these genes in response to various microbial pathogens in different populations. Moreover, HLA genes exhibit linkage disequilibrium,

Jörg Reinders (ed.), *Proteomics in Systems Biology: Methods and Protocols*, Methods in Molecular Biology, vol. 1394, DOI 10.1007/978-1-4939-3341-9_14, © Springer Science+Business Media New York 2016

meaning that they are often inherited together in blocks of genes. This dictates that studies take into account the role of HLA molecules both in isolation and in the context of their naturally occurring combinations or haplotypes.

HLA class I molecules, and the murine H-2 equivalent, are expressed on all nucleated cells and bind short peptides (typically 8–11 amino acids in length) derived from both self and foreign antigens. These peptide ligands are primarily generated in or transported into the cytoplasm and subsequently translocated into the endoplasmic reticulum (ER) where they assemble with nascent MHC class I molecules. These mature, peptide-loaded complexes are transported to the cell surface where they are scrutinized by CD8+ cytotoxic T lymphocytes (CTL). Should the peptide ligand be derived from a pathogen and be recognized as foreign in an immunocompetent host, the cell is killed via the cytotoxic armoury of the CTL.

The expression of MHC class II molecules is confined to a small subset of highly specialized cells called professional antigen-presenting cells (APCs). MHC class II molecules associate with longer peptides (9–25 amino acids in length) than class I molecules and this association occurs in late endosomal compartments, a distinct and separate cellular compartment to the ER-Golgi route inhabited by assembling MHC class I molecules. MHC class II molecules are recognized by CD4+ T helper cells and functional recognition of these complexes is intimately involved in both the humoral and cellular immune response. MHC class I and class II molecules form membrane-distal structures that comprise a cleft in which the antigenic peptide ligands reside [1–3]. The T cell receptor (TCR) on CD8+ or CD4+ T cells recognizes MHC molecules in the context of both the polymorphic class I or class II molecules, respectively, and the peptide antigen presented in the antigen-binding groove of these cell surface molecules [4]. The class III region-encoded molecules have quite different and diverse biochemical properties and are involved in inflammation and other immune activities. As such they include components of the complement system, cytokines (such as tumor necrosis factor and lymphotoxin), and heat-shock proteins.

Approaches that facilitate the direct isolation and identification of peptide antigens associated with class I or class II molecules have defined the ligand specificity of different MHC molecules. Moreover they have allowed direct identification of naturally processed and presented antigens derived from infectious microorganisms as well as self-peptides associated with autoimmune disorders and cancers. Several different approaches have been used to isolate MHC-bound peptides from cells, including analysis of acidified cell lysates [5–7], elution of peptides from the cell surface [8, 9], and immunoaffinity purification of the MHC-peptide complexes from detergent-solubilized cell lysates [10–14].

Table 1
Commonly used monoclonal antibodies for MHC-peptide immunoaffinity chromatography

Hybrid	Specificity	Isotype	Ref	ATCC	Comments
Anti-human					
L243	DRα	IgG2a	[23, 24]	Y	Lower peptide yield than LB3.1
LB3.1	DRα	IgG2a	[25]	Y	Use in preference to L243
SPV-L3	DQ	IgG2a	[23]		
B721	DP (DP1-5)	IgG3	[26]		
BB7.2	A2, A69	IgG2b	[27]		
ME1	B7, Bw22, B27	IgG1	[28]	Y	Cross-reacts with HLA B14 and Bw46
W632	A, B, C	IgG2a	[29]		
DT9	C, E	IgG2b	[30, 31]		Marginally lower yield of peptides in comparison to W6/32
Anti-mouse					
Y-3	K^b	IgG2b	[32]	Y	Cross-reacts with H-2k, p, q, r and s
28-14-8	D^b L^b		[33]		Higher yield of peptides for D^b than 28.8.6s
SF1.1.1.10	K^d	IgG2a	[14]	Y	
28.8.6 s	K^b, D^b	IgG2a	[34]	Y	Lower peptide numbers for D^b than 28-14-8s
MKD6	IA^d	IgG2a	[35]	Y	
Y-3P	IA	IgG2a	[36]	Y	Weak reactivity with I-A^k
10.2.16	IA ^{g7,k,r,f,s}	IgG2b	[37]	Y	

The use of immunoaffinity chromatography dramatically improves the specificity of the peptide extraction process. A single MHC allele can be isolated by the use of an appropriate monoclonal antibody (*see* Table 1) and some antibodies can even select a sub-population of MHC molecules with defined molecular or functional properties [15, 16]. The use of immunoaffinity chromatography to isolate specific MHC molecules provides the most appropriate material for identifying individual peptide ligands restricted by a defined MHC allele. Furthermore, the complexity of the eluates/lysates can be decreased by using cell lines that express limited numbers of HLA alleles, for example homozygous lymphoblastoid or mutant cell lines such as C1R which express very low levels of endogenous class I molecules but support high-level expression of transfected class I molecules [17]. This property makes these cells very attractive for examining endogenous peptides presented by individual class I alleles under normal physiological conditions [18–20] or during infection [21, 22].

In order to understand immunity at a systems level several elements must be examined in parallel. One crucial step is defining which peptides are selected by particular MHC molecules or combinations of MHC molecules for presentation on the surface of cells; how this changes both qualitatively and quantitatively at

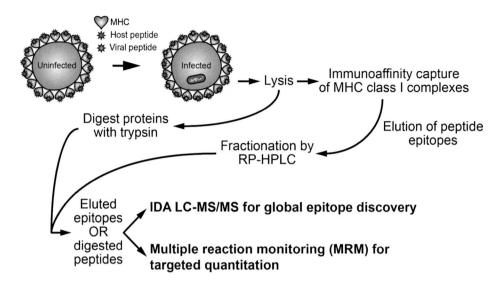

Fig. 1 A systems approach to antigen presentation using mass spectrometry. In this workflow, cells are infected with virus in vitro. At various time points, cells are harvested and lysed. A sample of lysate is taken and subjected to tryptic digestion. The remaining lysate is used to affinity purify MHC peptide complexes. Both tryptic peptides and non-tryptic MHC peptides are subjected to RP-HPLC before mass spectrometric interrogation. Mass spectrometric analysis involves both a global discovery approach of LC-MS/MS and the targeted method of LC-MRM, where a set of known peptides are detected and quantified. In this workflow it is possible to simultaneously quantify the presentation of virus or host peptide-MHC complexes, and the levels of their source antigens at multiple times during infection to develop a comprehensive picture of antigen presentation

different stages of development, during inflammation or infection, will critically inform systems immunology studies. These parameters may also be correlated with the host and pathogen proteome, which can be analyzed and quantitated from the same sample. The combination of peptidome and proteome data allows the relationship between antigen expression kinetics, abundance, and epitope liberation to be correlated with immune outcomes in animal models or humans [38]. This chapter explores methods for studying antigen presentation at a systems level by identifying peptides isolated from specific MHC class I or class II molecules from various types of APCs. It focuses on the use of serial immunoaffinity chromatography to study peptide determinant selection by different HLA haplotypes, the parallel determination of the proteome of the APC, and the quantitation of specific MHC-peptide complexes on the cell surface using targeted proteomics approaches (Fig. 1).

2 Materials

Prepare all solutions using ultrapure H_2O (18 mΩ cm at 25 °C, freshly drawn) and use MS-grade solvents, reagents, glassware, and plasticware.

2.1 Generation of MHC Eluate

All solutions for cross-linking with the exception of triethanolamine and DMP should be filtered through a 0.2 μM filter.

2.1.1 Preparation of Cross-Linked Immunoaffinity Column

1. Phosphate-buffered saline (PBS): 137 mM NaCl, 2.7 mM KCl, 10 mM Na_2HPO_4, and 2 mM KH_2PO_4, pH 7.4.

2. Purified monoclonal antibody (mAb) at 1–10 mg/ml in PBS: Ideally the mAb should only recognize the class I or class II allele of interest, although affinity and specificity issues frequently require a compromise (Table 1, *see* **Note 1**).

3. Suitable column (e.g., disposable plastic Econo-Column from BioRad).

4. Protein A resin (e.g., Repligen CaptivA™ PriMAB).

5. Borate wash buffer: 0.05 M Borate buffer pH 8.0. For 100 ml of buffer, add 3.97 ml of 0.1 N NaOH to 50 ml of 0.1 M boric acid/0.1 M KCl stock solution and make up to 100 ml with MS-grade H_2O.

6. Protein A wash buffer: 0.2 M Triethanolamine, pH 8.2 at RT. Prepare this solution fresh and pH just prior to use.

7. Dimethyl pimelimidate (DMP) cross-linker: 40 mM DMP-2HCl in 0.2 M triethanolamine pH 8.3. Prepare DMP by dissolving 250 mg DMP-2HCl (Sigma) in 22 ml 0.2 M triethanolamine pH 8.2. Adjust pH to 8.3 with NaOH, and bring to 24.1 ml. Prepare this solution fresh and pH just prior to use (*see* **Note 2**).

8. Termination buffer: Ice-cold 0.2 M Tris, pH 8.0.

9. Stripping buffer: 0.1 M Citrate, pH 3.0.

2.1.2 Generation of Cell Lysate

MHC-bound peptides may be affinity purified from whole tissue, isolated primary cells, or transformed cells grown in culture (*see* **Note 3**). The amount of material required is dependent on the downstream application (LC-MS/MS vs. LC-MRM) and will be highly dependent on the expression levels of MHC on the cells contained within the sample (*see* **Note 4**). Transformed cells grown in culture are the simplest sample type for MHC/peptide isolation with cell numbers ranging between 5×10^7 and 1×10^9 per isolation. Cells may be expanded, washed in PBS, and the pellets snap frozen in liquid nitrogen for storage at -80 °C for up to 6 months.

1. 2× Lysis buffer: 0.5 % NP-40 (IGEPAL 630 from Sigma is the equivalent), 50 mM Tris, pH 8.0 (from 1 M stock solution), 150 mM NaCl, protease inhibitor cocktail (cOmplete protease inhibitor from Roche or equivalent, should be made up fresh each time) in MS-grade H_2O. Prepare lysis buffer just prior to use and keep on ice.

2.1.3   Immunoaffinity
Purification of MHC Class I/
Class II Molecules

MHC class I and II molecules can be eluted from the same sample by using tandem columns. For human cell lines, we routinely pass the cell lysate through a class I column, followed sequentially by a column specific for HLA DR, then HLA DQ, and finally HLA DP.

1. mAb cross-linked protein A resin from Subheading 2.1.1.

2. Pre-column (non-cross-linked protein A sepharose in suitable column): The bed volume should be half that of the cross-linked column; that is, for a 1 ml protein A-mAb column, use a 0.5 ml pre-column.

3. Pepstatin A: 1 mg/ml stock in isopropanol, aliquot, and store at −20 °C.

4. PMSF: 0.1 M stock in absolute ethanol, aliquot, and store at −20 °C.

5. Wash buffer 1: 0.005 % IGEPAL 630, 50 mM Tris, pH 8.0, 150 mM NaCl, 5 mM EDTA, 100 μM PMSF, 1 μg/ml pepstatin A in MS-grade H_2O.

6. Wash buffer 2: 50 mM Tris, pH 8.0, 150 mM NaCl, in MS-grade H_2O.

7. Wash buffer 3: 50 mM Tris, pH 8.0, 450 mM NaCl, in MS-grade H_2O.

8. Wash buffer 4: 50 mM Tris, pH 8.0, in MS-grade H_2O.

9. Elution buffer: 10 % acetic acid in MS-grade water (use best grade glacial acetic acid, e.g., Sigma ACS grade).

2.1.4   Separation of MHC
Eluate by RP-HPLC

1. Peptide separation prior to loading on mass spectrometer is performed by reversed-phase (RP) chromatography using a C18 column in a high-pressure liquid chromatography (HPLC) system (e.g., 4.6 internal diameter × 50 mm long reversed-phase C18 endcapped HPLC column, Chromolith Speed Rod, Merck). LC mobile phases: Buffer A is 0.1 % trifluoroacetic acid (TFA) in MS-grade water and buffer B is 0.1 % TFA in MS-grade acetonitrile.

2. Low-protein-binding 1.5 ml tubes for fraction collection, e.g., Eppendorf LoBind tubes.

2.2   LC-MS/MS
Analysis
of MHC Eluate

1. Buffer A: 0.1 % formic acid (FA) in MS-grade H_2O.

2. Buffer B: 0.1 % FA in 80 % MS-grade acetonitrile.

3. Autosampler vials for mass spectrometry.

2.3   Targeted Mass
Spectrometric
Analysis
of MHC Eluate

In cases where peptide epitopes are known, isotopically labeled (AQUA) peptides can be used for absolute quantitation by targeted LC-MRM (multiple reaction monitoring) [38, 39]. AQUA peptides are composed of the same amino acid sequence as the natural equivalent but bearing one or more heavy amino acids.

1. Buffer A: 0.1 % FA in MS-grade H_2O.

2. Buffer B: 0.1 % FA in 98 % MS-grade acetonitrile.

3. Synthetic peptides at >98 % purity: Native peptide and isotopically labeled AQUA peptide (optimal monoisotopic shift between the two species of 6–8 Da) solubilized in appropriate buffer, e.g., DMSO (*see* **Note 5**).

4. Autosampler vials for mass spectrometry.

2.4 Analysis of Cellular Proteome Using FASP Digestion of MHC Lysate

In order to correlate the cellular proteome with the immunopeptidome, a small sample of the MHC lysate is subjected to tryptic digestion and mass spectrometric analysis [40]. Samples for tryptic digestion may be taken immediately after lysis or from the flow through of the MHC-affinity column. Retaining the flow through from the affinity column and freezing at –80 °C is advised so that if required, more tryptic digestions can be performed. We utilize the Expedeon FASP Protein Digestion Kit as a convenient way of generating tryptic peptides. This kit is compatible with a number of reducing agents; however we recommend TCEP.

1. Expedeon FASP Protein Digestion Kit (reagents except trypsin and reducing agent are provided in the FASP kit).

2. Ammonium bicarbonate (in kit): 50 mM.

3. NaCl (in kit): 0.5 M.

4. Urea sample solution (in kit): 1 ml of Tris solution to one tube of urea, vortex until powder is dissolved. Prepare just before use.

5. 10× Iodoacetamide solution (in kit): Add 100 µl of urea sample solution to one tube of iodoacetamide. Mix and dissolve by pipetting up and down. Transfer into clean microfuge tube, wrap in foil, and keep on ice. Prepare just before use.

6. Digestion solution (in kit): Dissolve one 1 µg tube of trypsin in 75 µl of 50 mM ammonium bicarbonate solution and keep on ice. Aim for 1:100 trypsin-to-protein ratio. Scale amount of trypsin according to how much lysate is used, i.e., if loading 400 µg of protein, use 4 µg trypsin. Prepare just before use.

7. Bradford reagent.

8. Trypsin (single shot 1 µg, Sigma).

9. 0.1 % FA.

10. TCEP: 0.5 M in 50 mM ammonium bicarbonate (Sigma).

3   Methods

3.1   Generation of MHC Eluate

3.1.1   Preparation of Cross-Linked Immunoaffinity Column

1. Remove required amount of protein A sepharose (supplied as a 50 % slurry in 20 % ethanol) and add to column. Generally cross-link 10 mg of antibody per 1 ml of resin. Allow to settle by gravity, check for air bubbles, and agitate the slurry if necessary to remove.

2. Wash with 10 column volumes (c.v.) of MS-grade H_2O followed by 10 c.v. of PBS.

3. Prepare mAb (ideally at 0.5–1 mg/ml in PBS) in 50 ml tube. Remove washed resin from column and add to antibody in tube. Rotate gently end-over-end at 4 °C for 1 h to allow binding (*see* **Note 6**).

4. Transfer resin and antibody back to column and allow antibody to flow through by gravity. Wash antibody-bound resin with 20 c.v. of borate buffer followed by 15 c.v. of freshly prepared 0.2 M triethanolamine, pH 8.2 at RT. Triethanolamine is used to ensure that there are no residual primary amines present that may interfere with the cross-linking reaction.

5. Flow 5 c.v. of freshly prepared DMP cross-linker through the column at RT leaving a meniscus just over the protein A column bed. Seal the bottom of column and allow to sit at RT for 1 h.

6. Terminate the cross-linking reaction by adding 10 c.v. of ice-cold termination buffer (0.2 M Tris, pH 8.0).

7. Remove unbound antibody by washing with 10 c.v of stripping buffer (0.1 M citrate pH 3).

8. Flow 10 c.v. of PBS pH 7.4 until pH of flow through is >7 (it may be convenient to stop here, wash, and store the column in PBS pH 7.4 supplemented with 0.02 % NaN_3). Generally columns are best used within a month; however this will vary depending on the specific antibody.

3.1.2   Generation of Cell Lysate

1. Cells (5×10^7 to 1×10^9) can be grown in spinner flasks, bioreactors, or roller bottles (*see* **Note 7**) to appropriate numbers, washed in PBS, and harvested by centrifugation ($2000 \times g$, 10 min at 4 °C). Pellets should be snap frozen in liquid nitrogen and may be harvested iteratively for storage at −80 °C for up to 6 months. If collecting tissues, they should be rinsed in cold PBS containing protease inhibitors and immediately snap frozen without liquid for later processing.

2. Prepare a 2× concentrated solution of lysis buffer. Lysis buffer is added at 2× to allow for the volume of the cell pellet to be taken into account prior to adjustment of the concentration of the lysate to 1×. Cells are lysed at 5×10^7–1×10^8 cells per ml

of 1× lysis buffer. If the volume of the cell pellet is close or over 50 % of the final volume required you may need to lyse at a lower cell density.

3. Add correct volume of 2× lysis buffer to the frozen cell pellets and thaw the pellets quickly in a bath of tepid (i.e., RT) water. The temperature of the material should remain cold to touch so do not let the material equilibrate, thaw until small ice clumps are left, and add ice-cold MS-grade water to a final volume so as the lysis buffer is now at 1× strength.

4. Briefly homogenize the lysate (e.g., using a Polytron Disperser) to disperse any left over ice pellets. For large cell pellets or tissue samples we recommend grinding cells under liquid nitrogen (rather than homogenizing) using a cryogenic mill (e.g., Mixer Mill MM 400, Retsch). If the mill is used, pellets are placed straight into the precooled mill pot and after grinding the powder is scraped into lysis buffer.

5. Rotate lysate end-over-end at 4 °C for 1 h. Retain a sample (100 μl to 1 ml depending on cell number and lysis volume) for FASP digestion; retained lysate may be stored at –80 °C.

6. Centrifuge lysate for 10 min at $2000 \times g$ (4 °C). This step removes the nuclei.

7. Take supernatant from previous step and spin for 75 min in an ultracentrifuge ($100,000 \times g$) at 4 °C. Multiple spins may be necessary to fully clarify the lysate.

8. Collect the supernatant. It should be clear. If there is an unclear layer at top of the tubes carefully remove this lipid-containing layer and filter through a 0.8 μm and a 0.45 μm filter.

3.1.3 Immunoaffinity Purification of MHC Class I/Class II

1. Using gravity flow or a peristaltic pump in a cold room, load cell lysate onto a protein A sepharose pre-column that has been pre-equilibrated in 10 c.v. wash buffer 1. Multiple pre-columns may be required depending on the size and type of sample and should be replaced upon clogging.

2. Collect pre-cleared lysate and slowly load onto the cross-linked mAb column(s). If gravity flow is too quick (<1 h for lysate to pass), use a peristaltic pump. For small samples (1–4 ml lysate), it is generally better to add lysate to several of 2 ml LoBind Eppendorf tubes containing resin and rotate slowly end over end at 4 °C for 1 h.

3. For maximal yield the lysate should be run through the column twice. Retain flow through and freeze at –80 °C for subsequent FASP analysis. If multiple columns are being run (i.e., class I and II elutions from the same lysate), we routinely run the lysate once only over each column. The columns can be set up on a retort stand in tandem so that the flow through from

one column runs directly to the next. It may be necessary to temporarily stop the flow of the different columns throughout the process, to ensure that they do not dry out (if the lysate is passing at slightly different flow rates through each). If binding has been performed in 2 ml LoBind Eppendorf tubes, after the 1-h incubation, spin resin gently and transfer supernatant to Eppendorf containing next mAb resin. Repeat incubation while washing and eluting from the first batch of mAb resin.

4. Wash the column(s) with 20 c.v. of each wash buffer in the following order: Wash buffer 1, wash buffer 2 (to remove detergent), wash buffer 3 (to remove nonspecifically bound material), wash buffer 4 (removes salt to prevent crystal formation).

5. Elute MHC molecules in 5 c.v. of elution buffer. *Add AQUA peptides here* if used, i.e., post-elution from column and pre-separation by RP-HPLC.

6. Empty column(s) can be discarded or soaked overnight in acetic acid, washed in MS-grade H_2O, and reused.

7. Progress to RP-HPLC fractionation of flow through. Alternatively the eluate can be frozen at −80 °C; however this will result in some sample loss.

3.1.4   Separation of MHC Eluate by RP-HPLC

A single RP-HPLC step may be used to isolate peptides if an immunological readout is used to assay peptide fractions. However biochemical analysis by mass spectrometry requires a minimum of two dimensions of RP-HPLC to achieve sufficient separation (*see* **Note 8**).

1. Peptides are separated from MHC heavy chain, β2m (for class I molecules), leached antibody, and contaminating detergent using a C18 reverse-phase column running on a mobile-phase buffer A of 0.1 % TFA and buffer B of 80 % acetonitrile/0.1 % TFA (*see* **Note 9**). We routinely use a 4.6 mm internal diameter × 50 mm (or 100 mm for greater separation *see* **Note 10**) long reversed-phase C18 endcapped HPLC column (Chromolith Speed Rod, Merck) on an ÄKTAmicro™ HPLC system (GE Healthcare).

2. Separate based on a rapid gradient of buffer A to B, which results in 10–30 peptide-containing fractions (e.g., 2–40 % B for 4 min, 40–45 % for 4 min, and a rapid 2 min increase to 100 % B; Fig. 2). Using this approach a small number of early fractions contain greater than 95 % of the peptides, whilst the later fractions contain IGEPAL 630 polymers which hamper MS analysis severely and β2-microglobulin (*see* **Note 11**).

3. Collect fractions (500 μl) into LoBind Eppendorf tubes. At this point fractions may be frozen at −80 °C.

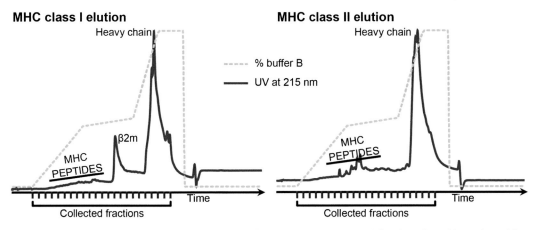

Fig. 2 Monolithic separation of affinity-purified MHC-peptide complexes and fractionation of bound peptides. Representative UV trace showing fractionation of the eluted mixture of peptides and heavy chains from both MHC class I and class II eluates. The early fractions contain MHC peptides and the later fractions contain heavy chains, detergent, and leached antibody. The β2-microglobulin peak is highlighted for the MHC class I eluate

4. Vacuum concentrate peptide-containing fractions (generally peptide-containing fractions contain up to 45 % buffer B) to reduce the concentration of acetonitrile. Typically dry to 10 μl and dilute to 15–25 μl in 0.1 % formic acid. Do not dry to completeness as this may result in sample loss due to adsorption to the plasticware.

3.2 LC-MS/MS Analysis of MHC Eluate

Two forms of mass spectrometry can be used to analyze fractions containing HLA-bound peptides (Fig. 3). This may consist of the traditional global LC-MS/MS analysis or more targeted methodologies such as multiple reaction monitoring. Global LC-MS/MS is recommended for samples where the MHC peptide repertoire composition is unknown. Peptide species are separated by an LC gradient and paired MS and MS/MS spectra are acquired by the mass spectrometer. Downstream analysis and identification of acquired spectra are facilitated by either manual sequencing or, preferably, the use of protein identification software algorithms (e.g., Mascot (MatrixScience), ProteinPilot™ (SCIEX)).

Numerous factors can affect the resulting number of peptide identifications from a global LC-MS/MS analysis, including initial starting cell number, expression level of the MHC molecules at the cell surface, efficiency of cell lysis, quantity of antibody used for immunoaffinity capture, appropriate and sufficient upstream HPLC fractionation, online LC gradient, and specific MS parameter settings. It is recommended to optimize these variables.

1. Place samples in a sonicating water bath for 5 min to detach peptides bound to plastic.

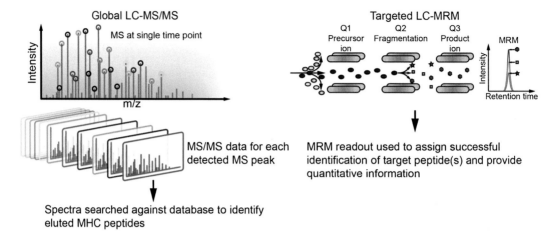

Fig. 3 A comparison of information-dependent acquisition (IDA) or LC-MS/MS and LC-MRM analysis. In global LC-MS/MS analysis a defined number of precursor ions are selected for fragmentation during each duty cycle of the mass spectrometer. The selection of precursors is a stochastic process typically based on abundance or ion intensity and limited to the top 10–50 most intense ions entering the instrument in that particular cycle. In contrast, LC-MRM involves the selection of specific, predefined, precursors that are targeted for analysis and detected based on a defined set of fragment ions. This allows relatively low-intensity ions to be selected in preference to more abundant co-eluting species. Area under the curve quantitation of the MRM transitions can then afford very accurate and specific quantitation

2. Centrifuge at $13,000 \times g$ for 10 min to pellet any particulates and transfer supernatant into an autosampler vial.

3. Add retention time standard peptides at appropriate concentrations, e.g., iRT peptide mix. This step is optional but highly recommended (*see* **Note 12**).

4. Load sample onto mass spectrometer and run optimized gradient. We routinely run MHC peptides on a 5600$^+$ TripleTOF (SCIEX) equipped with a Nanospray III ion source, where samples are loaded onto a microfluidic trap column packed with ChromXP C18-CL 3 μm particles (300 Å nominal pore size; equilibrated in 0.1 % FA/2 % acetonitrile) at 5 μl/min using a NanoUltra cHiPLC system. An analytical (75 μm × 15 cm ChromXP C18-CL 3 μm, 120 Å, Eksigent) microfluidic column is switched in line and peptides separated using linear gradient elution of 0–80 % acetonitrile over 90 min (300 nl/min). MS/MS switch criteria includes ions of $m/z > 200$ amu, charge state +2 to +5, intensity >40 cps, and the top 20 ions meeting this criteria are selected for MS/MS per cycle (*see* **Note 13**). Some examples of MS/MS spectra of class I and II peptides are shown in Fig. 4.

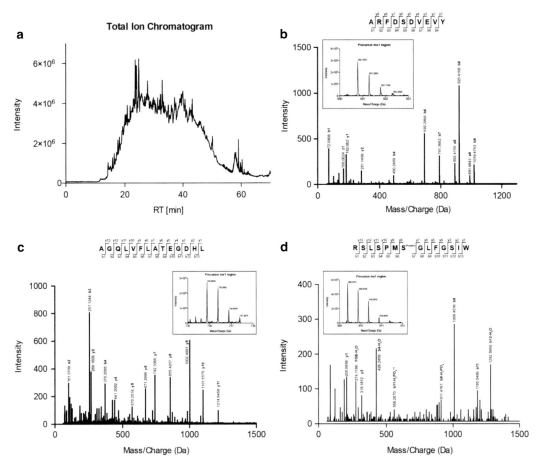

Fig. 4 Examples of biochemical analysis of peptides eluted from MHC class I and class II molecules. (**a**) A typical total ion chromatogram (TIC) of LC-separated MHC-bound peptides analyzed on a 5600⁺ TripleTOF mass spectrometer (SCIEX). (**b**–**d**) Annotated MS/MS spectra of various MHC-bound peptides. ARFDSDVEVY (*panel b*) and phosphorylated RSLSPMS*GLFGSIW (*panel d*) were eluted from the MHC class I molecules HLA-B*27:05 and HLA-B*57:01, respectively. AGQLVFLATEGDHL (*panel c*) was eluted from human MHC class II molecules. The *insets* of each panel show the corresponding precursor MS1 regions

3.3 Targeted Mass Spectrometric Analysis of MHC Eluate

Targeted methodologies, e.g., LC-MRM, are of use when specific peptide epitopes are known and a qualitative and/or quantitative readout is desired. Here, MS instrument parameters are set to target only the peptides of interest, ignoring the rest of the sample allowing low-abundance peptides to be detected in complex samples (*see* **Note 14**). LC-MRM also negates the need for high cell numbers as such targeted methods are, by their nature, highly specific and more sensitive than global LC-MS/MS approaches. For LC-MRM, synthetic peptides corresponding to those of interest can be used to design and optimize MRM parameters [41] or alternatively MS/MS data from discovery-based experiments can be used to identify optimal transition parameters. The detection of spiked AQUA peptides by LC-MRM allows integration of the area under the curve of the light and heavy peptides, providing quantitation of the light peptides. These values can be related back to the starting number of cells, to give the most accurate assessment possible of epitope copy number per cell [38, 39].

1. Prepare AQUA peptide stock, we generally reconstitute peptides at 5 mM in 100 % DMSO (choice of buffer will be user dependent, *see* **Note 5**). Run approximately 50 fmol of peptide diluted in 0.1 % FA on mass spectrometer to determine the dominant precursor ion. Peptides are then run targeting the dominant precursor ion and fragmented across a range of collision energies (CE) in order to determine the optimal energy to generate the most intense fragment ions [41]. Note that each MRM transition can use its own CE value, so different fragment ions will sometimes require different CE values. Typically the top four most intense fragment ions are used to build the MRM for a given peptide.

2. For each AQUA peptide perform a standard curve and choose a concentration in the linear part of the curve that will be used to spike MHC samples. Typically this will be between 1 and 100 fmol. Determine which fraction AQUA peptides elute during RP-HPLC for initial separation on the C18 column (prior to loading onto mass spec). By doing this it will be possible to run only fractions that contain the peptide of interest which will substantially reduce instrument time depending on the number of AQUA peptides being analyzed. Moreover, co-elution of the AQUA peptide provides further confidence in the detection of the target peptide.

3. Add isotopically labeled (AQUA) peptides of known concentration into the MHC eluate immediately following elution from the antibody affinity column. This will allow for the co-elution of the light (i.e., endogenous) and heavy (i.e., AQUA) peptide during RP-HPLC.

4. We routinely run LC-MRM experiments on a 5500 QTRAP (SCIEX) with similar LC conditions as above. MRM transitions are used with a dwell time between 5 and 40 ms, optimized to result in a cycle time that will lead to at least eight data points across a detected peptide. MS parameters are unit resolution for Q1 and Q3, with the MRM experiment coupled to an information-dependent acquisition (IDA) criterion set to trigger an EPI scan (10,000 Da/s; rolling CE; unit resolution) following any MRM transition exceeding 500 counts.

5. To quantitate the amount of the peptide of interest present in each HPLC fraction, the area under each MRM transition peak is calculated (using for example MultiQuant 2.0, SCIEX). After peak integration, the area value for all MRM transitions for the peptide is combined and compared against the area

value for the combined transitions of the AQUA peptide. The area ratio between the light and heavy peptides is used to determine the molar amount of the peptide of interest. Multiplying the molar amount of peptide by Avogadro's number and dividing by the cell number will give the number of peptide copies per cell.

3.4 Analysis of Cellular Proteome Using FASP Digestion of MHC Lysate

Tryptic digestion can be performed on samples taken before or after loading onto affinity column. If quantitation of the lysate cannot be performed, we would recommend digesting 100 μl (from a lysate made of 1×10^8 cells) with 1 μg of trypsin as a starting point.

1. If lysate has been frozen, thaw sample on ice.

2. Centrifuge at $16,000 \times g$ for 5 min, and collect supernatant.

3. Quantitate lysate by using Bradford reagent.

4. Treat up to 400 μg protein lysate with TCEP at 5 mM final concentration for 30 min at 60 °C to reduce sample.

5. Add 200 μl of urea sample solution, transfer to spin filter, and centrifuge $14,000 \times g$ for 15 min (maximum volume of the spin filter is 300 μl; if more volume of lysate is required to reach capacity, multiple loads through the spin filter can be performed).

6. Pass flow-through through the column again (spin at $14,000 \times g$ for 15 min).

7. Add 200 μl of urea sample solution to spin filter (*see* **Note 15**), centrifuge at $14,000 \times g$ for 15 min, and discard flow-through from collection tube.

8. Add 10 μl of 10× iodoacetamide solution and 90 μl of urea sample solution to spin filter and vortex for 1 min.

9. Incubate without mixing for 20 min in the dark.

10. Centrifuge for 10 min $14,000 \times g$ to remove iodoacetamide.

11. Add 100 μl of urea sample solution to the spin filter and centrifuge at $14,000 \times g$ for 10 min; repeat this step twice and discard flow-through.

12. Add 100 μl of 50 mM ammonium bicarbonate solution to spin filter and centrifuge at $14,000 \times g$ for 15 min; repeat this step twice.

13. Change collection tube and add 75 μl digestion solution (1:100 enzyme:protein) and vortex for 1 min.

14. Wrap tubes with parafilm to minimize evaporation, and incubate spin filters at 37 °C overnight.

15. Add 40 μl of 50 mM ammonium bicarbonate solution.

16. Centrifuge spin filter at $14,000 \times g$ for 10 min; repeat this step once.

17. Add 50 µl 0.5 M NaCl and centrifuge spin filter at $14,000 \times g$ for 10 min.

18. Transfer to LoBind tube, and spin down at $16,000 \times g$ for 10 min to remove debris.

19. Transfer to new LoBind tube and store peptides at –20 °C or –80 °C until ready for reversed-phase fractionation as per MHC eluate (Subheading 3.1.4, *see* **Note 16**).

4   Notes

1. The choice of monoclonal antibody is closely allied to the choice of cell line and the specificity and efficacy of the monoclonal antibody/antibodies used in the immunoaffinity isolation of MHC molecules. Monoclonal antibodies with specificity towards classes of MHC molecules, families of MHC molecules, individual alleles of MHC molecules, and even subsets of molecules of an individual allotype have been generated over the years and many hybridomas are readily accessible commercially through bodies such as the ATCC (www.atcc.org). Those we have successfully used for MHC/peptide elution experiments are shown in Table 1. This highlights that some antibodies are better suited for MHC elution than others. For example, although both L243 and LB3.1 antibodies affinity purify HLA-DR, LB3.1 yields at least double the number of MHC peptides. It is also important to determine whether the DMP cross-linker interferes with the ability of the antibody to immunoprecipitate (in which case other cross-linking methods may be required).

2. DMP can be bought in larger amounts; however the 250 mg vials are relatively cheap and work well for single use. This eliminates the possibility of decreased efficiency of cross-linking due to long-term storage of larger quantities of DMP at –20 °C. Generally one 250 mg vial is used to cross-link up to 20 mg of antibody to 2 ml of resin.

3. Choice of cell type: In order to maximize the yield of MHC class I or class II molecules the cell line used must be given serious consideration. Epstein–Barr virus-transformed B cell lines that express high levels of HLA A, B, or C class I molecules or HLA DR, DQ, or DP molecules are easily sourced from depositories such as ATCC. These cells can be grown to high density in cell culture and used to great effect in biochemical studies of bound ligands. Homozygous cell lines for most common class I or class II alleles are well documented and often express haplotypes of interest. The use of B-LCLs dictates the use of a discriminating antibody should a single

allele be required to be purified. In the absence of such an antibody, the ideal cell type for these experiments would express a single MHC molecule and have intact antigen processing and presentation pathways. Several mutant cell lines have been generated that approximate such a cell type. The B-lymphoblastoid cell line Hmy2.C1R was generated by gamma irradiation of LICR.LON.Hmy2 [42] and selected with antibodies against HLA A and HLA B alleles and complement. This resulted in a cell line with no detectable HLA A or B gene products, yet with intact antigen processing and presentation pathways [43]. Thus, these cells are able to support high-level expression of individually transfected HLA A, B, or C gene products [43]. Similar cell lines exist for class II elution studies. For example, the murine cell line M12.C3 lacks endogenous Ia and functional I-A^k expression can be restored by introduction of I-A^K α and β chains via transfection [44]. It should be noted however that not all cells express class II molecules endogenously, thus restricting the array of APCs amenable for the creation of appropriate cell lines. The same considerations apply to tissue samples, where MHC levels may vary and where there may be a mix of different cell types.

4. The number of cells required will vary depending on the application and the MHC levels expressed by the cells. Global LC-MS/MS analysis requires the largest number of cells. For a basic repertoire analysis of a cell line expressing high levels of MHC we would generally use 1×10^9 cells and expect to sequence between 1000 and 3000 peptides depending on the allele. During LC-MS/MS analysis the mass spectrometer is limited to the number of ions per second that can be fragmented to obtain MS/MS data. This dictates that it is the most abundant ions that are sequenced such that increasing the cell number does not necessarily increase the number of peptide identifications. Increasing the number of peptides requires a combination of adequate cell number, and further fractionation of the eluate prior to mass spectrometric analysis to decrease the complexity of the peptide mix. LC-MRM analysis requires fewer cells but can only be utilized when the sequence of the peptides to be targeted is known. We generally start with pellets of 1×10^8 for MRM analysis. Cell numbers can be lower than 1×10^8; however detection and quantitation of peptides from smaller cell numbers are dependent on how well the target peptide ionizes and how much of the given peptide/MHC complex is expressed in the surface of the cell.

5. AQUA peptides should be synthesized to the highest purity (>98 %) and solubilised in an appropriate buffer (e.g., buffer A, DMSO). The amino acid composition of the peptide will infer optimal buffer to use; a number of guides are available such as

http://www.genscript.com/peptide_solubility_and_stablity. html. Peptides should be quantified to an exact concentration (amino acid analysis; DirectDetect (Merck Millipore)). AQUA peptides are spiked into the sample immediately post-acid elution from the affinity column. The amount of each AQUA peptide to add requires optimization on a per-peptide basis, but typically ranges from 1 to 100 fmol per sample.

6. For each new antibody it is advisable to check the efficiency of the cross-linking reaction by SDS-PAGE. Take a sample of the original antibody (1) and a sample of the flow through (2) following incubation with the resin. Take aliquots of resin (25 μl) before the addition of DMP (3) and after incubation with DMP (4). It is also advisable to concentrate the flow through from the citric acid wash (5) using a 15–30 kDa cut-off concentrator to ~500 μl (this is done to see how much antibody has been left unbound), and take 25 μl to add to 2× SDS-PAGE loading dye. Samples should be run on a reducing 12 % SDS-PAGE gel and Commassie stained. Heavy and light chains of the antibody should be seen in samples 1 and 3. There should be little antibody in samples 2, 4, and 5. If the antibody has not been used for DMP cross-linking before, perform a small-scale purification to test that the antibody still retains its binding affinity.

7. Cells can be expanded by standard tissue culture techniques; however particular care should be taken with adherent cells. Removing cells with trypsin can strip coatings from treated tissue culture plastics, which may interfere with subsequent mass spectral analysis. We routinely passage cells using trypsin; however when harvesting for snap freezing we remove cells using 5–10 mM EDTA in PBS.

8. Column and mobile phase choice for multidimensional RP-HPLC are dictated by sample composition but considerations should include altered ion pair agent, altered mobile phase pH, altered stationary-phase ligand or mode, and column dimensions.

9. A common practice to separate peptides from heavy chain and β2m in the case of class I molecules is to use a low-protein-binding spin filter. In our experience, this results in significant loss of peptides, irrespective of filter brand or pre-blocking with BSA. We find that separation using RP-HPLC increases the yield of peptides at least tenfold.

10. We routinely separate β2m from heavy chains using a 4.6 mm internal diameter × 50 or 100 mm long reversed-phase C18 endcapped HPLC column (Chromolith Speed Rod, Merck). The choice between 50 and 100 mm is dependent on the cell number and whether the samples are to be used for LC-MS/MS

or LC-MRM. For LC-MS/MS where larger cell numbers are used and the goal is to sequence as many peptides as possible, we find that the 100 mm column gives better separation of the MHC peptides and greater peptide numbers (two- to three-fold over the 50 mm column). Peptide numbers may also be increased by using orthogonal separation methods such as SCX or HILIC. For MRM analysis, which is normally performed on smaller samples, the 50 mm rod is sufficient; however the impact of sample complexity on any given peptide of interest should be determined to optimize the amount of fractionation required.

11. The number of fractions collected needs to be a compromise between reducing the complexity of the sample for mass spectrometric analysis and keeping the number of samples to a practical number. We routinely collect 500 μl fractions, and then pool these into a single concatenated sample consisting of four to six fractions spread well across the gradient separation. In this way, the number of samples for LC-MS/MS is reduced, and the full gradient on the column attached to the mass spectrometer is utilized. Caution should be exercised with fractions coming off in high acetonitrile; although these may contain peptides of interest, the highly hydrophobic nature of the peptides in these samples may interfere with detection of peptides eluting earlier in the gradient and may be better run as individual fractions.

12. In addition to AQUA peptides, it is often desired for both LC-MS/MS and LC-MRM experiments to spike in a set of well-characterized peptides, which can be used (1) to normalize retention times across different experiments independent of the chromatographic system and/or (2) to precisely predict retention times of known target peptides. The indexed retention time (iRT) peptide mix, a set of 11 peptides derived from *Leptospira interrogans*, was specifically designed for this purpose [45].

13. Although a time-of-flight (TOF) mass spectrometer such as the 5600+ TripleTOF (SCIEX) is calibrated every three to five LC runs using a standard such as glu-fibrinopeptide, we also recommend preparing a standard sample of MHC peptides (i.e., non-tryptic) to run as a quality control for instrument performance. We often see a drop in the number of MHC peptides identified using this standard before a decrease in the intensity of the glu-fibrinopeptide calibration peak is detected.

14. Although the sample complexity is not particularly a problem for MRM analysis, it is our experience that some peptides are harder to detect when present in an increasingly complex sample or a sample containing highly hydrophobic sequences. For this reason fractionation is still recommended for LC-MRM analysis.

15. The FASP kit comes with 30 kDa cutoff spin filters; we however prefer to use 10 kDa cutoff spin filters from Pall.

16. Mass spectrometric analysis can be performed on the FASP digest without fractionation; however we find that this dramatically reduces the number of protein identifications. We recommend an off-line separation by RP-HPLC; alternatively other separation methods such as SCX may be used.

References

1. Madden DR (1995) The three-dimensional structure of peptide-MHC complexes. Annu Rev Immunol 15:587–622

2. Stern LJ, Brown JH, Jardetzky TS et al (1994) Crystal structure of the human class II MHC protein HLA-DR1 complexed with an influenza virus peptide. Nature 368(6468): 215–221

3. Fremont DH, Hendrickson WA, Marrack P et al (1996) Structures of an MHC class II molecule with covalently bound single peptides. Science 272(5264):1001–1004

4. Garcia KC, Teyton L, Wilson IA (1999) Structural basis of T cell recognition. Annu Rev Immunol 17:369–397

5. Falk K, Rötzschke O, Deres K et al (1991) Identification of naturally processed viral nonapeptides allows their quantification in infected cells and suggests an allele-specific T cell epitope forecast. J Exp Med 174:425–434

6. Sijts AJ, Neisig A, Neefjes J et al (1996) Two Listeria monocytogenes CTL epitopes are processed from the same antigen with different efficiencies. J Immunol 156(2):683–692

7. Rotzschke O, Falk K, Deres K et al (1990) Isolation and analysis of naturally processed viral peptides as recognized by cytotoxic T cells. Nature 348(6298):252–254

8. Storkus WJ, Zeh HJ, Maeurer MJ et al (1993) Identification of human melanoma peptides recognized by class I restricted tumor infiltrating T lymphocytes. J Immunol 151(7): 3719–3727

9. Storkus WJ, Zeh HJ, Salter RD et al (1993) Identification of T-cell epitopes: rapid isolation of class I-presented peptides from viable cells by mild acid elution. J Immunother 14(2): 94–103

10. Falk K, Rötzschke O, Stevanovic S et al (1991) Allele-specific motifs revealed by sequencing of self-peptides eluted from MHC molecules. Nature 351:290–296

11. Rammensee HG, Falk K, Rotzschke O (1993) Peptides naturally presented by MHC class I molecules. Annu Rev Immunol 11:213–244

12. Scull KE, Dudek NL, Corbett AJ et al (2012) Secreted HLA recapitulates the immunopeptidome and allows in-depth coverage of HLA A*02:01 ligands. Mol Immunol 51(2):136–142

13. Illing PT, Vivian JP, Dudek NL et al (2012) Immune self-reactivity triggered by drug-modified HLA-peptide repertoire. Nature 486(7404):554–558

14. Dudek NL, Tan CT, Gorasia DG et al (2012) Constitutive and inflammatory immunopeptidome of pancreatic beta-cells. Diabetes 61(11):3018–3025

15. Urban RG, Chicz RM, Lane WS et al (1994) A subset of HLA-B27 molecules contains peptides much longer than nonamers. Proc Natl Acad Sci U S A 91(4):1534–1538

16. Purcell AW, Kelly AJ, Peh CA et al (2000) Endogenous and exogenous factors contributing to the surface expression of HLA B27 on mutant APC. Hum Immunol 61(2):120–130

17. Storkus WJ, Howell DN, Salter RD et al (1987) NK susceptibility varies inversely with target cell class I HLA antigen expression. J Immunol 138(6):1657–1659

18. Macdonald WA, Purcell AW, Mifsud NA et al (2003) A naturally selected dimorphism within the HLA-B44 supertype alters class I structure, peptide repertoire, and T cell recognition. J Exp Med 198(5):679–691

19. Zernich D, Purcell AW, Macdonald WA et al (2004) Natural HLA class I polymorphism controls the pathway of antigen presentation and susceptibility to viral evasion. J Exp Med 200(1):13–24

20. Scally SW, Petersen J, Law SC et al (2013) A molecular basis for the association of the HLA-DRB1 locus, citrullination, and rheumatoid arthritis. J Exp Med 210(12):2569–2582

21. Ringrose JH, Yard BA, Muijsers A et al (1996) Comparison of peptides eluted from the groove of HLA-B27 from Salmonella infected and non-infected cells. Clin Rheumatol 15 Suppl 1:74–78

22. van Els CA, Herberts CA, van der Heeft E et al (2000) A single naturally processed

measles virus peptide fully dominates the HLA-A*0201-associated peptide display and is mutated at its anchor position in persistent viral strains. Eur J Immunol 30(4):1172–1181

23. Nepom BS, Nepom GT, Coleman M et al (1996) Critical contribution of beta chain residue 57 in peptide binding ability of both HLA-DR and -DQ molecules. Proc Natl Acad Sci U S A 93(14):7202–7206

24. Lampson LA, Levy R (1980) Two populations of Ia-like molecules on a human B cell line. J Immunol 125(1):293–299

25. Gorga JC, Knudsen PJ, Foran JA et al (1986) Immunochemically purified DR antigens in liposomes stimulate xenogeneic cytolytic T cells in secondary in vitro cultures. Cell Immunol 103(1):160–173

26. Watson AJ, DeMars R, Trowbridge IS et al (1983) Detection of a novel human class II HLA antigen. Nature 304(5924):358–361

27. Parham P, Brodsky FM (1981) Partial purification and some properties of BB7.2. A cytotoxic monoclonal antibody with specificity for HLA-A2 and a variant of HLA-A28. Hum Immunol 3(4):277–299

28. Ellis SA, Taylor C, McMichael A (1982) Recognition of HLA-B27 and related antigen by a monoclonal antibody. Hum Immunol 5(1):49–59

29. Schittenhelm RB, Dudek NL, Croft NP et al (2014) A comprehensive analysis of constitutive naturally processed and presented HLA-C*04:01 (Cw4)-specific peptides. Tissue Antigens 83(3):174–179

30. Braud VM, Allan DS, Wilson D et al (1998) TAP- and tapasin-dependent HLA-E surface expression correlates with the binding of an MHC class I leader peptide. Curr Biol 8(1):1–10

31. Thomas R, Apps R, Qi Y et al (2009) HLA-C cell surface expression and control of HIV/AIDS correlate with a variant upstream of HLA-C. Nat Genet 41(12):1290–1294

32. Hammerling GJ, Rusch E, Tada N et al (1982) Localization of allodeterminants on H-2Kb antigens determined with monoclonal antibodies and H-2 mutant mice. Proc Natl Acad Sci U S A 79(15):4737–4741

33. Straus DS, Stroynowski I, Schiffer SG et al (1985) Expression of hybrid class I genes of the major histocompatibility complex in mouse L cells. Proc Natl Acad Sci U S A 82(18): 6245–6249

34. Ozato K, Sachs DH (1981) Monoclonal antibodies to mouse MHC antigens. III. Hybridoma antibodies reacting to antigens of the H-2b haplotype reveal genetic control of isotype expression. J Immunol 126(1):317–321

35. Kappler JW, Skidmore B, White J et al (1981) Antigen-inducible, H-2-restricted, interleukin-2-producing T cell hybridomas. Lack of independent antigen and H-2 recognition. J Exp Med 153(5):1198–1214

36. Janeway CA Jr, Conrad PJ, Lerner EA et al (1984) Monoclonal antibodies specific for Ia glycoproteins raised by immunization with activated T cells: possible role of T cellbound Ia antigens as targets of immunoregulatory T cells. J Immunol 132(2):662–667

37. Oi VT, Jones PP, Goding JW et al (1978) Properties of monoclonal antibodies to mouse Ig allotypes, H-2, and Ia antigens. Curr Top Microbiol Immunol 81:115–120

38. Croft NP, Smith SA, Wong YC et al (2013) Kinetics of antigen expression and epitope presentation during virus infection. PLoS Pathog 9(1), e1003129

39. Tan CT, Croft NP, Dudek NL et al (2011) Direct quantitation of MHC-bound peptide epitopes by selected reaction monitoring. Proteomics 11(11):2336–2340

40. Wisniewski JR, Zougman A, Nagaraj N et al (2009) Universal sample preparation method for proteome analysis. Nat Methods 6(5):359–362. doi:10.1038/nmeth.1322

41. Lange V, Picotti P, Domon B et al (2008) Selected reaction monitoring for quantitative proteomics: a tutorial. Mol Syst Biol 4:222

42. Edwards PA, Smith CM, Neville AM et al (1982) A human-hybridoma system based on a fast-growing mutant of the ARH-77 plasma cell leukemia-derived line. Eur J Immunol 12(8):641–648

43. Alexander J, Payne JA, Murray R et al (1989) Differential transport requirements of HLA and H-2 class I glycoproteins. Immunogenetics 29(6):380–388

44. Nelson CA, Roof RW, McCourt DW et al (1992) Identification of the naturally processed form of hen egg white lysozyme bound to the murine major histocompatibility complex class II molecule I-Ak. Proc Natl Acad Sci U S A 89(16):7380–7383

45. Escher C, Reiter L, MacLean B et al (2012) Using iRT, a normalized retention time for more targeted measurement of peptides. Proteomics 12(8):1111–1121

Chapter 15

Profiling of Small Molecules by Chemical Proteomics

Kilian V.M. Huber and Giulio Superti-Furga

Abstract

Chemical proteomics provides a powerful means to gain systems-level insight into the mode of action of small molecules and/or natural products. In contrast to high-throughput screening efforts which only interrogate selected subproteomes such as kinases and often only consider individual domains, the methodology presented herein allows for the determination of the molecular targets of small molecules or drugs in a more relevant physiological setting. As such, the compound of interest is exposed to the entire variety of cellular proteins considering all naturally occurring posttranslational modifications and activation states. Samples prepared according to the procedures described in this protocol are compatible with lysates from cultured cell lines, primary cells, or samples from biopsies. In combination with state-of-the-art mass spectrometry techniques this approach grants access to a comprehensive view of small molecule-target protein interactions.

Key words Drug discovery, Target deconvolution, Chemical proteomics, Mode of action

1 Introduction

Phenotypicscreening represents an interesting strategy for identifying new potential therapeutics [1]. The fact that the compounds discovered by this approach exhibit a desired phenotype in living cells makes them attractive candidates for further development. However, understanding the molecular target and thus the mode of action can pose a significant challenge [2]. There are also a number of approved drugs which have proven efficacious in the treatment of human disease over decades for which it is not clear by which mechanism they work [1]. Knowledge of the relevant cellular target(s) would allow for the development of more potent and selective drugs with probably less side effects. Even for drugs whose mode of action is known a comprehensive target profile can assist further patient stratification both in terms of applicability and managing side effects [3]. Moreover, it can also reveal new potential indications due to previously unknown off-targets. Common "high-throughput" in vitro approaches to determine the targets of

Jörg Reinders (ed.), *Proteomics in Systems Biology: Methods and Protocols*, Methods in Molecular Biology, vol. 1394,
DOI 10.1007/978-1-4939-3341-9_15, © Springer Science+Business Media New York 2016

small molecules or natural products are often both time and cost intensive and usually only cover a selected range of protein classes, e.g., kinases or histone deacetylases. Furthermore, these assays do not provide any information if the targets identified in the screen are "real-and-relevant" interactors as such issues as differential tissue expression or even posttranslational modifications are not considered.

Several target deconvolution approaches have been developed covering a wide range of biochemical and genetic techniques as well as in silico methods [2, 4]. Among all those affinity-based approaches have contributed to the discovery of a number of important drug classes such as immunosuppressants [5, 6] and histone deacetylase (HDAC) inhibitors [7]. The combination of this methodology with protein mass spectrometry has been termed "chemical proteomics" and has lately been applied to diverse areas of research including the elucidation of the molecular mechanism of thalidomide teratogenicity [8] as well as the identification of novel regulators of necroptosis signaling [9] and potential anticancer targets [10]. In a typical chemical proteomic experiment, the small molecule or compound of interest is incubated with a relevant cell lysate which can be prepared from tissue culture cell lines or even primary cells and biopsy samples. Using the small-molecule compound as bait by means of a drug matrix, cellular interactors are captured and purified by affinity enrichment. After washing, the eluted proteins are digested to peptides and can subsequently be identified by mass spectrometry followed by bioinformatic analysis (Fig. 1). This procedure requires the compound of interest itself or a corresponding analogue to be amenable to chemical derivatization in order to be immobilized on the solid phase. The design of a suitable analogue is facilitated by prior knowledge of structure-activity relationships (SAR) or co-crystal structures of annotated targets. Alternatively, the so-called east–west approach may be applied to compounds devoid of those data [11]. The concept of this strategy is to prepare two coupleable analogues of the compound of interest of which each is modified at a different, preferably most distant site.

An alternative chemical proteomic approach is activity-based probe profiling (ABPP) which does not require chemical modification of the query compound by taking advantage of a reactive probe which binds covalently to a given class of proteins, e.g., serine hydrolases [12–14]. In this case, comparison of treated versus untreated sample yields the putative interactors. Recently, another proteomic target identification strategy based on ligand-induced thermal stabilization of proteins has been established [15]. This methodology termed thermal stability profiling does also not depend on chemical derivatization and instead uses the unmodified original compound of interest allowing for the detection of protein-ligand interactions in intact living cells. For a general overview of

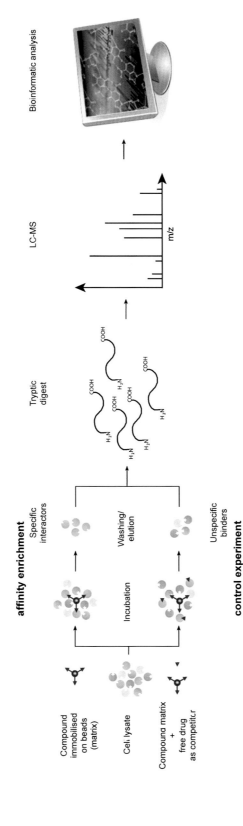

Fig. 1 Schematic outline of a chemical proteomic experiment. A drug matrix consisting of the immobilized small molecule on a solid phase is incubated with lysates from either cell lines, primary cells, or biopsy samples. After incubation, contaminants are removed by extensive washing and finally target proteins are eluted and analyzed by mass spectrometry followed by bioinformatic processing

recent proteomic technologies relevant to the field of drug and target discovery the reader is directed to the literature [14].

The procedures presented herein assume that the investigative compounds bear a nucleophilic handle suitable to react with *N*-hydroxy-succinimide (NHS) esters, e.g., a functional group such as a primary or secondary aliphatic amine. However, this protocol can easily be adapted to using biotinylated drugs by changing the solid phase accordingly.

---

2 Materials

2.1 Reagents

2.1.1 Lysis Buffer for Preparation of Whole-Cell Lysates

1. Nonidet P-40 (NP-40).
2. Tris–HCl pH 7.5.
3. Glycerol.
4. $MgCl_2$.
5. NaCl.
6. NaF.
7. Na_3VO_4.
8. Phenylmethylsulfonyl fluoride (PMSF).
9. Dithiothreitol (DTT).
10. *N*-p-Tosyl-L-phenylalanine chloromethyl ketone (TPCK).
11. Inhibitor cocktail containing leupeptin, aprotinin, soybean trypsin inhibitor.
12. Bradford reagent.

Lysis buffer (LB): 0.20 % NP-40, 50 mM Tris–HCl pH 7.5, 5 % glycerol, 1.5 mM $MgCl_2$, 100 mM NaCl, 25 mM NaF, 1 mM Na_3VO_4, 1 mM PMSF, 1 mM DTT, 10 μg/mL TPCK, 1 μg/mL leupeptin, 1 μg/mL aprotinin, 10 μg/mL soybean trypsin inhibitor; prepare freshly, keep on ice.

2.1.2 Compound Coupling Protocol

1. Coupleable compound (drug) (*see* **Note 1**).
2. NHS-activated sepharose beads (50 % slurry in isopropanol).
3. Dimethyl sulfoxide (DMSO) (abs.).
4. Triethylamine (TEA).
5. Ethanolamine.
6. Isopropanol.

2.1.3 Drug Pull-Down Protocol

1. Drug matrix (compound immobilized on beads).
2. DMSO (abs.).
3. Lysis buffer (LB).
4. HEPES.

5. EDTA.

6. NaCl.

7. NaOH.

8. Purified water.

9. Formic acid.

HEPES-NaOH-EDTA buffer pH 7.5: 50 mM HEPES, 0.5 µM EDTA pH 8.0, 100 mM NaCl; adjust pH to 7.5 using NaOH, prepare freshly, keep on ice.

2.2 Equipment

1. Needles and syringes.

2. Pipettes.

3. Filter tips.

4. Polycarbonate ultracentrifuge tubes.

5. Ultracentrifuge.

6. Photometer.

7. Tabletop centrifuge.

8. Rotoshaker.

9. HPLC with MS and/or UV detector.

10. Chromatography columns.

11. Mass spectrometry glass vials.

3 Methods

3.1 Preparation of Whole-Cell Lysates (Timing 2 h) (See Notes 2 and 3)

1. (a) For *cell pellets*, thaw pellets on ice and resuspend in lysis buffer (depending on pellet size, rule-of-thumb 1:1 ratio), transfer into homogenizer/Dounce apparatus (e.g., 0.9 mm syringe) and homogenize sample ten times (optional).

   (b) For *tissues*, transfer the sample into a tissue homogenizer/Dounce apparatus, wash with lysis buffer, and adjust to desired volume; homogenize sample ten times.

2. Transfer the homogenate to a Falcon tube and incubate on ice for 30 min.

3. Transfer homogenate to polycarbonate ultracentrifuge tubes, balance tubes, and centrifuge lysate for 10 min at 4 °C at $20,000 \times g$.

4. Transfer supernatant to fresh polycarbonate ultracentrifuge tubes, balance tubes, and centrifuge for 1 h at 4 °C at $90,000 \times g$.

5. Transfer supernatant (remove most of lipid layer, if possible) to a fresh Falcon centrifuge tube, and keep on ice.

6. Determine protein concentration (e.g., Bradford).

7. Prepare lysate aliquots (e.g., 5–10 mg total protein) or use directly for pull-downs (*see* **Note 4**).

Pausing point: After shock-freezing in liquid nitrogen the lysate aliquots may be stored at −80 °C until use.

3.2 Coupling of Compounds to Sepharose Beads (Timing 3 Days)

1. Pipet 100 μL of slurry (≈50 μL settled bed volume) in a 1.5 mL Eppendorf tube. *Caution*: *Always use filter tips and cut pipet tips for pipetting beads*!

2. Centrifuge beads for 3 min at room temperature at $75 \times g$, and remove supernatant.

3. Add 50 μL of DMSO (abs) to beads, suspend gently by inverting several times, centrifuge (as before), and discard supernatant.

4. Add 500 μL DMSO (abs), resuspend beads gently by inverting several times, transfer beads in a 2 mL Eppendorf tube, centrifuge (as before), and discard supernatant; repeat wash step another two times.

5. Resuspend beads in 50 μL DMSO (abs).

6. Add 0.025 μmol of coupleable compound to 50 % bead slurry; add 0.75 μL triethylamine, and mix carefully.

7. Incubate on roto-shaker for 16–24 h at RT with 10 rpm.

8. To check coupling efficiency, centrifuge beads (as before) and remove 10 μL (≈5 nmol) from supernatant.

9. Check for remaining unreacted compound by HPLC; if there are still significant amounts detected in the supernatant go back to **step 7** and extend coupling reaction time. Repeat **steps 8** and **9**.

10. Add 2.5 μL ethanolamine to drug-bead mixture in order to block unreacted NHS-ester groups.

11. Incubate on roto-shaker for at least 8 h at room temperature with 10 rpm.

12. Centrifuge beads (*see* above), remove supernatant, and wash twice with 500 μL of DMSO (abs).

13. Proceed directly with pull-down and wash with lysis buffer.

Pausing point: Alternatively, the drug-bead matrix can be stored for up to 2 weeks using the following procedure: Remove supernatant and add 50 μL of isopropanol, resuspend beads gently by inverting several times, centrifuge (as before), and discard supernatant. Add 500 μL isopropanol, resuspend beads gently by inverting several times, centrifuge (as before), and discard supernatant; repeat wash step once. Add 50 μL isopropanol to beads and resuspend gently; store coupled beads at 4 °C (away from light) until further use.

3.3 Pull-Down Procedure (Timing 6 h) (See Note 5)

1. Centrifuge beads for 3 min at $75 \times g$, and remove supernatant.

2. Add 1 mL lysis buffer and wash beads gently by resuspending and inverting several times.

3. Centrifuge beads (*see* above), remove supernatant, and repeat wash step three times.

4. Remove supernatant.

5. Dilute cell lysate with lysis buffer to a final protein concentration of 15 mg/mL.

6. Transfer to ultracentrifuge tube, balance tubes, and centrifuge for 20 min at 4 °C at $90,000 \times g$.

7. Remove 200 μg protein of whole-cell lysate from supernatant (as input control for Western blot).

8. Decant remaining supernatant directly onto compound-beads.

9. Gently resuspend washed compound-beads in cell lysate.

10. Incubate on roto-shaker for 2 h at 4 °C with 10 rpm.

11. After the incubation centrifuge beads for 3 min at $75 \times g$.

12. Wash disposable chromatography column twice with 1 mL lysis buffer.

13. *Caution: Perform the following steps at 4°C!* Resuspend beads gently by pipetting up and down and transfer to plugged columns. Let beads settle by gravity, unplug column, and then drain remaining buffer by gravity flow.

14. Add 5 mL lysis buffer, and let buffer drain by gravity flow.

15. Add 2.5 mL 50 mM HEPES-NaOH buffer, and let buffer drain by gravity flow.

16. Wash column tip twice with 0.5 mL HEPES buffer, place in centrifuge tube, and spin down for 1 min with $100 \times g$.

17. For MS analysis place a suitable glass vial under the column to collect the sample. Add 250 μL formic acid, let the sample drain by gravity, and remove remaining liquid from matrix by plugging/unplugging the column lid (3×). Submit sample to MS processing.

3.4 Concluding Remarks

The described procedures provide an effective means to reveal the interactors of small molecules and natural compounds. However, due to the high complexity of the samples and varying protein abundance it is recommendable to include control pull-downs which can be either the unreacted bead matrix itself or an unrelated compound matrix to estimate random and unspecific binding. Alternatively, lysates can be preincubated with the original, unmodified compound of interest to determine competitive binding (Fig. 1). If coupled with quantitative MS technologies such as iTRAQ or TMT this approach also allows for the determination of K_ds.

4   Notes

1. To maximize pull-down efficacy, coupleable compound analogues should be evaluated in vitro prior to the pull-down experiment.

2. This procedure is in general applicable for most protein targets of small molecules; however, some nuclear and transmembrane proteins may require optimized lysis conditions.

3. Depending on the stability of certain proteins it may be advisable to perform cell lysis and compound pull-down on the same day to avoid detrimental freeze-thawing cycles.

4. The required amount of total protein per pull-down depends largely on the sensitivity of the subsequent analytical method. As a rule of thumb, 5–10 mg should suffice for both applications described herein.

5. For the preparation of mass spectrometry samples it is crucial to avoid any potential contamination with keratin. Also, materials which are not sensitive to the chemicals used in this procedure should be considered preferably in order to avoid contamination with degradation products and/or polymers.

References

1. Swinney DC, Anthony J (2011) How were new medicines discovered? Nat Rev Drug Discov 10(7):507–519

2. Schenone M, Dancik V, Wagner BK et al (2013) Target identification and mechanism of action in chemical biology and drug discovery. Nat Chem Biol 9(4):232–240

3. Sillaber C, Herrmann H, Bennett K et al (2009) Immunosuppression and atypical infections in CML patients treated with dasatinib at 140 mg daily. Eur J Clin Invest 39(12):1098–1109

4. Terstappen GC, Schlupen C, Raggiaschi R et al (2007) Target deconvolution strategies in drug discovery. Nat Rev Drug Discov 6(11):891–903

5. Brown EJ, Albers MW, Bum Shin T et al (1994) A mammalian protein targeted by G1-arresting rapamycin-receptor complex. Nature 369(6483):756–758

6. Harding MW, Galat A, Uehling DE et al (1989) A receptor for the immuno-suppressant FK506 is a cis-trans peptidyl-prolyl isomerase. Nature 341(6244):758–760

7. Taunton J, Hassig CA, Schreiber SL (1996) A mammalian histone deacetylase related to the yeast transcriptional regulator Rpd3p. Science 272(5260):408–411

8. Ito T, Ando H, Suzuki T et al (2010) Identification of a primary target of thalidomide teratogenicity. Science 327(5971):1345–1350

9. Sun L, Wang H, Wang Z et al (2012) Mixed lineage kinase domain-like protein mediates necrosis signaling downstream of RIP3 kinase. Cell 148(1–2):213–227

10. Huber KVM, Salah E, Radic B et al (2014) Stereospecific targeting of MTH1 by (S)-crizotinib as an anticancer strategy. Nature 508(7495):222–227

11. Rix U, Superti-Furga G (2009) Target profiling of small molecules by chemical proteomics. Nat Chem Biol 5(9):616–624

12. Nomura DK, Dix MM, Cravatt BF (2010) Activity-based protein profiling for biochemical pathway discovery in cancer. Nat Rev Cancer 10(9):630–638

13. Savitski MM, Reinhard FBM, Franken H et al (2014) Tracking cancer drugs in living cells by thermal profiling of the proteome. Science 346(6205):1255784

14. Huber KVM, Olek KM, Müller AC et al (2015) Proteome-wide drug and metabolite interaction mapping by thermal-stability profiling. Nat Meth 12(11):1055–1057

15. Schirle M, Bantscheff M, Kuster B (2012) Mass spectrometry-based proteomics in preclinical drug discovery. Chem Biol 19(1):72–84

Chapter 16

Generating Sample-Specific Databases for Mass Spectrometry-Based Proteomic Analysis by Using RNA Sequencing

Toni Luge and Sascha Sauer

Abstract

Mass spectrometry-based methods allow for the direct, comprehensive analysis of expressed proteins and their quantification among different conditions. However, in general identification of proteins by assigning experimental mass spectra to peptide sequences of proteins relies on matching mass spectra to theoretical spectra derived from genomic databases of organisms. This conventional approach limits the applicability of proteomic methodologies to species for which a genome reference sequence is available. Recently, RNA-sequencing (RNA-Seq) became a valuable tool to overcome this limitation by de novo construction of databases for organisms for which no DNA sequence is available, or by refining existing genomic databases with transcriptomic data. Here we present a generic pipeline to make use of transcriptomic data for proteomics experiments. We show in particular how to efficiently fuel proteomic analysis workflows with sample-specific RNA-sequencing databases. This approach is useful for the proteomic analysis of so far unsequenced organisms, complex microbial metatranscriptomes/metaproteomes (for example in the human body), and for refining current proteomics data analysis that solely relies on the genomic sequence and predicted gene expression but not on validated gene products. Finally, the approach used in the here presented protocol can help to improve the data quality of conventional proteomics experiments that can be influenced by genetic variation or splicing events.

Key words Mass spectrometry, Metaproteomics, Proteogenomics, Proteomics informed by transcriptomics, Gene expression

---

1 Introduction

Improvements in high-throughput liquid chromatography-coupled tandem mass spectrometry (LC-MS/MS) and the broad availability of quantitative techniques, such as stable isotope labeling of amino acids in cell culture (SILAC) [1], dimethyl-labeling [2], isobaric tags for relative and absolute quantitation (iTRAQ) [3], absolute quantification (AQUA) peptides [4], and label-free methods like intensity-based absolute quantification (iBAQ) [5] amongst others, paved the way for comprehensive relative protein

Jörg Reinders (ed.), *Proteomics in Systems Biology: Methods and Protocols*, Methods in Molecular Biology, vol. 1394,
DOI 10.1007/978-1-4939-3341-9_16, © Springer Science+Business Media New York 2016

expression profiling in complex samples over different conditions [6, 7]. Typically, in a shotgun discovery proteomics workflow, proteins are digested to peptides by sequence-specific proteases, for instance with trypsin or Lys-C. For in-depth analysis, samples are often pre-fractionated prior to (e.g., by SDS-PAGE [8]) or after the digestion step (e.g., by isoelectric focusing or strong anion-exchange chromatography [9]) and subsequently submitted to LC-MS/MS.

The characteristic mass-to-charge ratios of the proteolytic peptides and their fragment ions, such as y and b ions that are typically recorded in data-dependent acquisition (DDA) mode by LC-MS/MS instruments [10], form distinct mass spectra that contain sequence information. To determine peptide sequences from fragment mass spectra and to infer proteins, sequence database search—by, e.g., Mascot [11], Sequest [12], X!Tandem [13], MyriMatch [14], Paragon [15], or Andromeda [16]—is the most commonly applied approach for peptide and protein identification [17]. Thereby experimental mass spectra are matched with in silico-generated mass spectra that are derived from available sequence databases. Thus, the chosen sequence database can directly affect search results because only peptides of which the amino acid sequences are deposited can be assigned. Retrieving protein sequence databases from public resources like Uniprot and ENSEMBL is a straightforward strategy for well-annotated model organisms with fully sequenced genomes, high-quality definition of transcriptional units, and predicted proteomes. But this approach represents the primary bottleneck to the widespread use of quantitative shotgun proteomics for the analysis of orphan organisms or metaproteomes. Furthermore, public databases are collections of all known and predicted proteins in a species but incomplete with respect to single-nucleotide variants (SNVs) and RNA-splice and -editing variants. Thus, they do not represent the real protein composition in a specific cell or tissue type [18].

With the increasing affordability of next-generation sequencing (NGS) technologies [19] many proteomic studies incorporated large-scale RNA expression analysis to correlate and verify protein by transcript expression data. A growing number of researchers start using this valuable resource to construct or refine their search space for improved protein identification and quantification [18, 20–30].

For instance, Wang et al. [18] investigated SNVs on protein level in human colorectal cancer cell lines SW480 and RKO. Therefore they performed RNA-sequencing (RNA-Seq), aligned short reads to the human genome, and added nonsynonymous protein-coding SNVs to the regular protein sequence database. Assuming that proteins with low-abundance transcripts are likely to be undetectable by shotgun proteomics, an RPKM (reads per kilobase of exon per million mapped reads) threshold for transcript abundance was applied to reduce the size of reference protein sequence database. This strategy improved sensitivity of peptide identification and reduced ambiguity in protein assembly. The same researchers incorporated their approach in the R package customProDB [31].

Recently, procedures have evolved that even do not need reference sequences for alignment. These emerging procedures are based on applying de novo assemblies of transcriptomes. Thus, these methods are in particular powerful when studying non-sequenced organisms or even communities of different species, as is often the case in the emerging field of metaproteomics.

The great potential of this technique has been benchmarked by Evans et al. [20], who used RNA-Seq to build a reference protein database by six-frame translation of the transcriptome of adenovirus-infected HeLa cells. This proteomics informed by transcriptomics (PIT) workflow identified more than 99 % of the proteins when using traditional protein sequence databases with annotated human and adenovirus proteins. Furthermore, we recently adopted the PIT approach to study the effect of "*Candidatus* Phytoplasma mali" strain AT infection on protein expression in *Nicotiana occidentalis* (tobacco) plants [32] and additionally applied Blast2GO [33] analysis for de novo annotation of identified proteins. Thereby, we could show that these resources can also be fruitfully applied for the construction of targeted proteomics assays. Also custom protein sequences derived from RNA-Seq might be incorporated in the analysis of mass spectrometry data acquired in data-independent (DIA) modes such as SWATH [34] and MSX [35].

In general, the PIT approach might be useful for various proteomics applications such as investigating metatranscriptomes/metaproteomes, and refining proteomics data analysis to cope with just predicted but non-validated gene/protein expression as well as incomplete annotation of genetic variation and splicing events. The general workflow for mass spectrometric analysis of protein samples by using RNA-Seq data is outlined in Fig. 1.

2   Materials

1. Users need to install Trinity software [36] in order to de novo assemble the transcriptome from short read RNA-seq data, available for unix-type operating systems from http://trinityrnaseq.github.io. It is best run in a high-performance computing environment with ~1 GB of RAM per 1 million paired-end reads. Alternatively, when lacking these resources, Trinity is also accessible on the Data Intensive Academic Grid (DIAG, http://diagcomputing.org/), a shared computational cloud for academic and nonprofit institutions.

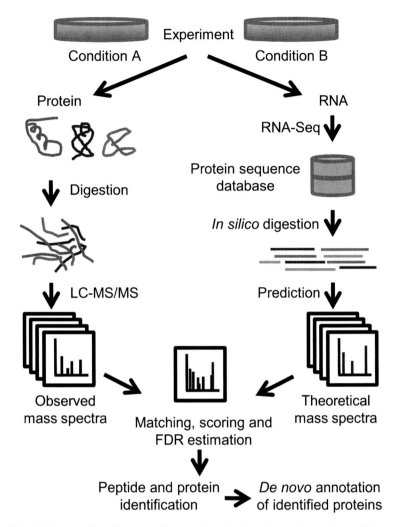

Fig. 1 Scheme of the bioinformatics workflow for the integrative analysis of shotgun mass spectrometry and RNA-Seq data. Proteins are subjected to LC-MS/MS analysis whereas RNA, ideally isolated from the same samples used for proteomics, is sequenced on next-generation sequencing (NGS) instruments [19]. Short read sequencing data is used to reconstruct protein sequences used as sequence database in the peptide search engine for identification of peptides and proteins from mass spectrometric raw data. Finally, identified proteins can be de novo annotated based on sequence homology search to known proteins

2. The "getorf" tool of the EMBOSS suite [37] is needed and available on the open-source, Web-based platform Galaxy (http://www.usegalaxy.org/).

3. Additional analysis steps take place in the statistical programming environment R [38], downloadable from http://cran.r--project.org/. Besides the base installation the Biostrings R package, available from http://www.bioconductor.org/, is needed.

4. Database searching of proteomic data is performed in the MaxQuant computational proteomics platform [16, 39] (http://www.maxquant.org/), which requires Windows operating system and the MS File Reader for accessing Thermo Scientific instrument files (Thermo Scientific, https://thermo.flexnetoperations.com/control/thmo/login).

5. For functional annotation of sequences and analysis of annotation data install Blast2GO [33] (http://www.blast2go.com/b2ghome).

3   Methods

The presented bioinformatics pipeline is a guide to the analysis of proteomic data generated by mass spectrometry with the help of a sample-specific protein sequence database for the identification and quantification of proteins. The workflow is composed of four major steps: RNA-Seq, protein sequence database construction, database searching of MS/MS data, and annotation of identified proteins.

3.1   RNA-Seq

Consult responsible NGS bioinformatics expert to define an appropriate experimental design matching your needs (*see* **Note 1**). We highly recommend isolating total RNA from the same samples used for protein extraction and proteomic analysis. In typical RNA-Seq workflows mRNA is enriched by Poly A+ selection prior to sequencing to increase the informative fraction in samples. However, when dealing with prokaryotes whole-transcriptome sequencing after depleting ribosomal RNAs with, e.g., RiboMinus-Kits (Invitrogen) might be the better choice. Alternatively, publicly available RNA-Seq data sets can be used as well (*see* **Note 2**).

3.2   Construction of Protein Sequence Database

The fundament of the protein sequence database construction is formed by the de novo transcriptome assembly from short read RNA-Seq data using Trinity software [36] (*see* **Note 3**). This assembly pipeline consists of the three consecutive modules Inchworm, Chrysalis, and Butterfly (Fig. 2). Briefly, the first module Inchworm generates transcript contigs from the RNA-Seq reads which are clustered in the second step by Chrysalis into regions that have probably originated from alternatively spliced transcripts or closely related gene families. Chrysalis encodes this structural complexity by building the Bruijn graphs for each cluster. Finally, Butterfly traces the RNA-Seq reads through the graphs and traverses supported graph paths to reconstruct full-length transcripts for alternatively spliced isoforms while teasing apart transcripts that correspond to paralogous genes. Trinity accepts pre-processed single- or paired-end short read data in either FASTQ or FASTA formats. Pre-processing involves removing barcodes used for multiplexing on the sequencing instrument

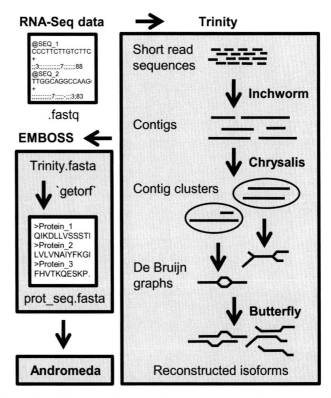

Fig 2 Overview of the reconstruction of protein sequences from short read RNA-sequencing (RNA-Seq) data. The Trinity pipeline [36] is used to build full-length transcripts. This software first assembles contigs (sets of overlapping sequences) in the Inchworm module. The Chrysalis module clusters contigs and processes each cluster to a de Bruijn graph component. Butterfly, the third module, extracts all probable sequences from the graph components and reports reconstructed sequences and their isoforms. Protein sequences are inferred by the EMBOSS tool "getorf" [37] and subjected to the Andromeda search engine [16] for peptide and protein identification from mass spectrometry raw data

and reads that probably contain sequencing errors (*see* **Note 4**). The reconstructed transcripts are translated to protein sequences using the EMBOSS tool "getorf." This file is subjected to the Andromeda search engine for peptide identification from mass spectrometry data (*see* Subheading 3.3).

1. Log in from your local working station to your Unix server where Trinity is installed. For instance, use the free SSH and telnet client PuTTY (or appropriate alternatives) when working on Windows operating system. Open terminal ("$" denotes the shell prompt) and copy raw or if necessary pre-processed sequencing data in FASTQ format to the working folder:

```
$ mkdir workingfolder
$    cp    rawdata/Illumina_singleread_*.fq    /
workingfolder
$ cd workingfolder
```

2. When multiple sequencing runs are conducted for a single experiment, concatenate RNA-Seq data into a single file to create a single reference Trinity assembly. When using paired-end data, combine all left and right reads into separate files:

```
$ cat Illumina_singleread_*.fq > combined_sin-
   gleread.fq
```

3. Assemble reads into transcripts using Trinity:

```
$ trinity_directory/Trinity.pl --seqType fq --JM 50G
   --single combined_singleread.fq --CPU 10
```

The "--seqType" option specifies input data format ("fq" for FASTQ, "fa" for FASTA), the "--JM" option controls amount of RAM used by processes of the Inchworm module, the "--single" option defines whether short reads are single or paired-end, and the "--CPU" option facilitates parallelization of processes. For more options see the Trinity documentation and choose appropriate parameters to fit your computational resources as well as RNA-Seq design like paired-end reads and strand specification of reads. Trinity per default outputs the de novo-assembled transcripts in FASTA format to "trinity_out_dir/Trinity.fasta" (*see* **Note 5**).

4. Copy "Trinity.fasta" to your local working station and upload the file to Galaxy (http://www.usegalaxy.org/) using the "Get Data" → "Upload File" function of the homepage.

5. Translate "Trinity.fasta" into proteins using the "EMBOSS" → "getorf" [37] function in Galaxy. Use transcripts with a minimum nucleotide length of 200 bp and output translated regions between start and stop codons. Save the resulting multiple FASTA file of protein sequences "Galaxy4-[getorf_on_data_1].fasta" to your local working station.

3.3   Searching MS/MS Data

Discovery proteomic raw data acquired on high-resolution MS instruments operating in data-dependent mode is analyzed in the MaxQuant computational proteomics platform [39] (Fig. 3). The integrated Andromeda peptide search engine [16] is used to search MS/MS spectra against the RNA-Seq-derived protein sequence database by taking the target decoy approach to control the false discovery rate (*see* **Note 6**).

1. Execute the "AndromedaConfig.exe" and switch to the Sequences tab (Fig. 4).

2. Select "Add new row" to create a new protein sequence database. Load the "Galaxy4-[getorf_on_data_1].fasta" file in the "Database" field.

3. Select an appropriate regular expression rule to correctly parse FASTA headers in the multiple FASTA file. The rule ">([^]*)" will output the whole header up to the first space in the MaxQuant search result files.

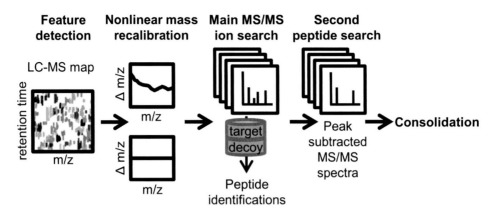

Fig 3 Main steps performed by the MaxQuant software [39] for the analysis of shotgun proteomic data acquired on high-resolution liquid chromatography-coupled tandem mass spectrometry (LC-MS/MS) instruments. The software suite contains algorithms to efficiently extract information from raw mass spectrometry data including peptide and protein identities as well as their high-accuracy quantification. First features are detected on MS1 level as three-dimensional objects in m/z, retention time, and intensity space. Isotope label partners are identified, features quantified and normalized, and their masses calculated precisely. To achieve mass accuracies in the p.p.b. range, mass errors ($\Delta m/z$) are estimated and mass offsets corrected by nonlinear mass recalibration. Furthermore MS/MS spectra are preprocessed and filtered prior to the main MS/MS ion search by the Andromeda search engine [16]. Peptide spectrum matches are scored and the false discovery rate (FDR) controlled on peptide and protein level by decoy database searching [17]. Additionally MaxQuant accounts for co-fragmented peptides occasionally observed due to the complexity of protein samples. Optionally, isotope-labeled peptide pairs are re-quantified if one label partner is missing and peptide identifications matched between different LC-MS/MS runs after their retention time alignment. In the consolidation phase proteins are assembled from peptide identifications, protein ratios calculated from peptide ratios by robust median averaging, and results compiled in multiple cross referenced tab-delimited .txt files that are used for further analysis steps in Perseus (www.maxquant.org) or R [38]

4. Save the rule in the "`Rule`" field. Afterwards click on "`File`" → "`Save`" → "`databases`". The RNA-Seq-derived protein sequence database has now been created.

5. Launch the "`MaxQuant.exe`" to analyze mass spectrometric raw data. Load the data and specify the experimental design of the experiment by assigning experiment and fraction to each file. Adjust parameters like isotope labels and their multiplicity, enzyme used for digestion of proteins to peptides, allowed missed cleavages, variable and fixed modifications, protein identification, and quantification criteria to fit your sample preparation methods and experimental design. In the field "`Sequences`" add the FASTA file you configured for Andromeda search engine using the "`AndromedaConfig. exe`". Check the box "`Include contaminants`", which will include a list of typically observed contaminants in samples like keratins. Select the number of threads you reserve for multithreading and speeding up the analysis. Start MaxQuant. The software exports result files in tab-delimited .txt format to the

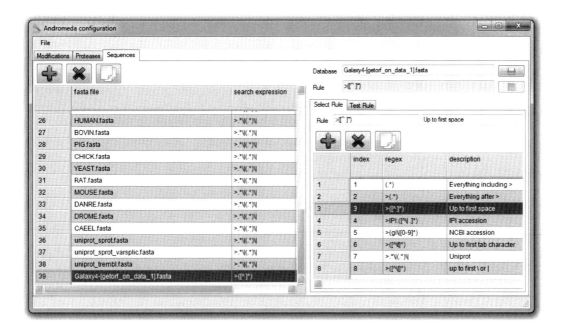

Fig. 4 Configuration of the RNA-Seq-derived protein sequences as search space in the Andromeda peptide search engine of the MaxQuant computational proteomics platform [16, 39]. After installation of the MaxQuant software package launch the "AndromedaConfig.exe" from the MaxQuant directory. The screenshot shows the graphical user interface of version 1.3.0.5 with the EMBOSS "getorf"- processed Trinity de novo assembly "Galaxy4-[getorf_on_data_1].fasta" selected as a new database

new folder "`txt`", a subfolder in the "`combined`" directory that was created when starting MaxQuant. For instance, the "`proteinGroups.txt`" contains information on the identified proteins in the processed raw files. Each single row contains the group of proteins that could be reconstructed from a set of peptides alongside with, e.g., quantitative information like normalized ratios between isotope label partners.

6. Inspect search results and perform secondary data analysis steps including testing for differential protein expression in Perseus (http://www.maxquant.org/) or R [38]. Apply appropriate statistics to identify proteins whose expression level changes between conditions.

3.4 Annotation of Sequence Data

In the last step the novel sequences identified to be present in the sample on RNA and protein level are to be annotated based on homology search to known sequence data. Annotating the whole de novo-assembled transcriptome is time consuming. However, taking the fraction that could be positively identified in the proteomics experiment will reduce data amount and speed up the analysis. Therefore the FASTA header of the protein sequences reported in the MaxQuant output files is used to parse the custom protein sequence database via regular expressions and extract the relevant protein sequences. This reduced multiple FASTA file of protein sequences is subjected to the Blast2GO software [33], a tool for the annotation of sequences and the analysis of annotation data (*see* **Note 7**).

1. Extract the amino acid sequences of the proteins that were identified in the proteomics experiment from the protein sequence database. For instance, this can be performed in R using the following commands ("> " denotes the R prompt).

   After starting R, download the Biostrings R Package or load the library if already installed:

   ```
   >source("http://bioconductor.org/biocLite.R")
   >biocLite("Biostrings")
   >library(Biostrings)
   ```

   Load the protein sequence database and the MaxQuant search result file "proteinGroups.txt":

   ```
   >proteinsequences=read.AAStringSet("C:/working-
       folder/MaxQuant_analysis/Galaxy4-[getorf_on_
       data_1].fasta")
   >proteingroups=        read.table("C:/workingfolder/
       MaxQuant_analysis/combined/txt/proteinGroups.
       txt", header=TRUE, sep="\t", na.strings=c("NaN"))
   ```

   Create a vector of the FASTA headers of identified proteins:
   ```
   >major_protein_id=as.vector(proteingroups    [,"Majority.
   protein.IDs"])
   ```
   Parse the protein sequence database via FASTA headers of identified proteins and save the new multiple FASTA file:

   ```
   >name=names(proteinsequences)
   >id=unique(grep(paste(major_protein_
       id,collapse="|"), name, value=FALSE))
   >identified_proteins_sequences=proteinsequences[id]
   >writeXStringSet(identified_proteins_sequences,
       file="C:/workingfolder/Blast2GO_analysis/pro-
       tein_sequences.fasta")
   ```

2. Load the file "protein_sequences.fasta" into the Blast2GO software by using the "File"➔ "Load Sequences (e.g.: .fasta)" option in the menu bar of the graphical user interface.

3. Start the Blast step by selecting "Blast" → "Run BLAST Step". This will use NCBI's BLAST service. You may choose the blast program and database to match your needs. For instance, use blastp to search protein sequences against the nr protein sequence database of NCBI.

4. The mapping step will link Blast hits to information stored in the Gene Ontology database. Select "Mapping" → "Run GO-Mapping Step".

5. Perform the annotation step. The mapped GO terms will now be assigned to the query sequences. Choose "Annotation" in the menu bar and "Run Annotation Step". Additionally run the "InterProScan" and "Merge InterProScan GOs to Annotation" which uses the functionality of InterPro annotations to retrieve motif and domain information. The corresponding GO terms are merged with already existing GO terms from the mapping step.

4   Notes

1. Take into account that sequencing depth and read length directly influence quality of de novo transcriptome assembly. A coverage of the protein-coding transcriptome as low as 11× might be sufficient to identify most of the present peptides in proteomics experiment [20], but underpowers other analyses like differential gene expression and SNV calling. Although short read sequencing on Illumina's Genome Analyzer for only 35 cycles has been used successfully in protein sequence database construction [25], keep in mind that longer reads reduce read ambiguity during transcriptome assembly, especially when dealing with metatranscriptomes. For instance, Adamidi et al. [23] combined long and relatively few reads obtained by 454 GS FLX (Roche) sequencing with shorter but many more reads from Illumina GAIIX (Illumina) sequencing to assemble a high-quality transcriptome of *Schmidtea mediterranea*.

2. Publicly available RNA-Seq data sets can be retrieved from repositories like ArrayExpress (https://www.ebi.ac.uk/arrayexpress/), Gene Expression Omnibus (GEO, http://www.ncbi.nlm.nih.gov/geo/), NCBI's Sequence Read Archive (SRA, http://www.ncbi.nlm.nih.gov/Traces/sra), European Nucleotide Archive (ENA, http://www.ebi.ac.uk/ena/), and the RNA-Seq Atlas (http://www.medicalgenomics.org/rna_seq_atlas).

3. Amongst Trinity several other software tools have been developed to assemble the transcriptome from RNA-Seq data without the need for a reference sequence. Alternatives are, e.g., DNAStar (http://www.dnastar.com/), Trans-ABySS (http://www.bcgsc.ca/platform/bioinfo/software/trans-abyss), SOAPdenovo-Trans (http://soap.genomics.org.cn/SOAPdenovo-Trans.html), Newbler (http://my454.com/products/analysis-software/index.asp), iAssembler (http://bioinfo.bti.cornell.edu/tool/iAssembler/), and Oases (https://www.ebi.ac.uk/~zerbino/oases/).

4. When the read length of the sequencing machine is longer than the molecule that is sequenced removing adapter sequences from reads becomes necessary. For data from

Illumina platforms these pre-processing steps can be performed with, e.g., Trimmomatic (http://www.usadellab.org/cms/index.php?page=trimmomatic) that is implemented as additional option in Trinity. A platform-independent solution is cutadapt (https://cutadapt.readthedocs.org/en/stable/).

5. Statistic parameters of the Trinity assembly can be examined using the "TrinityStats.pl" script to be found in the "utilities" folder in the Trinity installation directory. The number of transcripts, components, and the transcript contig N50 value will be reported. The contig N50 is a weighted median statistic such that 50 % of the entire assembly is contained in contigs or scaffolds equal to or larger than this value. Thus this value should be expected to be near to the average transcript length and can be used to confirm success of assembly whenever information about transcript length from reference sequences or assemblies are available.

6. MaxQuant currently enables the analysis of high-resolution MS data from, e.g., Thermo Scientific Orbitrap, and from Bruker's maXis qTOF and FT-ICR instruments only. Note that similar alternative data analysis tools, covering detection and quantification of peaks, isotope clusters and isotope-labeled peptide pairs, as well as MS/MS ion search, validation, and scoring of peptide identifications and their assembly to protein identifications including determination of protein ratios, can in principle be applied as well. For instance when using the Mascot search engine as implemented in the Proteome Discoverer software (Thermo Scientific) make sure that you have a local Mascot server installed to configure your RNA-Seq-based protein sequence database as search space.

7. Sequence data may be annotated with other tools as well. Mercator together with MapMan (http://mapman.gabipd.org/web/guest), GOanna as part of the AgBase (http://agbase.msstate.edu/cgi-bin/tools/GOanna.cgi), the KEGG Automatic Annotation Server (KAAS, http://www.genome.jp/kegg/kaas/), the KEGG Orthology Based Annotation System (KOBAS, http://kobas.cbi.pku.edu.cn/home.do), and the Trinotate pipeline for transcriptome functional annotation and analysis (http://trinotate.github.io) represent valuable replacement options for Blast2GO.

Acknowledgement

Our work was supported by the German Ministry for Education and Research (BMBF, grant number 0315082, 01EA1303 to S.S.), the European Union (FP7/2007-2013), under grant agreement n° 262055 (ESGI), and the Max Planck Society. This work is part of the Ph.D. thesis of T.L.

References

1. Ong S-E, Blagoev B, Kratchmarova I et al (2002) Stable isotope labeling by amino acids in cell culture, SILAC, as a simple and accurate approach to expression proteomics. Mol Cell Proteomics 1:376–386

2. Hsu J, Huang S, Chow N ct al (2003) Stable isotope dimethyl labeling for quantitative proteomics. Anal Chem 75:6843–6852

3. Wiese S, Reidegeld KA, Meyer HE et al (2007) Protein labeling by iTRAQ: a new tool for quantitative mass spectrometry in proteome research. Proteomics 7:340–350

4. Gerber SA, Rush J, Stemman O et al (2003) Absolute quantification of proteins and phosphoproteins from cell lysates by tandem MS. Proc Natl Acad Sci U S A 100:6940–6945

5. Schwanhäusser B, Busse D, Li N et al (2011) Global quantification of mammalian gene expression control. Nature 473:337–342

6. Mann M, Kulak NA, Nagaraj N et al (2013) The coming age of complete, accurate, and ubiquitous proteomes. Mol Cell 49:583–590

7. Gstaiger M, Aebersold R (2009) Applying mass spectrometry-based proteomics to genetics, genomics and network biology. Nat Rev Genet 10:617–627

8. Freiwald A, Weidner C, Witzke A et al (2013) Comprehensive proteomic data sets for studying adipocyte-macrophage cell-cell communication. Proteomics 13:3424–3428

9. Meierhofer D, Weidner C, Hartmann L et al (2013) Protein sets define disease states and predict in vivo effects of drug treatment. Mol Cell Proteomics 12:1965–1979

10. Mann M, Hendrickson RC, Pandey A (2001) Analysis of proteins and proteomes by mass spectrometry. Annu Rev Biochem 70:437–473

11. Perkins DN, Pappin DJC, Creasy DM et al (1999) Probability-based protein identification by searching sequence databases using mass spectrometry data. Electrophoresis 20:3551–3567

12. Eng JK, McCormack AL, Yates JR (1994) An approach to correlate tandem mass spectral data of peptides with amino acid sequences in a protein database. J Am Soc Mass Spectrom 5:976–989

13. Craig R, Beavis RC (2004) TANDEM: matching proteins with tandem mass spectra. Bioinformatics 20:1466–1467

14. Tabb DL, Fernando CG, Chambers MC (2007) MyriMatch: highly accurate tandem mass spectral peptide identification by multi-variate hypergeometric analysis. J Proteome Res 6:654–661

15. Shilov IV, Seymour SL, Patel AA et al (2007) The Paragon Algorithm, a next generation search engine that uses sequence temperature values and feature probabilities to identify peptides from tandem mass spectra. Mol Cell Proteomics 6:1638–1655

16. Cox J, Neuhauser N, Michalski A et al (2011) Andromeda: a peptide search engine integrated into the MaxQuant environment. J Proteome Res 10:1794–1805

17. Nesvizhskii AI (2010) A survey of computational methods and error rate estimation procedures for peptide and protein identification in shotgun proteomics. J Proteomics 73:2092–2123

18. Wang X, Slebos RJC, Wang D et al (2012) Protein identification using customized protein sequence databases derived from RNA-Seq data. J Proteome Res 11:1009–1017

19. Metzker ML (2010) Sequencing technologies—the next generation. Nat Rev Genet 11:31–46

20. Evans VC, Barker G, Heesom KJ et al (2012) De novo derivation of proteomes from transcriptomes for transcript and protein identification. Nat Methods 9:1207–1211

21. Lopez-Casado G, Covey PA, Bedinger PA et al (2012) Enabling proteomic studies with RNA-Seq: the proteome of tomato pollen as a test case. Proteomics 12:761–774

22. Sheynkman GM, Shortreed MR, Frey BL et al (2013) Discovery and mass spectrometric analysis of novel splice-junction peptides using RNA-Seq. Mol Cell Proteomics 12:2341–2353

23. Adamidi C, Wang Y, Gruen D et al (2011) De novo assembly and validation of planaria transcriptome by massive parallel sequencing and shotgun proteomics. Genome Res 21:1193–1200

24. He R, Kim M-J, Nelson W et al (2012) Next-generation sequencing-based transcriptomic and proteomic analysis of the common reed, Phragmites australis (Poaceae), reveals genes involved in invasiveness and rhizome specificity. Am J Bot 99:232–247

25. Song J, Sun R, Li D et al (2012) An improvement of shotgun proteomics analysis by adding next-generation sequencing transcriptome data in orange. PLoS One 7, e39494

26. Romero-Rodríguez MC, Pascual J, Valledor L et al (2014) Improving the quality of protein identification in non-model species. Characterization of Quercus ilex seed and

Pinus radiata needle proteomes by using SEQUEST and custom databases. J Proteomics 105:85–91

27. Wu X, Xu L, Gu W et al (2014) Iterative genome correction largely improves proteomic analysis of nonmodel organisms. J Proteome Res 13:2724–2734

28. Woo S, Cha SW, Merrihew G et al (2014) Proteogenomic database construction driven from large scale RNA-seq data. J Proteome Res 13:21–28

29. Armengaud J, Trapp J, Pible O et al (2014) Non-model organisms, a species endangered by proteogenomics. J Proteomics 105:5–18

30. Wang X, Zhang B (2014) Integrating genomic, transcriptomic, and interactome data to improve peptide and protein identification in shotgun proteomics. J Proteome Res 13:2715–2723

31. Wang X, Zhang B (2013) customProDB: an R package to generate customized protein databases from RNA-Seq data for proteomics search. Bioinformatics 29:3235–3237

32. Luge T, Kube M, Freiwald A et al (2014) Transcriptomics assisted proteomic analysis of Nicotiana occidentalis infected by "Candidatus Phytoplasma mali" strain AT. Proteomics 14:1882–1889

33. Conesa A, Götz S, García-Gómez JM et al (2005) Blast2GO: a universal tool for annota-

tion, visualization and analysis in functional genomics research. Bioinformatics 21:3674–3676

34. Gillet LC, Navarro P, Tate S et al (2012) Targeted data extraction of the MS/MS spectra generated by data-independent acquisition: a new concept for consistent and accurate proteome analysis. Mol Cell Proteomics 11:O111.016717

35. Egertson JD, Kuehn A, Merrihew GE et al (2013) Multiplexed MS/MS for improved data-independent acquisition. Nat Methods 10:744–746

36. Haas BJ, Papanicolaou A, Yassour M et al (2013) De novo transcript sequence reconstruction from RNA-seq using the Trinity platform for reference generation and analysis. Nat Protoc 8:1494–1512

37. Rice P (2000) The European Molecular Biology Open Software Suite EMBOSS: the European Molecular Biology Open Software Suite. Trends Genet 16:2–3

38. R Development Core Team R (2011) R: a language and environment for statistical computing. R Found Stat Comput 1:409

39. Cox J, Matic I, Hilger M et al (2009) A practical guide to the MaxQuant computational platform for SILAC-based quantitative proteomics. Nat Protoc 4:698–705

Chapter 17

A Proteomic Workflow Using High-Throughput De Novo Sequencing Towards Complementation of Genome Information for Improved Comparative Crop Science

Reinhard Turetschek, David Lyon, Getinet Desalegn, Hans-Peter Kaul, and Stefanie Wienkoop

Abstract

The proteomic study of non-model organisms, such as many crop plants, is challenging due to the lack of comprehensive genome information. Changing environmental conditions require the study and selection of adapted cultivars. Mutations, inherent to cultivars, hamper protein identification and thus considerably complicate the qualitative and quantitative comparison in large-scale systems biology approaches. With this workflow, cultivar-specific mutations are detected from high-throughput comparative MS analyses, by extracting sequence polymorphisms with de novo sequencing. Stringent criteria are suggested to filter for confidential mutations. Subsequently, these polymorphisms complement the initially used database, which is ready to use with any preferred database search algorithm. In our example, we thereby identified 26 specific mutations in two cultivars of *Pisum sativum* and achieved an increased number (17 %) of peptide spectrum matches.

Key words Proteomics, De novo sequencing, Polymorphism, Crop science, Cultivars, Mass spectrometry, *Pisum sativum*

1 Introduction

In recent decades entire genome sequences of many organisms were acquired and the amount of sequence information is continuously expanding at an increasing rate. Advanced functional annotation of genomic data in model organisms facilitates interpretation of newly generated data. Despite the fact that specific sequence information is unavailable for non-model organisms, a growing number and a broad range of phylogenetic diverse species, reaching from snake venoms [1–6] to whole microbial communities [7–11], are being subjected to proteomic studies. A great evolutionary distance to well-characterized species considerably complicates the compilation of comprehensive databases (DBs),

Jörg Reinders (ed.), *Proteomics in Systems Biology: Methods and Protocols*, Methods in Molecular Biology, vol. 1394,
DOI 10.1007/978-1-4939-3341-9_17, © Springer Science+Business Media New York 2016

which is crucial for every bottom-up proteomic approach. Prevalently this hurdle can be overcome by combining a relatively large unspecific database (e.g., viridiplantae, NCBI) with a custom-built specific database consisting of translated RNA-sequences (e.g., 6-frame translation of nucleotide to amino acid sequences, BLAST homology searches for functional annotation) [12]. In most of the cases, such a composite database is sufficient to gain new insights into the protein level. However, conclusions are often hampered when it comes to comparison of cultivars within the same species. Among cultivar-specific sequence polymorphisms match to a greater or lesser extent with the aforementioned database and consequently result in distinct identifications, not related to functional differences. Thus, differentiation of cultivars in the proteomic domain may be composed of both sequence variation and molecular processes. However, the identification of molecular adaptations of cultivars upon environmental constraints is a major focus of crop science. With the use of shotgun proteomics, comparative crop science not only aims to identify homologues but moreover quantifies differences among cultivars, thus supporting the development of breeding strategies [13]. Yet, most common database search algorithms (e.g., SEQUEST) require a good match with in silico-generated spectra (PMF and fragment ion series) and fail to identify polymorphisms derived from cultivar-specific sequences. Hence, the amino acid sequence is required for detailed DB comparison.

The sequence may be acquired de novo, by deriving the amino acid composition from fragment ions of peptides.

The idea of determining peptide primary structure via mass spectrometry without prior knowledge of the sequence was already developed in the 1970s by studying penicillinase [14]. In the 1980s, first tandem MS scans were manually sequenced with 2200 mass resolution [15]. Today, various automated de novo sequencing algorithms are available enabling high-throughput processing of MS/MS data [16–19]. Still, the reliability scoring, inherent to all de novo algorithms, remains an ongoing issue which highly influences accuracy and computation time [20]. Once confident sequence tags are obtained, these can be matched to a database with the help of various search engines [21–24] to retrieve homologue proteins.

After assigning homologues, comparing de novo tags with an adequate database is just one more step to extract sequence differences in order to determine mutations. This additional step, however, requires special care as de novo sequencing is prone to specific errors (e.g., the inability to distinguish between K and E in low mass accuracy measurement). Such errors are taken into account by a few programs, such as SPIDER [21], TagRecon [22], and OpenSea [23], that correct de novo tags and additionally allow inexact matches to DB sequences. By matching de novo tags

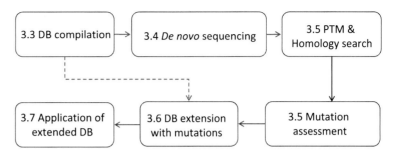

Fig. 1 Workflow with main steps from initial DB compilation and de novo sequence analysis to the application of the new organism aligned DB, explained in detail as follows (cf. Subheadings 3.2–3.6)

inexactly to the DB these algorithms show the capacity to more or less accurately—depending on the spectra quality and the search algorithm—identify posttranslational modifications, homologies, and mutations. With the use of such an automated search for homologies and mutations, our workflow (Fig. 1) aims to amend the initial DB with newly identified sequences (*see* Subheading 3.4). In case of cultivars, mutated sequences are not replacing original entries in the database, but are added with a header corresponding to the cultivar. Identified homologies to DB entries from different organisms are as well amending the new DB, but additionally the name of the organism must remain in the header not to confuse a homology with a mutation. Using this (extended) DB with any conventional algorithm (e.g., SEQUEST, Mascot) facilitates high-throughput MS/MS data analysis (compared to de novo sequencing) and additionally increases the confidence- and probability-based peptide identification (e.g., Xcorr) as well as protein sequence coverage, which results in more accurate quantification of cultivar-specific proteins (*see* Subheading 3.6). However, the determination of mutations via de novo sequencing is delicate and requires a few criteria. Therefore, particular attention has to be paid to reliable identification of mutations by critically taking mismatches with PTMs into account (*see* Subheading 3.2) and setting further criteria for stringent consideration of mutations (*see* Subheading 3.3).

2   Materials

2.1   Plant Material

1. Seeds from *P. sativum* ssp. cultivar Messire were provided by the Institute for Sustainable Agriculture CISC (Department of Plant Breeding, Cordoba, Spain). Cultivar Protecta was obtained from Probstdorfer Saatzucht GmbH & Co KG (Probstdorf, Austria).

2.2 Protein Extraction: Materials	1. Lyophilized plant material.
	2. TRIzol Reagent® RNA Isolation Reagent.
	3. Precipitation solution: 0.5 % β-Mercaptoethanol in acetone.
	4. Protein digestion: Endoproteinase LysC Sequencing grade, Poroszyme® Immobilized Trypsin Bulk Media.
2.3 LC-MS/MS Instrumentation	1. One-dimensional nano-flow LC (Dionex UltiMate 3000; Thermo Scientific, USA).
	2. EASY-Spray column, 15 cm × 75 μm ID, PepMap C18, 3 μm (Thermo Scientific, USA).
	3. LTQ Orbitrap Elite (Thermo Scientific, USA).
2.4 LC-MS/MS Analysis: Materials	1. Mobile-phase solvent A: 0.1 % Formic acid; solvent B: 90 % acetonitrile, 0.1 % formic acid.
2.5 Software	1. mEMBOSS 6.5 (European Molecular Biology Open Software Suite).
	2. PEAKS 7.0 (Bioinformatics Solutions Incorporation, Canada).
	3. SEQUEST (Proteome Discoverer 1.3; Thermo Scientific, USA).

3 Methods

3.1 Protein Extraction

Leaves of 4-week-old plants were sampled, immediately quenched, and ground in liquid N_2. Protein from lyophilized material was extracted in TRIzol® according to Lee et al. [25] with a few modifications: 3 ml of β-mercaptoethanol in acetone was used for precipitation overnight at –20 °C. The protein pellet was washed and digested with LysC and trypsin according to Staudinger et al. [26].

3.2 LC-MS/MS Analysis

Peptide digests (1 μg) were applied to a one-dimensional nano-flow LC. The peptides were separated using a 95-min nonlinear gradient from 98 % of solvent A to 45 % of solvent B at a flow rate of 300 nl/min. The nLC-ESI-MS/MS analysis was optimized for standard high-throughput analysis at a resolution of 120,000 (FTMS) with 20 MS/MS scans in the LTQ at the following settings: rapid scan mode, minimum signal threshold counts 1000, prediction of ion injection time, repeat count 1, repeat duration 30 s, exclusion list size 500, exclusion duration 60 s, exclusion mass width 5 ppm relative to reference mass, early expiration enabled (count 1, S/N threshold 2), monoisotopic precursor selection enabled, rejected charge state: 1, normalized collision energy: 35, and activation time 30 ms.

3.3 Custom Database Design

A composite protein-fasta file was created by merging the following six databases:

Uniprot UniRef100 (all identical sequences and subfragments with 11 or more residues are placed into a single record—http://www.uniprot.org/help/uniref) sourced at 15-05-2013 from the following Taxa:

1. *Pisum sativum*

2. *Rhizobium leguminosarum*

3. *Glomus*

4. *Mycosphaerella*

5. Legume-specific protein database (LegProt) [27] including information from the following organisms: *Pisum sativum, Lotus japonicus, Medicago sativa, Glycine max, Lupinus albus, Phaseolus vulgaris.*

6. Processed dbEST NCBI sourced from http://www.coolseasonfoodlegume.org/

Pisum Sativum Unigene v1, P. Sativum Unigene wa1, Pisum Sativum Unigene v2;

Nucleotide sequences were six-frame-translated using mEMBOSS. For each accession number the longest continuous amino acid sequence (longest ORF) within a frame was chosen. If multiple sequences (of different frames) were of the same maximum length, all of them were kept (each with a different accession number, including the frame number).

The 6 fasta files described above were combined, producing a new fasta containing 135,754 entries. Protein sequences 100 % identical in sequence and length were combined by subsequently adding one header after the other, separating them by the following characters " __***__ " (no matter if the redundancies originated from one or multiple fasta files). All other entries were simply added to the end of the new file. The first accession number of the header was repeatedly written at the very beginning of the header line, separated by a " | " in order to consistently view and parse the accession numbers.

3.4 De Novo Sequencing and Homology Search

Several automated software solutions for de novo sequencing are available to date. The outcome very much depends on the quality of processed spectra and the selected algorithm [28]. Higher spectra quality can be achieved by adaptation of the fragmentation (*see* **Note 1**). Here, the de novo search was performed with PEAKS [18] employing settings according to the resolution and mass accuracy of the mass spectrometer used (*see* Subheading 2.3). By calculating the narrowest possible mass error tolerance most occurring PTMs cannot be mistaken for a mutation. Hence, a mass error of 5 ppm from the monoisotopic precursor was allowed with fragment ion mass error of 0.5 Da (*see* **Note 2**). De novo tags with a minimum average local confidence (ALC) of ≥15 were subjected to PTM identification to determine the most frequent modifications in order to recalculate an adequate mass error tolerance (*see* **Note 2**). Accordingly, de novo tags were matched against the designed database (*see* Subheading 3.1) and searched for mutations with the implemented SPIDER tool. Maximum three of the previously searched PTMs were allowed. The peptide spectrum matching score (−10 lgP) was set to 20. A maximum of two missed cleavages per peptide and nonspecific cleavage at one end of the peptide were allowed—this apparently loose restriction facilitates the identification of mutated K or R residues.

3.5 Evaluation of Homology and Mutation

Various automated programs [21, 23, 24] crucially simplify the identification of sequence variance. However, these identifications must be critically filtered to obtain only confident sequence amendments to the original database. First, exclusively proteotypic peptides are added to the original database, because a variation in any other peptide cannot be specifically attributed to one protein. However, as a result of using a merged DB containing sequence information of several RNA data (*see* Subheading 3.1), variations are sometimes assigned to multiple protein entries with the same function but slightly different sequences. In such case, the protein entry with the most assigned peptides and highest coverage is chosen for further processing. If the number of assigned peptides and coverage is similar for several DB entries, all of them are chosen for further processing. Second, the mutated peptide must not have a non-mutated counterpart: if a mutated and a non-mutated peptide are attributed to the same sequence in the database, the identified mutation is likely to be a false positive. Third, the sequence of the mutation must be confirmed by at least two MS/MS spectra. Thereby, a de novo error—identifying a mutation—caused by a low-quality spectrum is largely avoided and mutations gain confidence. A typical quantitative shotgun proteomics experiment requires the measurement of several replicates, which usually acquire enough MS/MS to confirm mutations.

3.6 Database Extension with Mutations

Confidently identified mutated amino acid residues are added to the database by copying the original fasta entries with the mutated sequence. Original entries must be kept, because the located mutations may be characteristic for just one cultivar. The mutated sequences are found in new entries with modified accession and header (*see* example below).

Accession: ACU20233.1_**m1**

Header: unknown [glycine max] [**Me_IV25**]

The accession shows a suffix ("_m1") indicating a sequence alteration. If cultivars show different mutations at the same protein entry, the number of the suffix is ascending. The header is supplemented with squared brackets including the information about the cultivar and the polymorphisms in the sequence (in the current example in cv. Messire isoleucine substitutes valine at position 25). Here, an entry of the LegProt database (NCBI) [27] is shown which is not yet annotated. After inserting the mutation, the sequence can again be blasted to achieve improved annotation, albeit a change in only one amino acid will result in a similar BLAST result. Additionally, when working with non-model organisms, it is worth considering re-annotation of the genome by use of proteomic data in a proteogenomics approach [29].

3.7 Application of Extended Database

The database, amended with mutated sequences, potentially enhances the identification of any preferred DB search algorithm and enables high-throughput processing of MS/MS data. By increasing peptide scores (e.g., Xcorr) and protein sequence coverage, proteins are more confidently identified. Moreover, inclusion of exclusively proteotypic peptides (*see* Subheading 3.5) expands the list of candidates for other proteomic approaches (e.g., SRM, MRM).

3.8 Iterated Search with DB Search Algorithm

The new extended DB was used for a standard DB search using the SEQUEST algorithm with the following settings: 5 ppm precursor mass tolerance, 0.5 Da fragment mass tolerance, acetylation of the N-terminus, and oxidation of methionine as dynamic modifications. Minimum peptide confidence was set to medium, and minimum Xcorr to 2. A minimum of two peptides per protein were required for identification.

In the present study of *P. sativum* with the cultivars Messire and Protecta we identified 48 variations to original DB entries, of which 26 are mutations showing high cultivar specificity. Both cultivars have five mutations in common. Messire showed 12 and Protecta 9 characteristic mutations. Furthermore, 22 homologues were identified from entries of different species (e.g., *G. max* from the LegProt DB). The ratio of replaced and substituting amino acids (Fig. 2) shows that most frequently valine and alanine are both replacing and substituting other amino acids in our experiment.

The number of peptide spectral matches (PSMs) shows how many of fragmented ions match to the applied DB. Thus, a rather complete DB will result in a higher number of PSMs compared to an imperfect DB. Here Fig. 3 shows that the number of PSMs increased significantly (17 %) for the two studied cultivars after amending the initial DB with sequence variations. Besides improving protein identification, the elevated number of PSMs crucially contributes to more accurate and confident quantification.

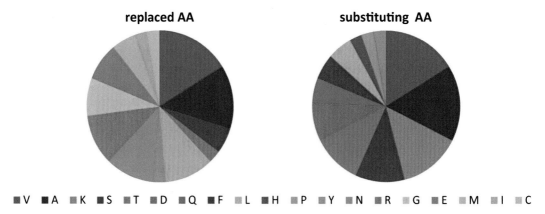

Fig. 2 Ratio of replaced (from the initial DB) and substituting amino acids

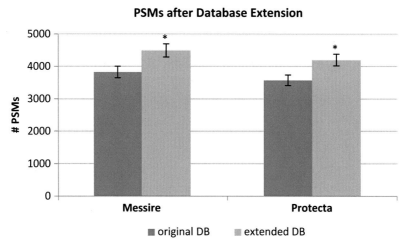

Fig. 3 The total number of PSMs is significantly increased (student's *t*-test, $p < 0.05$) in both cultivars when the extended DB is applied; $n = 24$, error bars indicate standard error at 95 % confidence intervals

4 Notes

1. Optimizing prerequisites for de novo sequencing

   De novo sequencing essentially depends on a complete series of fragment ions which are generated according to the applied fragmentation technique. Thus, the quality of de novo sequencing can be enhanced crucially by applying different fragmentations (CID/HCD/ETD) to the same precursor and subsequently merging the spectra in order to achieve a great number of fragment ions [30]. The choice of instrumentation often points to high speed and sensitivity (e.g., LTQ using CID) with the drawbacks of reduced mass resolution and accuracy, resulting in an impossibility to resolve fragment ions' charge state. This deficit causes considerable difficulties to de novo sequencing because of additional ambiguity. Through performance of MS/MS scans at high resolution (i.e., FT) the precision of de novo sequencing benefits especially for precursors of higher charge states ($\geq$3+). However, high resolution requires higher AGC values, which increases acquisition time remarkably. The trade-off between spectra-quality (FT) and speed (LTQ) can be accounted for by a data-dependent decision tree [31], where the site of the fragment scan depends on the charge state and the m/z of the peptide. Consequently MS/MS scans from $\geq$3+ charged precursors crucially improve in quality.

2. Definition of mass error tolerance for accurate de novo sequencing

The calculation of the mass error tolerance comprises the comparison of an AA mass with a PTM to another AA. This gap is the mass difference which needs to be resolved to distinguish a PTM from a mutation. Since a database search with more than 200 possible PTMs [32] requires infeasible computing capacity, the calculation refers to the seven most observed PTMs in this experiment. The most frequently occurring PTMs were determined by peaks with 5 ppm precursor mass error tolerance and 0.6 Da fragment mass error tolerance: oxidation of methionine (+15.99), sodium adduct (+21.98, on D-, E-, and C-term), carbamylation (+43.01 on N-term), methyl ester (+14.02 on D-, E-, and C-term), deamidation (+0.98 on N and Q), acetylation (+42.01 on N-term), and replacement of two protons by calcium (+37.95 on D-, E-, and C-term). The mass error tolerance can be calculated as follows: The mass of a possible modification is added to the AA's mass. When modifications affect the C- or N-term, the PTM's mass is added to each AA's mass. These values are subtracted from the masses of each amino acid. Thus, a mass error window to differentiate between an AA and a modified AA is calculated. For MS/MS scans in an ion trap a mass accuracy of 100–200 ppm must be additionally subtracted from this mass difference. Accordingly, a bulk of false positives (modified peptides identified as mutations) can already be excluded by setting the mass error tolerance to 0.5 Da in an MS/MS scan, that is, e.g., the identification of valine as asparagine which mass differs in 0.94 Da from a peptide with methyl ester at the C-terminus. Considering a mass accuracy of 200 ppm in an ion trap (0.4 Da at 2000 m/z) the mass difference narrows to 0.54 Da. For an FT MS/MS scan the mass error tolerance is preferably set a lot lower (e.g., 0.05 Da). Additionally, setting the precursor mass tolerance to $\leq$5 ppm critically minimizes false positives.

References

1. Fox JW, Ma L, Nelson K et al (2006) Comparison of indirect and direct approaches using ion-trap and Fourier transform ion cyclotron resonance mass spectrometry for exploring viperid venom proteomes. Toxicon 47(6):700–714

2. Mackessy SP (2002) Biochemistry and pharmacology of colubrid snake venoms. J Toxicol Toxin Rev 21(1–2):43–83

3. Nawarak J, Sinchaikul S, Wu CY et al (2003) Proteomics of snake venoms from Elapidae and Viperidae families by multidimensional chromatographic methods. Electrophoresis 24(16):2838–2854

4. OmPraba G, Chapeaurouge A, Doley R et al (2010) Identification of a novel family of snake venom proteins Veficolins from Cerberus rynchops using a venom gland transcriptomics and proteomics approach. J Proteome Res 9(4):1882–1893

5. Serrano SMT, Shannon JD, Wang D et al (2005) A multifaceted analysis of viperid snake venoms by two-dimensional gel electrophoresis: an approach to understanding venom proteomics. Proteomics 5(2):501–510

6. Wong ESW, Morgenstern D, Mofiz E et al (2012) Proteomics and deep sequencing comparison of seasonally active venom glands in the platypus reveals novel venom peptides and distinct expression profiles. Mol Cell Proteomics 11(11):1354–1364

7. Abraham PE, Giannone RJ, Xiong W et al (2014) Metaproteomics: extracting and mining proteome information to characterize metabolic activities in microbial communities. Curr Protoc Bioinformatics 46:13.26.1–13.26.14

8. Becher D, Bernhardt J, Fuchs S et al (2013) Metaproteomics to unravel major microbial players in leaf litter and soil environments: challenges and perspectives. Proteomics 13(18–19):2895–2909

9. Hao C, Liu Q, Yang J et al (2008) Metaproteomics: exploration of the functions of microbial ecosystems. Chin J Appl Environ Biol 14(2):270–275

10. Siggins A, Gunnigle E, Abram F (2012) Exploring mixed microbial community functioning: recent advances in metaproteomics. FEMS Microbiol Ecol 80(2):265–280

11. Wang H-B, Zhang ZX, Li H et al (2011) Characterization of metaproteomics in crop rhizospheric soil. J Proteome Res 10(3):932–940

12. Romero-Rodríguez MC, Pascual J, Valledor L et al (2014) Improving the quality of protein identification in non-model species. Characterization of Quercus ilex seed and Pinus radiata needle proteomes by using SEQUEST and custom databases. J Proteomics 105:85–91

13. Vanderschuren H, Lentz E, Zainuddin I et al (2013) Proteomics of model and crop plant species: status, current limitations and strategic advances for crop improvement. J Proteomics 93:5–19

14. Morris HR, Williams DH, Ambler RP (1971) Determination of the sequences of protein-derived peptides and peptide mixtures by mass spectrometry. Biochem J 125(1):189–201

15. Johnson RS, Biemann K (1987) The primary structure of thioredoxin from Chromatium vinosum determined by high-performance tandem mass spectrometry. Biochemistry 26(5):1209–1214

16. Frank A, Pevzner P (2005) PepNovo: de novo peptide sequencing via probabilistic network modeling. Anal Chem 77(4):964–973

17. Fischer B, Roth V, Roos F et al (2005) NovoHMM: a hidden Markov model for de novo peptide sequencing. Anal Chem 77(22):7265–7273

18. Ma B, Zhang K, Hendrie C et al (2003) PEAKS: powerful software for peptide de novo sequencing by tandem mass spectrometry. Rapid Commun Mass Spectrom 17(20):2337–2342

19. Taylor JA, Johnson RS (1997) Sequence database searches via de novo peptide sequencing by tandem mass spectrometry. Rapid Commun Mass Spectrom 11(9):1067–1075

20. Ma B, Johnson R (2012) De novo sequencing and homology searching. Mol Cell Proteomics 11(2):O111.014902

21. Han Y, Ma B, Zhang K (2005) SPIDER: software for protein identification from sequence tags with de novo sequencing error. J Bioinform Comput Biol 3(3):697–716

22. Dasari S, Chambers MC, Slebos RJ et al (2010) TagRecon: high-throughput mutation identification through sequence tagging. J Proteome Res 9(4):1716–1726

23. Searle BC, Dasari S, Turner M et al (2004) High-throughput identification of proteins and unanticipated sequence modifications using a mass-based alignment algorithm for MS/MS de novo sequencing results. Anal Chem 76(8):2220–2230

24. Tabb DL, Saraf A, Yates JR 3rd (2003) GutenTag: high-throughput sequence tagging via an empirically derived fragmentation model. Anal Chem 75(23):6415–6421

25. Lee FW, Lo SC (2008) The use of Trizol reagent (phenol/guanidine isothiocyanate) for producing high quality two-dimensional gel electrophoretograms (2-DE) of dinoflagellates. J Microbiol Methods 73(1):26–32

26. Staudinger C, Mehmeti V, Turetschek R et al (2012) Possible role of nutritional priming for early salt and drought stress responses in *Medicago truncatula*. Front Plant Sci 3:285

27. Lei Z, Dai X, Watson BS et al (2011) A legume specific protein database (LegProt) improves the number of identified peptides, confidence scores and overall protein identification success rates for legume proteomics. Phytochemistry 72(10):1020–1027

28. Pevtsov S, Fedulova I, Mirzaei H et al (2006) Performance evaluation of existing de novo sequencing algorithms. J Proteome Res 5(11): 3018–3028

29. Armengaud J, Trapp J, Pible O et al (2014) Non-model organisms, a species endangered by proteogenomics. J Proteomics 105:5–18

30. Guthals A, Clauser KR, Frank AM et al (2013) Sequencing-grade de novo analysis of MS/MS triplets (CID/HCD/ETD) from overlapping peptides. J Proteome Res 12(6):2846–2857

31. Frese CK, Altelaar AF, Hennrich ML et al (2011) Improved peptide identification by targeted fragmentation using CID, HCD and ETD on an LTQ-Orbitrap Velos. J Proteome Res 10(5):2377–2388

32. Gooley AA, Packer NH (1997) The importance of protein co- and post-translational modifications in proteome projects. In: Wilkins MR et al (eds) Proteome research: new frontiers in functional genomics. Springer, Berlin, pp 65–91

Chapter 18

From Phosphoproteome to Modeling of Plant Signaling Pathways

Maksim Zakhartsev, Heidi Pertl-Obermeyer, and Waltraud X. Schulze

Abstract

Quantitative proteomic experiments in recent years became almost routine in many aspects of biology. Particularly the quantification of peptides and corresponding phosphorylated counterparts from a single experiment is highly important for understanding of dynamics of signaling pathways. We developed an analytical method to quantify phosphopeptides (pP) in relation to the quantity of the corresponding non-phosphorylated parent peptides (P). We used mixed-mode solid-phase extraction to purify total peptides from tryptic digest and separated them from most of the phosphorous-containing compounds (e.g., phospholipids, nucleotides) which enhances pP enrichment on TiO_2 beads. Phosphoproteomic data derived with this designed method allows quantifying pP/P stoichiometry, and qualifying experimental data for mathematical modeling.

Key words Phosphopeptide enrichment, Mixed-mode solid-phase extraction, Metal oxide affinity chromatography, Mathematical modeling

1 Introduction

Mathematical modelingand dynamic simulation of signal transduction pathways is an important topic in systems biology [1, 2]. One of the purposes of the dynamic modeling in plant physiology is to evaluate the degree of involvement of different signaling pathways in plant responses to external perturbations [3, 4] or to explain phenotypic appearances of plant mutants. Protein phosphorylation is one of the fastest posttranslational modifications (PTM) that is an intrinsic mechanism of the signal transduction in some signaling pathways (e.g., MAPK cascades). Phosphorylation of signaling proteins traced in time allows revealing involvement of corresponding pathways into adaptive responses [5, 6]. Signaling pathways are organized in cascades of counteracting (e.g., cyclic) irreversible reactions [7], which generate and amplify the cellular signal (Fig. 1). Phosphorylation (by kinases)/de-phosphorylation (by phosphatases) of substrate-proteins is an elemental event in many signal transduction pathways [1, 3, 7]. Normally, proteins in

Jörg Reinders (ed.), *Proteomics in Systems Biology: Methods and Protocols*, Methods in Molecular Biology, vol. 1394,
DOI 10.1007/978-1-4939-3341-9_18, © Springer Science+Business Media New York 2016

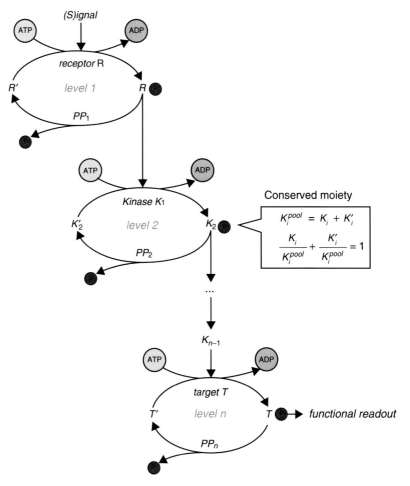

Fig. 1 A simplified hypothetical scheme of a signal (*S*) transfer from receptor (*R*) to target (*T*) through a linear signal transduction pathway consisting of different kinases (*K*). At each cascade level, only two interconvertable forms are shown: active and inactive (marked by ′). The mechanism of activation is through phosphorylation by means of group transfer from phosphate-donor (e.g., ATP, GTP). The active form of a species from a preceding level activates species at a subsequent level. *PPi*—*i*th phosphatase. The figure was compiled from [1, 7]

cyclic reactions function under assumption of a conserved moiety (Fig. 1): a total pool of a protein remains unchanged in short time, but its degree of phosphorylation shows a fast response to the environmental stimuli. Therefore, it is crucial to measure the total pool of the signaling proteins as well as the degree of their phosphorylation, which, when measured in a time course, can provide one with sufficient information for parameter estimation of mathematical models based on kinetic expressions of individual reactions.

1.1 MS-Based Proteomics

Shotgun mass spectrometry (MS)-based proteomics offers a unique opportunity for identification and quantification of thousands of peptides (P) in a single analysis, and in combination with computational methods provides information on expression of complete proteome of a research object at a given state [8, 9]. The shotgun MS-based proteomics also allows identifying most PTMs of proteins, such as phosphorylation, acetylation, methylation, glycosylation, and ubiquitination [10, 11], which drive signaling pathways, and their dynamics can be used for decoding signaling networks. The common workflow of sample preparation for MS-based proteomics relies on different analytical techniques: enzymatic protein digest, sample fractionation based on various chromatographic techniques, sample enrichment or desalting by solid-phase extraction (SPE), etc. [12]. Since the fraction of phosphopeptides is relatively low (<1 %) regarding to total peptides in trypsinized protein digests, usually an enrichment step is required for their confident identification and quantification [12–14].

1.2 Total Peptide Purification

The classical approach to purify total peptides from complex digest mixes exploits hydrophobic interactions between side chains of hydrophobic amino acids and C_{18}-groups of reverse-phase SPE sorbent [15]. However, this approach results in co-purification of a variety of nonpolar components, such as lipids, pigments, or sterols. This often is particularly problematic when working with strongly pigmented plant tissues. At low pH (e.g., <3.0) the zwitterion of a peptide becomes fully protonated (i.e., a weak cation or weak base) which allows the use of strong cation exchanger (SCX) for their purification, but it also results in co-purification of charged impurities such as nucleotides. Thus, a combination of the reverse-phase and SCX modes provides enhanced peptide purification and additionally a better removal of nonpolar and charged impurities. Oasis® MCX (mixed-mode cation-exchange and reversed-phase sorbent) provides such dual modes of retention by (1) strong sulfonic-acid-cation exchanger and (2) reversed-phase interactions in combination with hydrophilic interactions on a single organic co-polymer. The sorbent is highly selective and sensitive for extraction of basic compounds from acidified biological matrices (such as plasma, urine, bile, and pigmented plant tissue) and demonstrates very good capacity for the peptide purification from tryptic protein digests. Sample preparation based on use of mixed-mode solid phases provides superior removal of nearly all phospholipids and weaker acids, achieving double-positive effects by (1) eliminating phosphorous-containing compounds (phospholipids, nucleic acids, etc.) that strongly interfere with phosphopeptide enrichment and (2) eliminating the major sources of matrix effects (e.g., neutrals), a known case of ion suppression, loss of sensitivity, and inaccurate quantification by liquid chromatography with a tandem mass analyzer (LC-MS/MS).

1.3 Phosphopeptide (pP) Enrichment

The typical pP enrichment protocols rely on several unit operations, such as total peptide purification/desalting after the digest, pP enrichment, and further desalting, in order to provide salt-free samples for the LC-MS/MS analysis. Usually, all desalting steps are carried out on C_{18} Stop-and-go-extraction tips (StageTips) [15]. Most common pP enrichment protocols are based on metal oxide affinity chromatography (MOAC) on titanium, aluminum, and zirconium oxides [5, 13, 16]. The MOAC of pP is based on the affinity of phosphates to metals, but this interaction is interfered by non-phosphorylated acidic peptides which also show affinity for the metals [14]. Therefore, enhancers of the phosphopeptide selectivity on MOAC are widely used (e.g., lactic or β-hydroxypropanoic acids [16], 2,5-dihydroxybenzoic or phthalic acids [17]). The enhancers reduce unspecific binding of non-phosphorylated acidic peptides, in this sense giving preferences for pP to bind to the metal oxides. Obviously, the second desalting step after the use of enhancers is extremely important, but it leads to an additional loss of phosphopeptides.

In some research, the phosphoprotein enrichment [18, 19] is used prior to the digest followed by the phosphopeptide enrichment procedure to increase yield of phosphopeptides. However, in the context of our approach it is meaningless since non-phosphorylated cognates will be removed.

Other phosphorous-containing compounds (phospholipids, nucleotides, etc.) also compete with pP for the metal centers to bind. Therefore, we have chosen a strategy of eliminating the phosphorous-containing compounds on the level of total peptide purification/desalting (mixed-mode SPE) to give the advantages for the pP at the enrichment step on conventional TiO_2 without enhancers. Our method describes pP enrichment from a microsomal fraction [20] of roots of wild-type *Arabidopsis thaliana* (Fig. 2).

1.4 Peptide (P) and Phosphopeptide (pP) Quantification

There are different methods of peptide quantification in quantitative proteomics, which implement either absolute or relative peptide quantifications. The quantification techniques are based on the use of stable isotope labeling (e.g., SILAC), isobaric labeling (e.g., TMT, iTRAQ), isotope-coded affinity tag (ICAT), internal peptide standards [21], and label-free quantification (e.g., iBAQ). The workflow we developed is based on label-free relative peptide quantification (iBAQ by MaxQuant [22]) before and after pP enrichment, using the amount (literally a mass in μg) of total peptides and its loss during pP enrichment for the iBAQ value normalization (Fig. 2). This approach allows us to estimate the degree of phosphorylation of the signaling protein pool (*see* **Note 1**).

However, peptides and their phosphorylated counterparts have significant differences in ionization/detection efficiency (so-called flyability) [23]. Determination of flyability ratio for a particular pP/P pair (based on the technical replicates and conserved moiety assumption) allows to correct the signal intensities of the corresponding species and to quantify the absolute phosphorylation stoichiometry in each obtained pP/P pair [23]. This aspect also has to be taken into account for the final quantification.

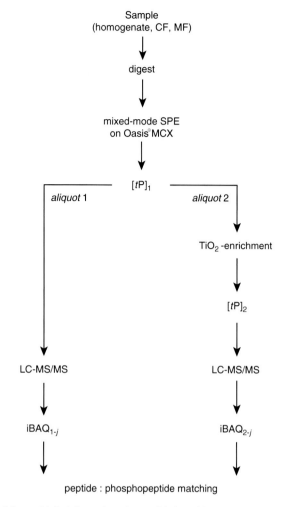

Fig. 2 Workflow of label-free phosphopeptide/peptide ratio quantification based on TiO$_2$ enrichment of phosphopeptides. Refer to **Note 1** for the calculation algorithm and notations. *CF* cytosolic fraction, *MF* microsomal fraction, *SPE* solid-phase extraction, *tP* total peptide concentration, *iBAQ* intensity-based absolute quantification is used for label-free quantitation (for further details please refer to MaxQuant manual instructions)

2   Materials

Prepare all solutions using double-deionized ultrapure water (0.055 µS/cm; *see* **Note 2**) and analytical grade reagents. Use pipette tips and microcentrifuge tubes made from low-binding-capacity plastics to minimize peptide loss by adsorption to the plastic. Take maximum care to avoid keratin contamination. Prepare and store all reagents at room temperature (unless indicated otherwise).

2.1 Protein Quantification

1. Protein quantification reagents according to manual instruction for use of NanoOrange® protein quantification kit (Molecular Probe).

2. Bovine serum albumin (BSA) standard solutions in water, 1 mL of 0, 0.1, 0.3, 0.6, 1, 3, 6, 10, 30, 60, 100, 300, 600, 1000, 1500, 2000 µg/mL each.

3. BSA standard solutions in UTU buffer, 1 mL of 0, 0.1, 0.3, 0.6, 1, 3, 6, 10, 30, 60, 100, 300, 600, 1000, 1500, 2000 µg/mL each.

4. BSA standard solutions in loading buffer for LC-MS/MS, 1 mL of 0, 0.1, 0.3, 0.6, 1, 3, 6, 10, 30, 60, 100, 300, 600, 1000, 1500, 2000 µg/mL each.

2.2 Total Protein Extraction

1. Hammer and aluminum foil.

2. Liquid nitrogen.

3. Miracloth® filter paper (Merk Millipore).

4. Potter® homogenizer (10 mL; VWR).

5. Homogenization buffer: 330 mM sucrose, 100 mM KCl, 1 mM EDTA, 50 mM Tris–HCl adjusted with MES to pH 7.5, 6.5 mM dithiothreitol (DTT; add freshly). Protease inhibitor cocktail (Sigma), phosphatase inhibitor cocktail 2 (Sigma), and phosphatase inhibitor cocktail 3 (Sigma) were added from stock solutions in 50 µL each/10 mL of the ice-cold homogenization buffer just before use.

6. UTU-buffer: 6 M Urea, 2 M thiourea, 50 mM Tris–HCl, pH 8.0.

7. Ultrasound bath (e.g., Ultrasonic Cleaner USC 300TH, VWR).

8. High-speed refrigerated benchtop centrifuge for max. speed $65,000 \times g$ (e.g., Sigma 3-30KS).

2.3 In-Solution Digest

1. Reduction buffer: 6.5 mM (or 1 µg/µL; w/v) DTT in water.

2. Alkylation buffer: 27 mM (or 5 µg/µL; w/v) iodoacetamide in water.

3. 10 mM Tris–HCl, pH 8.0.

4. Lysyl endopeptidase (LysC) stock solution: 0.5 µg/µL (Promega).

5. Trypsin stock solution: 0.5 µg/µL (Promega).

6. 2 % Trifluoroacetic acid (TFA) in water (v/v).

2.4 Peptide Purification

1. Oasis® MCX 1 cc Vac Cartridge, 30 mg per Cartridge, 30 µm particle size (Waters).

2. 100 % Methanol.

    3. Water.

    4. 2 % Formic acid in water (v/v).

    5. 5 % Ammonium hydroxide in 80 % methanol (pH 11.0) (v/v).

    6. Thermo-Strips and Caps, 8×0.2 mL (ThermoScientific).

2.5 C_8-TiO_2-StageTip Column Preparation

The 200 µL C_8-StageTips are commercially available (e.g., ThermoScientific; or elsewhere) or can be in-house manufactured according to [15, 24].

1. C_8-StageTip: 200 µL Pipet tip packed with two C_8 disks.

2. Titanium dioxide (TiO_2) beads: Titansphere® beads (5–10 µm particle size, 100 Å pore size, spherical particle shape, TiO_2 crystals; GL-Sciences). Store under dry conditions to keep specificity against phosphopeptides (*see* **Note 3**).

3. 100 % Methanol.

2.6 Phosphopeptide Enrichment on C_8-TiO_2-StageTip

1. 5 % Acetonitrile, 0.2 % TFA (pH 2.0) in water (v/v).

2. 80 % Acetonitrile, 0.2 % TFA (pH 2.0) in water (v/v).

3. 5 % Ammonium hydroxide (pH 11.0) in water (v/v).

4. 5 % Piperidin in water (v/v).

5. 20 % Phosphoric acid in water (v/v).

2.7 C_{18}-StageTip Column Preparation

The 200 µL C_{18}-StageTips are commercially available (e.g., ThermoScientific; or elsewhere) or can be in-house manufactured according to [15, 24].

1. C_{18}-StageTip: 200 µL Pipet tip packed with two C_{18} disks.

2. 80 % Acetonitrile, 0.2 % TFA.

3. 0.5 % Acetic acid in water (v/v).

4. 80 % Acetonitrile, 0.5 % acetic acid (v/v).

5. 5 % Acetonitrile, 0.2 % TFA (v/v).

2.8 LC-MS/MS

2.8.1 Liquid Chromatography

1. Chromatographic system: Easy-nLC 1000 (ThermoScientific).

2. Column: EASY-Spray column, PepMap® RSLC, C18 (ThermoScientific); particle size 2 µm; pore size 100 Å; column dimensions 75 µm × 50 cm (I.D. × L).

3. Loading buffer: 5 % Acetonitrile, 0.2 % TFA, pH 2.0.

4. Buffer A: 0.5 % Acetic acid, pH 2.0.

5. Buffer B: 0.5 % Acetic acid, 80 % acetonitrile, pH 2.0.

2.8.2 Mass Spectrometry Equipment

1. Mass analyzer: Q Exactive Plus (ThermoScientific).

3 Methods

3.1 Protein Quantification

1. Use 3 μL of proteins/ peptides contenting solution for quantification, which is performed according to the manual instruction for use of NanoOrange® protein quantification kit (*see* **Note 4**). Choose the BSA standard (in water, in UTU buffer, or in loading buffer for LC-MS/MS) in accordance with the matrix of the protein/peptide-contenting solution (*see* **Note 5**).

3.2 Total Protein Extraction

This protocol is adapted from [20]. All protein extraction steps must be done on ice.

1. Weigh root samples (g of fresh weight; gFW), wrap them individually in aluminum foil, and freeze them immediately in liquid nitrogen (*see* **Note 6**).

2. Break the frozen samples into small pieces with the hammer while keeping them wrapped in aluminum foil. Make sure that the samples are constantly frozen after harvesting to avoid rapid dephosphorylation.

3. Transfer the cell material into ice-cold homogenization buffer in a ratio of 5 mL homogenization buffer per 1 gFW.

4. Resuspend the cell material thoroughly by gentle stirring or shaking to get rid of clots.

5. Grind the sample manually in a Potter® homogenizer on ice, 50 smooth strokes per 7 mL sample.

6. Filter the homogenate through four layers of Miracloth® to get rid of cell wall material and tissue debris.

7. Centrifuge the homogenate at $7.5 \times 10^3 \times g$ for 15 min at 4 °C to get rid of unbroken cells and organelles.

8. Collect the supernatant.

9. Centrifuge the supernatant at $48 \times 10^3 \times g$ for 80 min at 4 °C to precipitate microsomal vesicles.

10. Collect the supernatant which represents the cytosolic fraction (i.e., water-soluble proteins, can be used for other experiments).

11. Resuspend the pellet, which represents the microsomal fraction (i.e., endomembranes and membrane-associated proteins), in 500 μL of ice-cold UTU-buffer.

12. Rigorously vortex the suspension and ultra-sonicate it for approximately 30 s.

13. Quantify total protein content in the microsomal fraction using NanoOrange ® protein quantification kit and BSA standards in UTU buffer.

3.3 In-Solution Protein Digest

1. Reduction: Add 1 μL of the reduction buffer per each 50 μg of the total protein and incubate for 30 min at 25 °C and 260 rpm.

2. Alkylation: Add 1 μL of alkylation buffer per each 50 μg of the total protein and incubate for 20 min at 25 °C in the dark and 260 rpm.

3. Digest 1: Add 0.5 μL of LysC stock solution per each 50 μg of the total protein and incubate for 3 h at 37 °C and 260 rpm.

4. Dilution: Dilute the sample by fivefold with 10 mM Tris–HCl, pH 8.0 (*see* **Note 7**).

5. Digest 2: Add 1 μL of trypsin stock solution per each 50 μg of the total protein and incubate overnight at 37 °C and 260 rpm.

6. Stop the digest: Acidify the digest to 0.2 % TFA final concentration (add 1/10 volume of 2 % TFA to reach pH 2.0).

7. Spin the sample on centrifuge at $20 \times 10^3 \times g$, 3 min, at room temperature to pellet any insoluble materials.

3.4 Peptide Purification

1. Conditioning: Condition the MCX cartridge with 1 mL of 100 % methanol (*see* **Note 8**).

2. Equilibrate the MCX cartridge with 1 mL of water.

3. Sample loading: Load the digest to the conditioned MCX cartridge (*see* **Note 9**). Collect the flow-through and re-load it two more times.

4. Wash 1: Wash the cartridge with 1 mL of 2 % formic acid (pH 2.0) to lock ionized compounds on MCX (*see* **Note 10**).

5. Wash 2: Wash the cartridge with 1 mL of 100 % methanol to remove interfering unionized weaker acids and neutrals. Completely expel the methanol from the cartridge.

6. Elution: Elute peptides from the cartridge with 5×200 μL of 5 % NH_4OH in 80 % methanol (pH 11) (*see* **Note 11**). The final total volume is 1000 μL.

7. Split the eluate into 15 % ("aliquot 1") and 85 % ("aliquot 2") of the sample volume (*see* Fig. 2).

8. Dry both aliquots of the eluate to complete dryness in a vacuum centrifuge (10 mbar, $235 \times g$, rotor temperature 37 °C, e.g., Christ® RVC 2–25 CD plus).

9. Redissolve the dried "aliquot 1" in 50 μL of LC-MS/MS loading buffer (e.g., 5 % acetonitrile, 0.2 % TFA).

10. Rigorously vortex the sample, sonicate it for 30 s, and centrifuge at $2 \times 10^3 \times g$ for 5 min.

11. Transfer the supernatant of "aliquot 1" into fresh micro-tubes (0.2 mL).

12. Quantify total peptide content in "aliquot 1" using NanoOrange®protein quantification kit and BSA standards in

loading buffer for LC-MS/MS. This is $[Peptides]_1$ according to notations at Fig. 2.

13. Reserve the "aliquot 1" for the further LC-MS/MS analysis.

3.5 C₈-TiO₂-StageTip Column Preparation

1. Activation of TiO_2 beads: Activate TiO_2 beads at 130 °C for 30 min prior to use (*see* **Note 3**).

2. Preparing TiO_2 bead stock suspension: Weigh 25 mg of TiO_2 beads, resuspend it in 500 μL of 100 % methanol, and vortex the suspension well.

3. Loading of TiO_2 bead stock suspension: Load 20 μL of the stock suspension (overall 1 mg; *see* **Note 12**) on top of the C_8 disk in the 200 μL C_8-StageTip (*see* **Note 13**).

4. Let the suspension settle down under the gravity force in order to distribute the beads evenly.

5. Spin (*see* **Note 14**) the C_8-TiO_2-StageTip to force the solution through.

3.6 Phosphopeptide Enrichment on C₈-TiO₂-StageTip

1. Redissolve the dried "aliquot 2" in 50 μL of 80 % acetonitrile and 0.2 % TFA.

2. Rigorously vortex the sample, sonicate it for 30 s, and centrifuge at $2 \times 10^3 \times g$ for 5 min.

3. Insert the 200 μL C_8-TiO_2-StageTip into a spin adapter and place it in a fresh microcentrifuge tube.

4. C_8-TiO_2-StageTip conditioning: Load 100 μL of 80 % acetonitrile and 0.2 % TFA to the C_8-TiO_2-StageTip and spin it to force the solution through.

5. Replace the waste microcentrifuge tube with a fresh tube.

6. Sample loading: Load the sample (**step 1**) onto conditioned C_8-TiO_2-StageTip (**step 4**) and spin it to force the sample through.

7. Sample reloading: Collect the flow-through, reload the sample again, and then spin it to force the sample through. Repeat this step twice.

8. Wash: Load 100 μL of 5 % acetonitrile and 0.2 % TFA to the C_8-TiO_2-StageTip and then spin it to force the sample through into waste microcentrifuge tube.

9. Add 50 μL of 20 % phosphoric acid into a fresh microcentrifuge tube where the phosphopeptides will be eluted in.

10. Elution 1: Elute the phosphopeptides with 50 μL of 5 % NH_4OH (pH 11.0) from the 200 μL C_8-TiO_2-StageTip into a microcentrifuge tube with 20 % phosphoric acid (**step 10**) (*see* **Note 11**).

11. Elution 2: Elute the remaining phosphopeptides with 50 μL of 5 % piperidine from the C_8-TiO_2-StageTip into the same tube. The final volume of the eluate is 150 μL.

3.7 Desalting on C₁₈-StageTip

1. Conditioning of C_{18}-StageTips: Load 100 µL of 80 % acetonitrile and 0.2 % TFA to the C_{18}-StageTip and spin it to force the solution through into a waste microcentrifuge tube.

2. Load 2×100 µL of 0.5 % acetic acid to the C_{18}-StageTip and spin it to force the solution through into a waste microcentrifuge tube.

3. Sample loading: Load the sample (Subheading 3.6, **step 11**) onto pre-conditioned C_{18}-StageTip and spin it to force the solution through into a waste microcentrifuge tube.

4. Washing: Load 2×100 µL of 0.5 % acetic acid to the C18-StageTip and spin it to force the solution through into a waste microcentrifuge tube.

5. Elution: Elute the phosphopeptide enriched fraction by 2×20 µL of 80 % acetonitrile and 0.2 % TFA into fresh microcentrifuge tube.

6. Spin down the eluate to dryness (10 mbar, $235 \times g$, rotor temperature 37 °C, e.g., Christ® RVC 2–25 CD plus).

7. Redissolve the phosphopeptides in 50 µL of loading buffer for LC-MS/MS (i.e., 5 % acetonitrile, 0.2 % TFA).

8. Rigorously vortex the sample, sonicate it for 30 s, and centrifuge at $2 \times 10^3 \times g$ for 5 min.

9. Transfer the supernatant of "aliquot 2" into a fresh micro-tube (0.2 mL).

10. Quantify the total peptide content in the sample using NanoOrange®protein quantification kit and BSA standards in loading buffer for LC-MS/MS. This is $[Peptides]_2$ according to notations at Fig. 2.

3.8 LC-MS/MS

3.8.1 Liquid Chromatography

1. Injection volume: 1–5 µL to achieve at least 2 µg of overall column load with the total peptides.

2. Flow rate: 250 nL/min.

3. Gradient %B: 0 min 5 %, 200 min 35 %, 240 min 60 %, 242 min 90 %, 257 min 90 %, 258 min 5 %, 263 min 5 %.

4. Operation column temperature: 50 °C.

5. Operation pressure: 500 bar.

3.8.2 Mass Spectrometry

1. Polarity: Positive.

2. Full MS: Resolution 70,000 (at $m/z = 200$ Th); AGC target 1e6; maximum IT 20 ms; scan range 300–1600 m/z.

3. dd-MS²: Resolution 17,500 (at $m/z = 200$ Th); AGC target 1e5; maximum IT 120 ms; TopN 5; isolation window 2.2 m/z; scan range 200–2000 m/z; NCE 25.

4. dd-Settings: Underfill ratio 0.1 %; dynamic exclusion 40 s.

3.9 Data Employment

This method is mainly designated for quantification of stoichiometry in pairs of phosphorylated peptide and its corresponding unmodified cognate, i.e., to measure phosphorylation stoichiometry. However, this method can also be applied for search/screen of gross phosphorylation sites or qualitative assessment of phosphorylation of proteins from certain signaling pathways, without detecting the unmodified cognates.

Stimulus response (i.e., dynamic perturbation) experimental approaches are widely used in systems biology to provoke dynamic responses of the studied system. The phosphorylation stoichiometry of signaling proteins is a state variable in mathematical models of the conserved moieties or cascade reactions of the signaling pathways (Fig. 1). Steady-state phosphorylation stoichiometry and its time-dependent dynamics in response to a perturbation event allow parameter estimation of the kinetic equations that describe the corresponding cascades in signaling pathways [1]. Measurements of steady-state phosphorylation stoichiometry under different signal strength allow quantification of local and global response coefficients of the signaling pathway, if the kinetic properties of the reaction cascades are known [7]. This type of modeling can be performed either in package programs like MATLAB (The MathWorks) and MATHEMATICA (Wolfram Research) or in specialized software like Simmune [25].

4 Notes

1. Please refer to Fig. 2 for the notations associated with corresponding workflow steps. The loss of peptides' amounts (l) during phosphopeptide enrichment can be estimated as

$$l = \frac{[tP]_1}{[tP]_2} \tag{1}$$

where $[tP]i$—concentration of total peptides before (1) and after (2) enrichment (μg/mL). The peptide quantification must be accurate; therefore please refer to Subheading 3.1.

The amount (i.e., mass in μg) of the injected peptides (mi) for LC-MS/MS analysis can be calculated as

$$m_i = x_i \times [tP]_i \tag{2}$$

where xi—injection volume used for the LC-MS/MS analysis ($i = 1,2$) (μL).

The mass-specific content of individual peptide (Pj) and its phosphorylated species (pPj) can be estimated from its label-free quantification (iBAQi,j) and normalized per corresponding mi ($j = N$):

$$\begin{cases} P_j = \dfrac{i\mathrm{BAQ}_{1.j}}{m_1} \\[2ex] pP_j = \dfrac{i\mathrm{BAQ}_{2.j}}{m_2} \times l \end{cases} \tag{3}$$

where the correction factor l is calculated in Eq. 1. The peptide content after phosphopeptide enrichment must be corrected with the loss of the total peptides mass (Eq. 1). The total pool of the particular peptide (P_j^{pool}) consists of non-phosphorylated and phosphorylated species:

$$P_j^{\mathrm{pool}} = P_j + pP_j \tag{4}$$

and correspondingly a part of each species is

$$\frac{P_j}{P_j^{\mathrm{pool}}} + \frac{pP_j}{P_j^{\mathrm{pool}}} = 1 \tag{5}$$

2. Hereinafter designated just as water.

3. The specificity of TiO_2 beads against phosphopeptides is reduced by water absorption when it is kept without desiccation. The specificity can be recovered by heating the beads in a drying oven at 130 °C for 30 min [16].

4. Accurate and highly specific quantification of protein/peptide is essential for this approach. We have selected NanoOrange®protein quantification kit (Molecular Probe) for this purpose, because it is very sensitive and specific to proteins/peptides, has a wide quantification range (0.1–2000 μg/mL), and is compatible with nucleic acids, reducing agents, and detergents.

5. UTU buffer or 5 % acetonitrile and 0.2 % TFA significantly quench the fluorescence and therefore they must be included into the solution matrix for compensation.

6. The frozen samples can be stored at −80 °C for further analysis.

7. The dilution step is required in order to get final 1.2 M urea, 0.4 M thiourea, and 10 mM Tris–HCl (pH 8.0), which is favorable for trypsin operation.

8. Do not let the cartridge dry; always expel one mobile phase with another.

9. The 1 cc cartridge from Waters has a load volume of a matrix with an analyst up to 50 mL. At this step, the sample is in 1.2 M urea, 0.4 thiourea, 10 mM Tris–HCl, and 0.2 % TFA, pH 2.0.

10. This step also removes undigested proteins and salts.

11. Phosphopeptides are not stable in alkaline conditions; therefore, in order to minimize the exposure time, it is advised to

dry the eluate immediately at vacuum centrifuge, as it is exemplified in Subheading 3.4, **steps 7** and **8**, or immediately neutralize the alkaline solution with strong acid as it is exemplified in Subheading 3.6, **steps 10** and **11**.

12. 1 mg TiO_2 beads per a single C_8-StageTip column have a binding capacity of ~100 μg of total peptides from *Arabidopsis* [16].

13. The choice of the C_8 material is based on the idea that the membrane disk is only used to retain the TiO_2 beads inside the tip, but the C_8-disk itself does not participate in the phosphopeptide enrichment. The C_8-StageTips can be stored at room temperature.

14. "Spin" hereinafter refers to centrifugation of a StageTip in a bench microcentrifuge (e.g., Mini Star Silverline, Galaxy Mini Centrifuge, VWR) at $2 \times 10^3 g$ (or 6×10^3 rpm) for 1 min at room temperature.

References

1. Klipp E, Liebermeister W (2006) Mathematical modeling of intracellular signaling pathways. BMC Neurosci 7(Suppl 1):S10. doi:10.1186/1471-2202-7-S1-S10

2. Mariottini C, Iyengar R (2013) Chapter 16—system biology of cell signaling. In: Walhout AJM, Vidal M, Dekker J (eds) Handbook of systems biology. Academic, San Diego, pp 311–327

3. Duan G, Walther D, Schulze W (2013) Reconstruction and analysis of nutrient-induced phosphorylation networks in Arabidopsis thaliana. Front Plant Sci 4:540. doi:10.3389/fpls.2013.00540

4. Niittylä T, Fuglsang AT, Palmgren MG et al (2007) Temporal analysis of sucrose-induced phosphorylation changes in plasma membrane proteins of arabidopsis. Mol Cell Proteomics 6(10):1711–1726. doi:10.1074/mcp.M700164-MCP200

5. Schulze WX (2010) Proteomics approaches to understand protein phosphorylation in pathway modulation. Curr Opin Plant Biol 13(3):279–286. doi:10.1016/j.pbi.2009.12.008

6. Blagoev B, Ong S-E, Kratchmarova I et al (2004) Temporal analysis of phosphotyrosine-dependent signaling networks by quantitative proteomics. Nat Biotechnol 22(9):1139–1145. doi:10.1038/nbt1005

7. Kholodenko BN, Hoek JB, Westerhoff HV et al (1997) Quantification of information transfer via cellular signal transduction pathways. FEBS Lett 414(2):430–434. doi:10.1016/S0014-5793(97)01018-1

8. Hein MY, Sharma K, Cox J et al (2013) Chapter 1—proteomic analysis of cellular systems. In: Walhout AJM, Vidal M, Dekker J (eds) Handbook of systems biology. Academic, San Diego, pp 3–25

9. Cox J, Mann M (2011) Quantitative, high-resolution proteomics for data-driven systems biology. Annu Rev Biochem 80(1):273–299. doi:10.1146/annurev-biochem-061308-093216

10. Choudhary C, Mann M (2010) Decoding signalling networks by mass spectrometry-based proteomics. Nat Rev Mol Cell Biol 11(6):427–439. doi:10.1038/nrm2900

11. Altelaar AFM, Munoz J, Heck AJR (2013) Next-generation proteomics: towards an integrative view of proteome dynamics. Nat Rev Genet 14(1):35–48. doi:10.1038/nrg3356

12. Olsen J, Macek B (2009) High accuracy mass spectrometry in large-scale analysis of protein phosphorylation. In: Lipton M, Paša-Tolic L (eds) Mass spectrometry of proteins and peptides. Humana Press, Totowa, NJ, pp 131–142

13. Schmelzle K, White FM (2006) Phosphoproteomic approaches to elucidate cellular signaling networks. Curr Opin Biotechnol 17(4):406–414. doi:10.1016/j.copbio.2006.06.004

14. Larsen MR, Thingholm TE, Jensen ON et al (2005) Highly selective enrichment of phosphorylated peptides from peptide mixtures using titanium dioxide microcolumns. Mol Cell Proteomics 4(7):873–886. doi:10.1074/mcp.T500007-MCP200

15. Wisniewski JR, Zougman A, Nagaraj N et al (2009) Universal sample preparation method for proteome analysis. Nat Methods 6(5): 359–362. doi:10.1038/nmeth.1322

16. Nakagami H (2014) StageTip-based HAMMOC, an efficient and inexpensive phosphopeptide enrichment method for plant shotgun phosphoproteomics. In: Jorrin-Novo JV et al (eds) Plant proteomics. Humana Press, New York, pp 595–607

17. Thingholm TE, Jorgensen TJD, Jensen ON et al (2006) Highly selective enrichment of phosphorylated peptides using titanium dioxide. Nat Protoc 1(4):1929–1935. doi:10.1038/nprot.2006.185

18. Beckers GM, Hoehenwarter W, Röhrig H et al (2014) Tandem metal-oxide affinity chromatography for enhanced depth of phosphoproteome analysis. In: Jorrin-Novo JV et al (eds) Plant proteomics. Humana Press, New York, pp 621–632

19. Colby T, Röhrig H, Harzen A et al (2011) Modified metal-oxide affinity enrichment combined with 2D-PAGE and analysis of phosphoproteomes. In: Dissmeyer N, Schnittger A (eds) Plant kinases. Humana Press, New York, pp 273–286

20. Pertl H, Himly M, Gehwolf R et al (2001) Molecular and physiological characterisation of a 14-3-3 protein from lily pollen grains regulating the activity of the plasma membrane H+ ATPase during pollen grain germination and tube growth. Planta 213(1):132–141. doi:10.1007/s004250000483

21. Pratt JM, Simpson DM, Doherty MK et al (2006) Multiplexed absolute quantification for proteomics using concatenated signature peptides encoded by QconCAT genes. Nat Protoc 1(2):1029–1043

22. Cox J, Mann M (2008) MaxQuant enables high peptide identification rates, individualized p.p.b.-range mass accuracies and proteome-wide protein quantification. Nat Biotechnol 26(12):1367–1372. doi:10.1038/nbt.1511

23. Steen H, Jebanathirajah JA, Springer M et al (2005) Stable isotope-free relative and absolute quantitation of protein phosphorylation stoichiometry by MS. Proc Natl Acad Sci U S A 102(11):3948–3953. doi:10.1073/pnas.0409536102

24. Rappsilber J, Mann M, Ishihama Y (2007) Protocol for micro-purification, enrichment, pre-fractionation and storage of peptides for proteomics using StageTips. Nat Protoc 2(8): 1896–1906

25. Meier-Schellersheim M, Xu X, Angermann B et al (2006) Key role of local regulation in chemosensing revealed by a new molecular interaction-based modeling method. PLoS Comput Biol 2(7), e82. doi:10.1371/journal.pcbi.0020082

Chapter 19

Interpretation of Quantitative Shotgun Proteomic Data

Elise Aasebø, Frode S. Berven, Frode Selheim, Harald Barsnes, and Marc Vaudel

Abstract

In quantitative proteomics, large lists of identified and quantified proteins are used to answer biological questions in a systemic approach. However, working with such extensive datasets can be challenging, especially when complex experimental designs are involved. Here, we demonstrate how to post-process large quantitative datasets, detect proteins of interest, and annotate the data with biological knowledge. The protocol presented can be achieved without advanced computational knowledge thanks to the user-friendly Perseus interface (available from the MaxQuant website, www.maxquant.org). Various visualization techniques facilitating the interpretation of quantitative results in complex biological systems are also highlighted.

Key words Quantification, Data interpretation, Perseus, Data post-processing

1   Introduction

Quantitative shotgun proteomics has become the method of choice for the description of large scale biological systems [1]. The approach relies on the proteome-wide quantification of proteins, often including screening for posttranslational modifications [2, 3]. Different quantification methods are available to the researcher, ultimately providing a list of relatively and/or absolutely quantified proteins [4–6]. Before inferring any biological sense from the quantitative results, the data must undergo several post-processing steps, typically including normalization, statistical evaluation, and functional enrichment—which can be challenging due to the amount of data, and its specificity and complexity [7].

In this chapter, we present a simple workflow for the post-processing of proteome-wide quantification results which can be applied without advanced knowledge in computer science thanks to the user friendly Perseus interface. The workflow is here applied to a freely available dataset used for illustrative purposes, consisting

Jörg Reinders (ed.), *Proteomics in Systems Biology: Methods and Protocols*, Methods in Molecular Biology, vol. 1394, DOI 10.1007/978-1-4939-3341-9_19, © Springer Science+Business Media New York 2016

of five cell lines derived from acute myeloid leukemia (AML) patients measured with a spiked-in internal standard (IS) obtained from the combination of the same five AML cell lines metabolically labeled with heavy isotopes [8] and analyzed using MaxQuant version 1.4.1.2 [9]. Subsequently, we exemplify the use of several visualization techniques allowing the critical interpretation of the data: volcano plots, principal component analysis (PCA), and hierarchical clustering.

2 Material

1. The dataset here used for illustrative purposes is freely available through the ProteomeXchange [10] consortium via the PRIDE [11] partner repository under the accession number PXD000441 (Results_MQ_5cell-line-mix.zip).

2. Perseus is a software tool freely available upon registration on the MaxQuant website (http://www.maxquant.org). It can be downloaded directly from http://www.perseus-framework. org. After downloading the software and extracting the zipped folder, open the Perseus program located in the Perseus folder. For Perseus v1.5.0.0 and newer, download the plugin PluginProteomicRuler.dll from http://perseus-framework. org/plugins and put it in the main Perseus folder. *See* **Note 1**.

3 Methods

3.1 Loading Data in Perseus

1. Open the *Perseus* program (annotated with the *Perseus* logo) located in the main *Perseus* folder.

2. Download annotations: Click the blue tab indicated in Fig. 1a and then *Annotation download* (Fig. 1b) in the menu thatwwshows up. Go to the provided Dropbox folder and download the appropriate *mainAnnot.txt.gz* file in the OrganismSpecific folder (in this example use *mainAnnot.homo_sapiens.txt.gz*), putthedownloadedannotationfileinto *Perseus\conf\annotation*. Extract the zipped file (*see* **Note 2**).

3. Import your output file: Press the green upwards-pointing arrow to load your data (Fig. 1c). When you hover over the arrow you will get the message: *Generic matrix upload*.

4. After clicking the green arrow, a matrix opens where you can upload your .txt file (Fig. 2), in this case use the "proteinGroups_5cell-line-mix.txt" file (*see* **Note 3**). Transfer quantitative data (ratios) into the *Expression* field (*see* **Note 4**). You can also choose other parameters of interest, such as "Number of proteins", "Unique peptides", etc. Put numerical

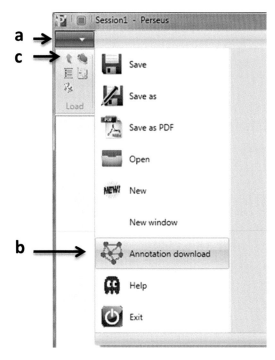

Fig. 1 When using Perseus for the first time, it is recommended to download annotations (**a**, **b**). To import a data set (Generic matrix upload), press the *green arrow* (**c**)

columns into the *Numerical* field, and columns containing text into the *Text* field. Some columns are preselected if you are working with MaxQuant output. Click *OK* and you will see the selected columns in the Perseus matrix.

3.2 Filtering and Rearranging Columns

1. Start by filtering on the categorical columns. Go to: *Filter rows→ Filter rows based on categorical column*. In the appearing window, select the categorical column you wish to remove. In this example, remove *contaminants, reverse hits* and *only identified by site*.

2. Remove the empty columns: *Rearrange→ Remove empty columns*.

3. Rename the columns: *Rearrange→ Rename columns*. In this dataset "1" is Molm-13, "2" is MV4-11, "3" is NB4, "4" is OCI-AML3 and "5" is THP-1.

4. Remove proteins without an expression value: *Filter rows→ Filter rows based on valid values*. Use default parameters for this dataset (*see* **Note 5**).

5. Have a look to the right in your Perseus window. All the steps you have performed this far appear as individual matrices. Here you can easily navigate between the matrices, and inspect the steps you have performed. If you want to delete a matrix,

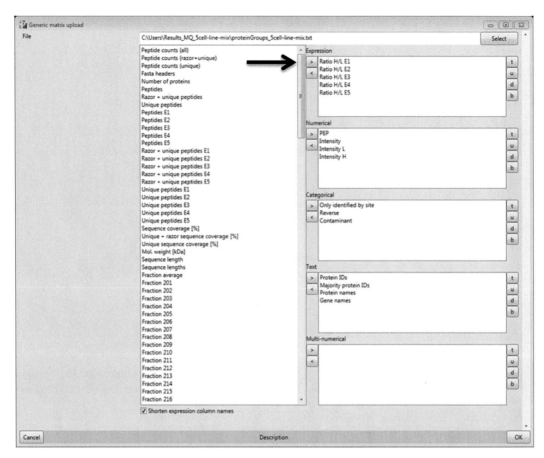

Fig. 2 Select the output file (.txt format) to upload (usually the ProteinGroups.txt file for MaxQuant). The unique column headers in the output are listed in the *left column*. Select the columns to analyze in Perseus and press the *arrows* to load them into the columns to the *right*. Be sure to load protein expression data into the *Expression* fields and numerical information—such as the number of peptides or sequence coverage—into the *Numerical* field

simply select the matrix and click the red cross. If you are using Perseus version 1.5.0.15 or newer, you can (re-)name the matrices by double clicking them (Fig. 3).

6. The MaxQuant output display the ratios as Heavy/Light, so in order to compare all light samples against the heavy sample (i.e., the internal standard) we have to invert the comparison to get Light/Heavy. Go to: *Basic → Transform*, and write "1/(x)" in the *Transformation* field.

7. Transform the expression values into log values (*see* **Note 6**). Go to: *Basic → Transform →* choose "log2(x)".

3.3 Normalization

1. The example dataset is now in $\log_2$ values; thus the normal distribution should be centered on zero. Visualize the distribution of the dataset by using a histogram. Go to: *Visualization → Histogram*. Accept the default settings by

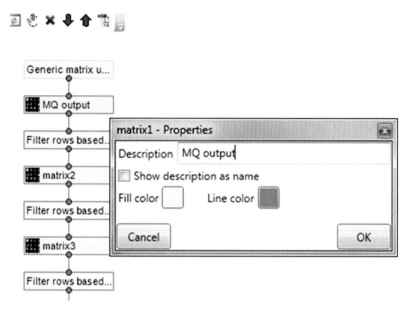

Fig. 3 Matrix overview. Each new task that changes the *Data* matrix will appear as a new matrix. It is possible to navigate between different matrices, delete paths and (re-) name matrices, thus allowing inspecting intermediate results

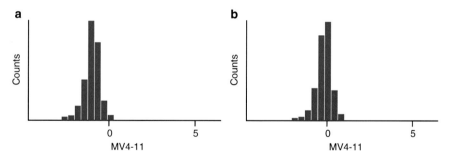

Fig. 4 Histogram displaying the protein ratio distribution before (**a**) and after (**b**) normalization, exemplified by two of the cell lines. The *Histogram* tab can be found next to the *Data* tab in the same matrix

clicking *OK*. The histogram will appear in the same matrix, you find it as a new tab next to the *Data* tab. You may notice that you ought to normalize the data (Fig. 4a).

2. To normalize, go to: *Normalization → Subtract*. Choose *Columns* in the *Matrix access* field and keep the default for subtracting the median. Click *OK*. Make a new histogram and check that the center of distribution has changed (Fig. 4b).

3.4 Gene Annotation

1. Add categorical annotations. Go to: *Annot. Columns → Add annotation*. Select the *mainAnnot.homo_sapiens.txt* file as the *Source* (Fig. 5). Choose from "GOBP name" down to "Keyword" in the *Annotations to be added* field and transfer the selected annotations over to the empty field using the arrow. Click *OK*.

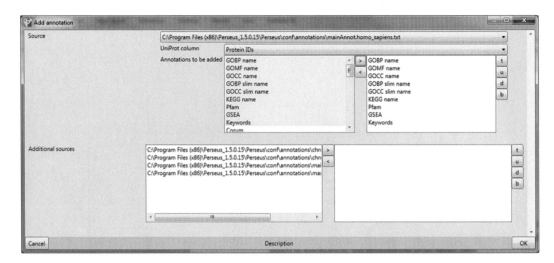

Fig. 5 Add annotation—a prerequisite is that annotations are already downloaded (described in Subheading 3.1.2 and Fig. 1)

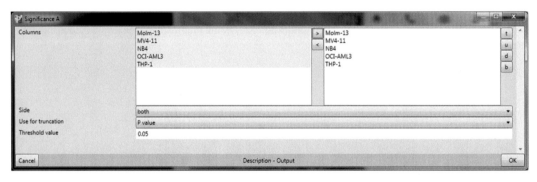

Fig. 6 Identify regulated proteins in each sample by using the *Significance A* test with the indicated parameters. Regulated proteins are marked with the symbol "+" in the *Data* matrix

3.5 Statistical Evaluation

1. Go to: *Basic → Significance A*. Select all cell lines and transfer them to the empty field using the arrow. In the *Use for truncation* field choose *P value* and use the default (0.05) as *Threshold value* (Fig. 6). Click *OK*. The protein ratios that are significant outliers relative to the sample population are now annotated with "+" in the matrix.

2. Compare the cell lines derived from patients at time of diagnosis (MV4-11 and OCI-AML3) to the other cells lines derived from patients during relapse (Molm-13, NB4 and THP-1). Start by making two groups: *Annot. rows → Categorical annotations rows* and specify the groups as "Diagnosis" or "Relapse" as shown in Fig. 7. Click *OK*.

3. Do a two-samples *t*-test to compare the groups: *Tests → Two samples t-test*. Select *P value* in the *Use for truncation* field and keep *0.05* as *Threshold value*, and default settings for other parameters (Fig. 8). Click *OK*.

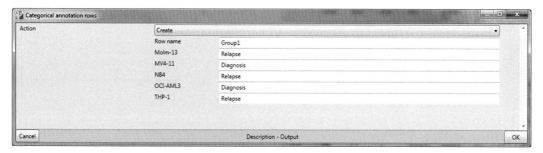

Fig. 7 *Categorical annotation rows* is used to create groups. In this case we create "Group1" (default name) and mark the samples with either "Diagnosis" or "Relapse"

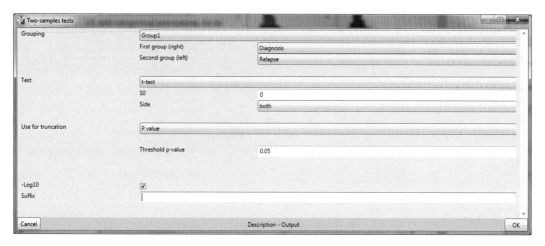

Fig. 8 Find significantly expressed proteins between the groups "Diagnosis" or "Relapse" by applying a *Two-samples tests*. Use the indicated parameters. The significantly expressed proteins are marked with the symbol "+" in the *Data* matrix

3.6 Volcano Plot

1. Compare the two groups. Go to: *Visualization → Scatter plot* and keep *Columns* in the *Matrix access* field. Click *OK*.

2. Find the *Scatter plot* tab next to the *Data* tab in the same matrix. Select *t-test difference* and *−Log t-test p value* in the tabs at right of the scatter plot (Fig. 9a).

3. To look for specific pathways or annotations go to the *Categories* tab and select terms related to apoptosis (Fig. 9b).

4. Alternatively, go to: *Misc. → Volcano Plot*. Keep the default parameters and click *OK*.

3.7 Principal Component Analysis (PCA)

1. To perform principal component analysis, valid expression values for each protein are required. To filter, go to: *Filter rows → Filter rows based on valid values* and write "5" in the *Minimum number of values* field. Click *OK*.

2. Create a PCA plot: *Clustering/PCA → Principal component analysis*. Tick in the tab for *Categoryenrichment in components* and keep the default settings (*see* **Note** 7). Click *OK*.

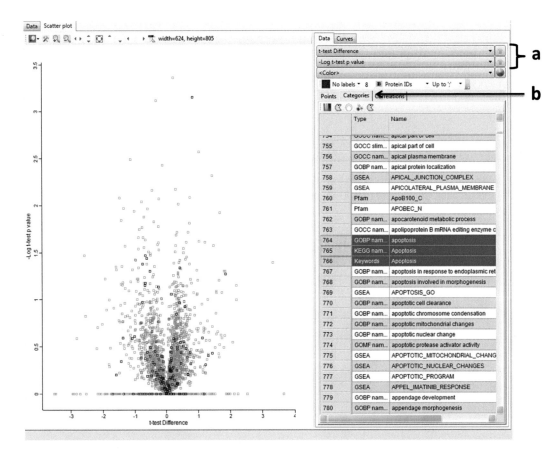

Fig. 9 Volcano plot. Start by creating a scatter plot and compare the *t-test difference* to the *-Log t-test p value* (**a**) to visualize the volcano plot. Inspect enriched terms, such as "apoptosis," by going to the *Categories* tab (**b**). The proteins annotated by the term appear in *red* in the volcano plot

3. Find the PCA plot next to the *Data* tab and change from *No labels* to *All labels* (Fig. 10a).

4. Color the groups "Diagnosis" and "Relapse" with two different colors by clicking on the color symbol (Fig. 10b).

5. Select the proteins responsible for the separation in the PCA plot by changing the mode from *zoom* (magnifying glass) to *select* (square) (Fig. 10c). The selected proteins in this example are marked in red in the PCA plot (Fig. 10d) and the protein name is marked in blue in the box indicated in Fig. 10e.

3.8 Hierarchical Clustering

1. Normalize on the protein level: Before clustering you need to normalize the protein values at the row level. Go to: *Normalization → Z-score* and select *Rows* in the *Matrix Access* field (*see* **Note 8**). Use default settings for the other parameters. Click *OK*.

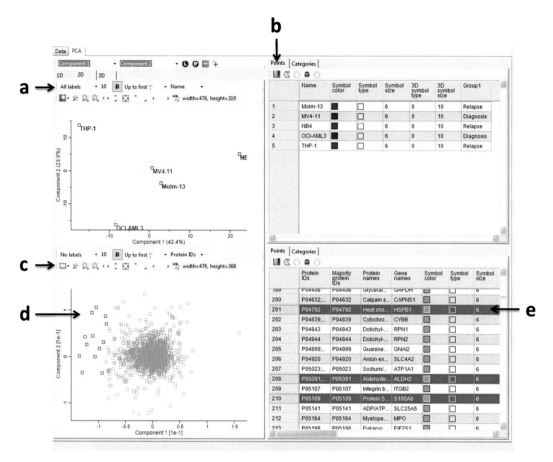

Fig. 10 Principal Component Analysis (PCA). Change from *No labels* to *All labels* (**a**) to reveal the sample names of the dots in the PCA plot. Color the samples belonging to each group in different colors (**b**) to see if the cell lines of one group cluster together. To look at the proteins that cause the separation of the samples, *see* the box at the *lower left*. Change from the *zoom* to *select* option (**c**) and select the proteins to the *left* (**d**). Note that the selected proteins appear with a *blue line* in the table at *lower right* (**e**)

2. We want to make a cluster of the proteins that are significantly regulated between the two groups "Diagnosis" and "Relapse". Filter to keep the significant proteins: *Filter rows→ Filter rows based on categorical column*, select *t-test Significant* in the *Column* field and *Keep matching rows* in the *Mode* field. Click *OK*.

3. To perform hierarchical clustering, go to: *Clustering/ PCA→ Hierarchical clustering*. Use the default settings and click *OK*.

4. Define clusters at the protein (row) and group (column) level as indicated in Fig. 11a. In this dataset *Number of clusters* was set to "2" at both the protein and group level. Accept with *Apply* and note that two clusters appear in the figure and click *OK*.

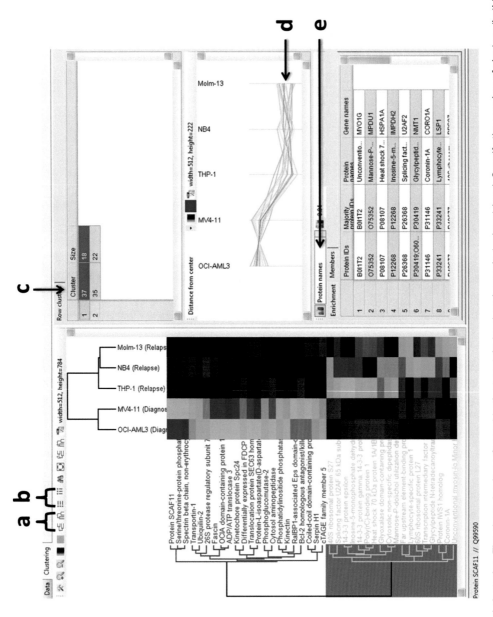

Fig. 11 Hierarchical clustering. The clustering is performed both at protein (*row*) and sample (*column*) level. Specify the number of clusters (in this example two) at both levels by clicking the symbols indicated in (**a**). Configure the row and column names by clicking the symbols indicated in (**b**). At the *upper right* there is a table of *Row clusters* (**c**), you can select a cluster by clicking the line. The proteins of the selected cluster are displayed in the profile plot (**d**). By selecting *Protein names* in (**e**), you can see the table of protein names in the *Members* tab

5. Configure the row and column names by clicking the symbols indicated in Fig. 11b. In the *Row names* field choose *Protein names* and in the *Column names* field select *Group1* under *Addtl. Column names*. Click *OK*.

6. Inspect the row clusters by selecting one of the clusters as indicated in Fig. 11c. The expression profiles of the proteins corresponding to the selected cluster appear in the plot below (Fig. 11d).

7. In order to explore which proteins that belong to a selected cluster, change the tab indicated in Fig. 11e from < *None*> to *Protein names* and look under the *Members* tab. If any annotation (i.e., KEGG pathway, GO term) is enriched, it will appear if you click the bar symbol next to the *Protein names* field.

4 Notes

1. For more help and discussion with other users, visit the Perseus Google Group at https://groups.google.com/forum/#!forum/perseus-list.

2. You need a program to extract zipped files. Here we used 7-Zip File Manager, downloaded freely from the internet. After installing 7-Zip, right click on the zipped folder and "extract" the folder.

3. You can upload tab separated files, such as the .txt files from a MaxQuant search, but also output files from other software and types of experiments (genomics, transcriptomics, metabolomics, etc.), as long as the file contains a unique column header and is in the .txt format. In this example we look at the protein expression data, but you can also explore and analyze peptide and modification data with this software.

4. The expression values will differ between ratios (in case of SILAC, TMT, iTRAQ, etc.) and intensities (in case of label free), depending on your experiment.

5. In general you should have expression values in minimum 50 % of your samples, this allows for better statistics and clustering. It is also possible to specify the minimum number of valid values in each group or in at least one group in the *Mode* field. This can be relevant if a protein is expressed in one group and absent in another group.

6. If you have intensity values you might use a base 10 logarithmic transformation, while base 2 logarithm is recommended for ratios and is used in this example.

7. By ticking off *Categoryenrichment in components* you can later explore the annotations the proteins are related to. Go to the *Categories* tab above the protein list and click one of the annotations. Sometimes you will see that the proteins that contribute most to the PCA plot are related to the same annotations, and this might have biological implications. Look for instance at the proteins related to "GTPaseactivation".

8. With Z-score normalization the mean of the protein row is subtracted from each protein and divided by the standard deviation of protein row. This should be performed at protein level before hierarchical clustering, as it allows visualizing the protein clusters differentially expressed between the samples analyzed, independently of average protein abundance. It is then important to note that, even though proteins in one cluster have the same expression profile, for instance lower in OCI-AML3 and MV4-11 than the three other cell lines, these proteins will not necessarily have the same expression intensity.

9. Even though Perseus was initially created for the analysis of MaxQuant results, its versatile input format also makes it compatible with most proteomic software, including OpenMS [12, 13], the TransProteomic Pipeline (TPP) [14], or PeptideShaker [15].

10. Although the methods presented here are generic and can be applied to most proteomics studies, it is important to critically tailor the post-processing workflow to the specificities of each experimental design.

11. Perseus is not limited to proteomic analyses, and can be operated on other types of omics datasets. It can also be used for multi-omic studies [16].

Acknowledgements

F.S.B. and F.S. acknowledge the support by the Norwegian Cancer Society. H.B. is supported by the Research Council of Norway.

References

1. Aebersold R, Mann M (2003) Mass spectrometry-based proteomics. Nature 422:198–207

2. Olsen JV, Mann M (2013) Status of large-scale analysis of post-translational modifications by mass spectrometry. Mol Cell Proteomics 12:3444–3452

3. Altelaar AF, Munoz J, Heck AJ (2013) Next-generation proteomics: towards an integrative view of proteome dynamics. Nat Rev Genet 14:35–48

4. Vaudel M, Sickmann A, Martens L (2010) Peptide and protein quantification: a map of the minefield. Proteomics 10:650–670

5. Bantscheff M, Lemeer S, Savitski MM et al (2012) Quantitative mass spectrometry in proteomics: critical review update from 2007 to the present. Anal Bioanal Chem 404:939–965

6. Bantscheff M, Schirle M, Sweetman G et al (2007) Quantitative mass spectrometry in proteomics: a critical review. Anal Bioanal Chem 389:1017–1031

7. Vaudel M, Sickmann A, Martens L (2014) Introduction to opportunities and pitfalls in functional mass spectrometry based proteomics. Biochim Biophys Acta 1844:12–20

8. Aasebo E, Vaudel M, Mjaavatten O et al (2014) Performance of super-SILAC based quantitative proteomics for comparison of different acute myeloid leukemia (AML) cell lines. Proteomics 14:1971–1976

9. Cox J, Mann M (2008) MaxQuant enables high peptide identification rates, individualized p.p.b.-range mass accuracies and proteome-wide protein quantification. Nat Biotechnol 26:1367–1372

10. Vizcaino JA, Deutsch EW, Wang R et al (2014) ProteomeXchange provides globally coordinated proteomics data submission and dissemination. Nat Biotechnol 32:223–226

11. Martens L, Hermjakob H, Jones P et al (2005) PRIDE: the proteomics identifications database. Proteomics 5:3537–3545

12. Kohlbacher O, Reinert K, Gropl C et al (2007) TOPP—the OpenMS proteomics pipeline. Bioinformatics 23:e191–e197

13. Bertsch A, Gropl C, Reinert K et al (2011) OpenMS and TOPP: open source software for LC-MS data analysis. Methods Mol Biol 696:353–367

14. Deutsch EW, Mendoza L, Shteynberg D et al (2010) A guided tour of the Trans-Proteomic Pipeline. Proteomics 10:1150–1159

15. Vaudel M, Burkhart JM, Zahedi RP et al (2015) PeptideShaker enables reanalysis of MS-derived proteomics data sets. Nature biotechnology 33:22–24

16. Cox J, Mann M (2012) 1D and 2D annotation enrichment: a statistical method integrating quantitative proteomics with complementary high-throughput data. BMC Bioinformatics 13(Suppl 16):S12

Chapter 20

A Simple Workflow for Large Scale Shotgun Glycoproteomics

Astrid Guldbrandsen, Harald Barsnes, Ann Cathrine Kroksveen, Frode S. Berven, and Marc Vaudel

Abstract

Targeting subproteomes is a good strategy to decrease the complexity of a sample, for example in body fluid biomarker studies. Glycoproteins are proteins with carbohydrates of varying size and structure attached to the polypeptide chain, and it has been shown that glycosylation plays essential roles in several vital cellular processes, making glycosylation a particularly interesting field of study. Here, we describe a method for the enrichment of glycosylated peptides from trypsin digested proteins in human cerebrospinal fluid. We also describe how to perform the data analysis on the mass spectrometry data for such samples, focusing on site-specific identification of glycosylation sites, using user friendly open source software.

Key words Glycoproteomics, Enrichment, Data interpretation

1 Introduction

Enrichment for subproteomes can help circumvent the challenge of a few high abundant proteins masking proteins of lower abundance in biological samples and body fluids [1]. An example of such a subproteome are the glycoproteins, proteins carrying one or more carbohydrates (glycans) of varying size and structure at particular amino acid residues in the protein sequence [2, 3]. When a glycan is attached to a protein it is referred to as a glycosylation—one of the most common post-translational modifications. Glycoproteins are most often secreted or membrane-attached [2], and are known to be involved in protein folding and protection from degradation [3–6]. They also play important roles in cell communication, signaling, aging and cell adhesion [7–9]. Several known clinical biomarkers, as well as therapeutic targets, are glycoproteins [10–17].

In this chapter, we present a simple protocol for glycopeptide enrichment, and subsequent protein identification after shotgun proteomic analysis. Note that the liquid-chromatography coupled to mass spectrometry (LC-MS) acquisition is not detailed as it does not differ from standard shotgun proteomics [18]. The glycopeptide enrichment described here is based on solid-phase extraction of N-linked glycopeptides, as described by Tian et al. [19] and Berven et al. [20], with minor modifications as described in detail in [18]. In the last steps of the protocol, the glycans are released from the glycopeptides by the enzyme PNGase F, rendering the identification of the glycans' chemical composition and structure not possible. The glycan release by PNGase F leads to a deamidation of the asparagine residue where the glycan was attached, converting it to an aspartic acid residue, resulting in a 1 Da mass increase when an amide group is exchanged for a hydroxyl group [21]. This mass shift allows for glycosylation site identification in the mass spectrometry data. However, in order to distinguish glycosylation sites from natural deamidation, it is necessary to (manually or automatically) validate the detected glycosylation sites, e.g., specify that the deamidated asparagine must satisfy the following glycosylation pattern [N X^P S/T] (where X^P is any amino acid except proline), see [18] Supplementary File 1C. For the experiment described here, the peptides were manually inspected and filtered as part of the post-processing. It should be noted that in this approach for identifying glycosylation sites, a deamidation of an asparagine residue in the glycosylation pattern is only an indicator of a likely former glycosylation site, i.e., there is no direct detection of the glycosylation.

The data interpretation protocol is here demonstrated using a dataset of human cerebrospinal fluid (CSF) obtained from the Cerebrospinal Fluid Proteome Resource (CSF-PR) [18] (www.probe.uib.no/csf-pr), and is conducted using SearchGUI [22] (http://searchgui.googlecode.com) and PeptideShaker [23] (http://peptide-shaker.googlecode.com), both freely available from their respective Web pages.

2   Materials

If not otherwise stated, all chemicals and products are purchased from Sigma-Aldrich (St. Louis, MO, USA). All solutions should be prepared with deionized water.

1. 0.1 % N-octyl-β-D-glucopyranoside (NOG).

2. 3 kDa ultracentrifugation filters (Amicon Ultra-4, Merck Millipore, Billerica, MA, USA).

3. Denaturation buffer: 8 M urea/0.4 M ammonium bicarbonate/0.1 % SDS (Bio-Rad Laboratories, Hercules, CA, USA).

4. 120 mM Tris(2-carboxyethyl)phosphine (TCEP).

5. 160 mM iodoacetamide (IAA), light sensitive.

6. 100 mM ammonium bicarbonate (Ambic).

7. Trypsin porcine (Promega).

8. Oasis™ HLB 10 mg (30 μm) plates (Waters, Milford, MA, USA).

9. Oasis™ HLB μElution plates (Waters, Milford, MA, USA).

10. 100 % formic acid (FA).

11. 0.1 % FA.

12. 0.1 % trifluoroacetic acid (TFA).

13. 50 % acetonitrile (ACN)/0.1 % TFA.

14. 80 % ACN–0.1 % TFA.

15. 80 % ACN–0.1 % FA.

16. 0.1 M sodium periodate, light sensitive.

17. BcMag® hydrazide-modified magnetic beads, 30 mg/mL (BioClone Inc. San Diego, CA, USA).

18. Dynal® magnetic bead separation rack (Life Technologies).

19. 100 % N,N-dimethylformamide (DMF, toxic).

20. PNGase F enzyme for proteomics 1 unit/μL.

21. 5 M hydrochloric acid (HCl).

22. Dataset: The dataset used here for illustrative purposes is freely available from the ProteomeXchange Consortium [24] via the PRIDE partner repository [25], with the identifiers PXD000651-PXD000657. The dataset can also be inspected (and proteins and peptides exported) via CSF-PR at http://probe.uib.no/csf-pr.

23. Software: SearchGUI [22] is an open source user friendly interface for simple use of multiple search engines (http://searchgui.googlecode.com).

24. Software: PeptideShaker [23] is an open source user friendly interface for the interpretation of results from multiple search engines (http://peptide-shaker.googlecode.com).

3 Methods

3.1 Glycopeptide Enrichment

Protocol modified from [19] and [20].

1. Purify and concentrate the CSF sample using 3 kDa ultracentrifugation filters, pre-cleansed with 1 mL NOG. Add CSF sample + 1 mL deionized water (MQ) and spin at 3000 ×g for 45 min at 4 °C. Add 1 mL MQ and spin for approximately 1 h, or until there is between 50 and 100 μL left in the filter. Transfer to Eppendorf tube and concentrate to ≈15 μL.

2. Add 135 µL denaturation buffer (for 100–1000 µg protein) and vortex.

3. Add TCEP to a final concentration of 10 mM. Incubate with shaking for 1 h at 37 °C.

4. Add IAA to a final concentration of 12 mM. Incubate with shaking for 30 min in the dark at 20 °C.

5. Add 1 mL 100 mM Ambic to get the urea concentration below 1 M.

6. Add trypsin in a 1:50 trypsin to protein ratio. Incubate with gentle shaking for 12–16 h at 37 °C.

7. Acidify sample by adding approximately 12 µL 100 % FA, drop by drop. Keep the lid open to avoid pressure building up inside the tube.

8. Perform cleanup at 4 °C using Oasis HLB 10 mg (30 µm) plates. Condition column with 1 mL 50 % ACN–0.1 % TFA, wash×2 with 1 mL 0.1 % TFA, all at 200 ×*g* for 1 min, before addition of sample and spinning at 150 ×*g* for 3 min. Wash again using 1 mL 0.1 % TFA×3 and elute with 200 µL 50 % ACN–0.1 % TFA×2, all at 200 ×*g* for 1 min.

9. Transfer the sample to a new Eppendorf tube and concentrate to dryness.

10. Reconstitute the sample in 400 µL 0.1 % TFA and vortex.

11. Add sodium periodate to a final concentration of 10 mM. Incubate with shaking for 1 h at 20 °C in the dark.

12. Repeat **step 8**, but use 80 % ACN–0.1 % TFA for conditioning and elution. Leave the sample in the Oasis collection plate.

13. Vortex the hydrazide modified magnetic beads and pipet 133 µL (4 mg) to a new tube. Wash with 1 mL 80 % ACN–0.1 % TFA for 5 min with extensive shaking (1200 rpm should be used for all bead incubations, *see* **Notes 1** and **2**).

14. Add the peptides from the Oasis well to the washed beads and incubate overnight at 20 °C.

15. Spin the sample and save supernatant (contains the unbound peptides) for future analysis.

16. Wash the beads for 5 min with 1 mL of the following:

    (a) 80 % ACN–0.1 % TFA×3

    (b) Denaturation buffer×3

    (c) 100 % DMF×3 (*see* **Note 3**)

    (d) 100 mM Ambic×3

17. Add 100 µL 100 mM Ambic to the beads.

18. Add 1.5 µL PNGase F enzyme. Incubate overnight at 37 °C.

19. Collect supernatant containing the released deglycosylated peptides into new Eppendorf tubes.

20. Add 200 μL 100 mM Ambic to the beads and wash for 5 min at 20 °C.

21. Collect the supernatant and pool with the peptides collected in **step 19**.

22. Acidify the sample by adding 7 μL 5 M HCl, drop by drop. Keep lid open to avoid pressure building up inside the tube.

23. Add 200 μL 0.1 % FA and vortex.

24. Perform clean-up at 4 °C as is described in **steps 8** and **12**, but this time use Oasis μElution plates (*see* **Note 4**) and 500 μL 80 % ACN–0.1 % FA for conditioning and elution and 500 μL 0.1 % FA for washing, and place the sample tube in the magnetic rack before transfer to the Oasis plate to remove any remaining beads. Elute with 150 μL ×2.

25. Transfer sample (300 μL) to a new Eppendorf tube, concentrate to dryness and freeze at –80 °C until LC-MS analysis.

26. Dissolve in appropriate solvent and volume for LC-MS analysis (*see* **Note 5**).

3.2 Data Interpretation

The data interpretation consists of two main parts: (1) match the spectra to a database, and (2) interpret the matches to infer proteins and glycosylation sites. For the first part, so-called search engines are used, retrieving a list of Peptide Spectrum Matches (PSMs). In the second, the PSMs are assembled into inferred peptides and proteins, the quality of the identification results is evaluated in order to limit the prevalence of false positive hits, and post-translational modification (PTM) sites are inferred. More details on the identification process can be found in the following reviews [26–28].

In this chapter, the above task is demonstrated using user friendly open source software, SearchGUI (version 1.20.8) and PeptideShaker (version 0.33.6). These tools notably present the advantage to support multiple search engines for PTM studies in a user friendly environment. For more details on how to operate these tools, please refer to the respective free tutorials [29] (http://compomics.com/bioinformatics-for-proteomics). Note that the concepts introduced here can be transposed to most proteomics applications, like OpenMS [30, 31], the TransProteomic Pipeline (TPP) [32], or MaxQuant [33].

3.3 Database Search

1. After starting SearchGUI, Click *Add* and select the spectrum files to search in the *Input & Output* panel at the top of the dialog shown in Fig. 1. Here, spectrum files consist of peak lists of the original MS2 spectra in the mgf format (http://www.matrixscience.com/help/data_file_help.html#GEN). To convert raw data to the mgf format it is recommended to use ProteoWizard [34]. In this experiment, fractionation has been performed, so there is a total of 20 files to be searched.

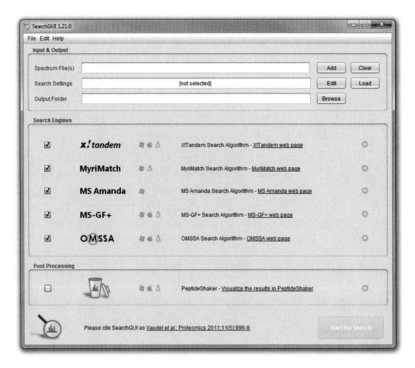

Fig. 1 SearchGUI main dialog. The main dialog appearing when the tool is started. Spectrum files, search settings, and output folder have to be loaded and specified. The search engines shown are all automatically selected, but can be unchecked. The search can be automatically processed in PeptideShaker if this option is selected

2. *Edit* the *Search Settings* or *Load* an already saved search settings file. As displayed in Fig. 2, the following input is required: (a) choose a *Database* (*FASTA*) file; here, the human complement of the UniProtKB/Swiss-Prot database [35] available from the UniProt website (www.uniprot.org). SearchGUI then proposes to add decoy sequences, press *yes*. (b) Add fixed and variable PTMs by selecting the relevant modifications and pressing '≪' to add to the appropriate list; here we use *carbamidomethyl c* as fixed, and *oxidation of m* and *deamidation of n* as variable modifications. (c) Set the *Protease & Fragmentation* settings; *Enzyme: Trypsin, Precursor Mass Tolerance: 10 ppm, Fragment Mass Tolerance: 0.7 Da* and *Max Missed Cleavages* (*by trypsin*): *2*, and for the rest keep the defaults. For more information on how to set the search parameters, please *see* [36]. Save the settings file for future reuse in other searches.

3. Select the *Output Folder* where the search output will be stored by pressing *Browse* and navigating to the desired folder.

4. (a) It is possible to directly open the project in PeptideShaker after the search. To do this, check the box under *Post Processing*. A window will appear allowing the setting of the PeptideShaker parameters. Under *Project Details* give a *Project Name*, a

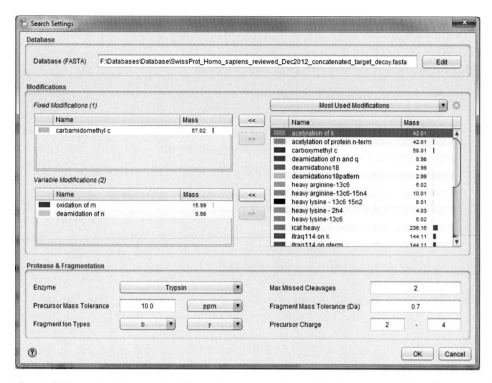

Fig. 2 SearchGUI search settings dialog. This dialog appears when editing the search settings, the parameters used in this experiment are selected. The database (FASTA file) is the human complement of the UniProtKB/Swiss-Prot database available from the UniProt website (www.uniprot.org)

Sample Name, and select the Ensembl [37] species by pressing *Edit*; in this case *Vertebrates* as species type and *Human* (*Homo Sapiens*) [*Ensembl 76*] as species, where 76 is the version of Ensembl. Then select the folder to save the output of the search by pressing *Browse* and browsing to the desired folder.

or

(b) It is also possible to first do the search and then later load the results in PeptideShaker. Then, leave the box under *Post Processing* unchecked.

5. Press *Start the Search*!.

6. While the search is running, a dialog shows updates on the progress of the search. When finished, the search output is written to the output folder in the form of a zipped file containing the result file of every search engine, and which can be loaded in PeptideShaker (*see* **Note 6**).

3.4 Glycosylation Site Identification

Steps 1–4 should be skipped if automatic post-processing in PeptideShaker was selected in the previous section.

1. If the data was not directly processed in PeptideShaker, start PeptideShaker, press *New Project* and give a name for *Project Reference*, *Sample Name*, and edit species (*see* previous section).

2. Browse to find and select the correct identification file(s), i.e., SearchGUI output file(s). If not automatically selected when identification files are loaded, also select spectrum file(s) (mgf) and database file (FASTA).

3. Edit *Search Settings* and *Import Filters* if not automatically loaded and leave *Preferences* to default.

4. Press *Load Data*! and a dialog appears showing updates on the progress of the project creation.

5. Upon completion, the main display of PeptideShaker allows the browsing of the proteomics dataset, as displayed in Fig. 3. Note that glycosylation sites are highlighted in the peptide and protein sequences using the PTM color coding set when editing the search parameters. An enlarged *Spectrum & Fragment Ions* window for a selected glycopeptide in the dataset is displayed in Fig. 4.

3.5 Data Export   Project features and result reports containing the possible glycosylation sites (at the protein, peptide, and PSMs level) can be exported by pressing *Export→ Identification Features*. In the *Export Features* window displayed in Fig. 5 the type of report to export can be chosen and custom reports created (*see* **Notes** 7 and **8**).

4 Notes

1. All following washes/incubations with beads must be done with extensive shaking (1200 rpm) to avoid beads depositing at the bottom of the tube. After incubation, spin down the tube briefly to collect beads that might have been stuck in the lid during shaking. Then use the magnetic rack for bead–supernatant separation. Wait for all beads to gather and get attached to the back of the tube where the magnet is before pipetting the supernatant gently without inducing stress on the beads.

2. If using non-magnetic Macroporous beads, pipet 50 μL, wash with deionized water and spin at $13,000 \times g$.

3. DMF is toxic and teratogenic (dangerous for the developing embryo/fetus), so it should be handled with care and suitable protection, and not by pregnant women.

4. Oasis plates with lower binding capacity (μElution) are used because the amount of peptides is substantially lower after glycopeptide enrichment.

5. It is suggested to reconstitute in 10 μL 3 % ACN–5 % FA, and to inject 5 μL for 250 μg starting material and 2 μL for 1 mg starting material. However, injection volume depends on the sensitivity of the instrument.

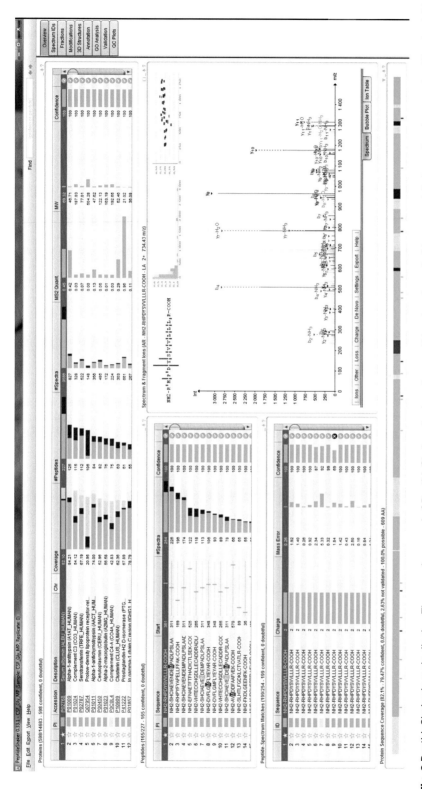

Fig. 3 PeptideShaker's main display. The overview panel of PeptideShaker displays the result in a top down fashion upon data import and processing. The *top table* lists all identified proteins and their details. Below, the *upper left table* lists all peptide matches identified for the selected protein, and under it a table lists the peptide-spectrum matches (PSMs) for the selected peptide. The spectrum annotated with fragment ions deduced from the selected PSM is displayed to the *bottom right*. At the *bottom* of the display, the protein sequence is displayed with the identified peptides color coded. Note that peptides can be navigated by clicking the sequence. Deamidated asparagines (glycosylation sites) are highlighted in the peptide and protein sequences using the color coding from the search parameters. More details on the project are available via the other tabs in the *upper right corner*

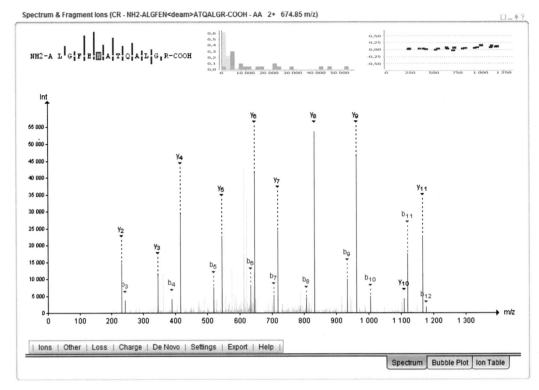

Fig. 4 PeptideShaker's *Spectrum & Fragment Ions* display. The *Spectrum & Fragment Ions* panel displays information on the annotation of the spectrum based on the selected peptide. At the *top left*, the intensity of the fragment ions annotated on the spectrum is illustrated with bars at the possible fragmentation site on the peptide sequence, where the modification is color coded. The intensities of b and y ions are in *blue* and *red*, respectively. Here, the deamidation of the asparagine on the ENAT motif is displayed in *brown* and clearly flanked with fragment ions. At the *top center*, a histogram shows the respective shares of annotated and not annotated peaks, in *green* and *grey*, respectively, using intensity bins. At the *top right*, the *m/z* deviation of every fragment ion is plotted against the fragment ion *m/z*. Below, the spectrum is displayed with the annotated peaks in *red*. It is possible to customize the annotation of the spectrum using the menu at the *bottom*

6. If following Subheading 3.3, **step 4b**, go to the folder where the zipped SearchGUI output file is stored and select the identification files in PeptideShaker.

7. Further (manual or automatic) validation is required to confirm if the detected sites are real glycosylation sites, *see* [18] Supplementary File 1C.

8. It is crucial to remember that this approach identifies deamidated asparagine residues in the glycosylation patterns. While these are highly likely to be former glycosylation sites, there is no direct measurement of the glycosylation event, and no information on the structure of the glycan.

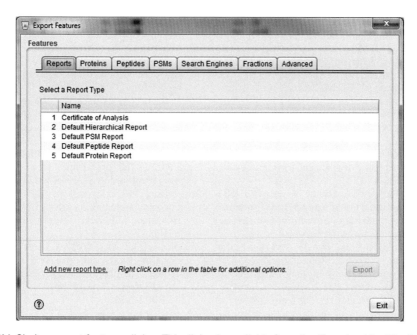

Fig. 5 PeptideShaker export features dialog. This dialog is available from the *Export→ Identification Features*. Under the panel *Reports*, several default reports can be selected and exported, while custom made reports can be created and edited by right-clicking the table

Acknowledgements

A.C.K. is supported by the Kristian Gerhard Jebsen Foundation. H.B. is supported by the Research Council of Norway.

References

1. Gevaert K, Van Damme P, Ghesquiere B et al (2007) A la carte proteomics with an emphasis on gel-free techniques. Proteomics 7:2698–2718
2. Zhang H, Li XJ, Martin DB et al (2003) Identification and quantification of N-linked glycoproteins using hydrazide chemistry, stable isotope labeling and mass spectrometry. Nat Biotechnol 21:660–666
3. Gamblin DP, Scanlan EM, Davis BG (2009) Glycoprotein synthesis: an update. Chem Rev 109:131–163
4. Shental-Bechor D, Levy Y (2008) Effect of glycosylation on protein folding: a close look at thermodynamic stabilization. Proc Natl Acad Sci U S A 105:8256–8261
5. Sola RJ, Rodriguez-Martinez JA, Griebenow K (2007) Modulation of protein biophysical properties by chemical glycosylation: biochem-ical insights and biomedical implications. Cell Mol Life Sci 64:2133–2152
6. Varki A (1993) Biological roles of oligosaccharides: all of the theories are correct. Glycobiology 3:97–130
7. Roth J (2002) Protein N-glycosylation along the secretory pathway: relationship to organelle topography and function, protein quality control, and cell interactions. Chem Rev 102:285–303
8. Sato Y, Endo T (2010) Alteration of brain glycoproteins during aging. Geriatr Gerontol Int 10 Suppl 1:S32–S40
9. Ruggeri ZM, Mendolicchio GL (2007) Adhesion mechanisms in platelet function. Circ Res 100:1673–1685
10. Berger MS, Locher GW, Saurer S et al (1988) Correlation of c-erbB-2 gene amplification and

protein expression in human breast carcinoma with nodal status and nuclear grading. Cancer Res 48:1238–1243

11. Hudziak RM, Schlessinger J, Ullrich A (1987) Increased expression of the putative growth factor receptor p185HER2 causes transformation and tumorigenesis of NIH 3T3 cells. Proc Natl Acad Sci U S A 84:7159–7163

12. Vogelzang NJ, Lange PH, Goldman A et al (1982) Acute changes of alpha-fetoprotein and human chorionic gonadotropin during induction chemotherapy of germ cell tumors. Cancer Res 42:4855–4861

13. Bosl GJ, Lange PH, Fraley EE et al (1981) Human chorionic gonadotropin and alphafetoprotein in the staging of nonseminomatous testicular cancer. Cancer 47:328–332

14. Thompson DK, Haddow JE (1979) Serial monitoring of serum alpha-fetoprotein and chorionic gonadotropin in males with germ cell tumors. Cancer 43:1820–1829

15. Catalona WJ, Richie JP, Ahmann FR et al (1994) Comparison of digital rectal examination and serum prostate specific antigen in the early detection of prostate cancer: results of a multicenter clinical trial of 6,630 men. J Urol 151:1283–1290

16. Canney PA, Moore M, Wilkinson PM et al (1984) Ovarian cancer antigen CA125: a prospective clinical assessment of its role as a tumour marker. Br J Cancer 50:765–769

17. Topol EJ, Byzova TV, Plow EF (1999) Platelet GPIIb-IIIa blockers. Lancet 353:227–231

18. Guldbrandsen A, Vethe H, Farag Y et al (2014) In-depth characterization of the cerebrospinal fluid proteome displayed through the CSF Proteome Resource (CSF-PR). Mol Cell Proteomics 13(11):3152–63

19. Tian Y, Zhou Y, Elliott S et al (2007) Solid-phase extraction of N-linked glycopeptides. Nat Protoc 2:334–339

20. Berven FS, Ahmad R, Clauser KR et al (2010) Optimizing performance of glycopeptide capture for plasma proteomics. J Proteome Res 9:1706–1715

21. Gonzalez J, Takao T, Hori H et al (1992) A method for determination of N-glycosylation sites in glycoproteins by collision-induced dissociation analysis in fast atom bombardment mass spectrometry: identification of the positions of carbohydrate-linked asparagine in recombinant alpha-amylase by treatment with peptide-N-glycosidase F in 18O-labeled water. Anal Biochem 205:151–158

22. Vaudel M, Barsnes H, Berven FS et al (2011) SearchGUI: an open-source graphical user interface for simultaneous OMSSA and X!Tandem searches. Proteomics 11:996–999

23. Vaudel M, Burkhart JM, Zahedi RP et al (2015) PeptideShaker enables reanalysis of MS-derived proteomics data sets. Nature biotechnology 33:22–24

24. Vizcaino JA, Deutsch EW, Wang R et al (2014) ProteomeXchange provides globally coordinated proteomics data submission and dissemination. Nat Biotechnol 32:223–226

25. Martens L, Hermjakob H, Jones P et al (2005) PRIDE: the proteomics identifications database. Proteomics 5:3537–3545

26. Vaudel M, Sickmann A, Martens L (2012) Current methods for global proteome identification. Expert Rev Proteomics 9: 519–532

27. Nesvizhskii AI (2010) A survey of computational methods and error rate estimation procedures for peptide and protein identification in shotgun proteomics. J Proteomics 73:2092–2123

28. Chalkley RJ, Clauser KR (2012) Modification site localization scoring: strategies and performance. Mol Cell Proteomics 11:3–14

29. Vaudel M, Venne AS, Berven FS et al (2014) Shedding light on black boxes in protein identification. Proteomics 14:1001–1005

30. Kohlbacher O, Reinert K, Gropl C et al (2007) TOPP—the OpenMS proteomics pipeline. Bioinformatics 23:e191–e197

31. Bertsch A, Gropl C, Reinert K et al (2011) OpenMS and TOPP: open source software for LC-MS data analysis. Methods Mol Biol 696: 353–367

32. Deutsch EW, Mendoza L, Shteynberg D et al (2010) A guided tour of the Trans-Proteomic Pipeline. Proteomics 10:1150–1159

33. Cox J, Mann M (2008) MaxQuant enables high peptide identification rates, individualized p.p.b.-range mass accuracies and proteome-wide protein quantification. Nat Biotechnol 26:1367–1372

34. Kessner D, Chambers M, Burke R et al (2008) ProteoWizard: open source software for rapid proteomics tools development. Bioinformatics 24:2534–2536

35. Apweiler R, Bairoch A, Wu CH et al (2004) UniProt: the Universal Protein knowledgebase. Nucleic Acids Res 32:D115–D119

36. Vaudel M, Burkhart JM, Sickmann A et al (2011) Peptide identification quality control. Proteomics 11:2105–2114

37. Flicek P, Amode MR, Barrell D et al (2014) Ensembl 2014. Nucleic Acids Res 42: D749–D755

Chapter 21

Systemic Analysis of Regulated Functional Networks

Luis Francisco Hernández Sánchez, Elise Aasebø, Frode Selheim,
Frode S. Berven, Helge Ræder, Harald Barsnes, and Marc Vaudel

Abstract

In biological and medical sciences, high throughput analytical methods are now commonly used to investigate samples of different conditions, e.g., patients versus controls. Systemic functional analyses emerged as a reference method to go beyond a list of regulated compounds, and identify activated or inactivated biological functions. This approach holds the promise for a better understanding of biological systems, of the mechanisms involved in disease progression, and thus improved diagnosis, prognosis, and treatment. In this chapter, we present a simple workflow to conduct pathway analyses on biological data using the freely available Reactome platform (http://www.reactome.org).

Key words Pathway analysis, Data interpretation, Functional proteomics

1 Introduction

In systems biology, thousands of compounds—metabolites, genes, transcripts, proteins, etc.—are studied in a global approach to investigate the biological processes differentially triggered between samples [1, 2]. Identified and quantified compounds are mapped against databases of known interactions and functions, and differentially expressed pathways extracted, providing the user with insight on the significantly regulated biological processes [3, 4]. Two factors are crucial for the success of this procedure [5, 6]: (1) the quality of the functional knowledgebase used as reference for the study, and (2) the accuracy of the matching of experimental data with the knowledge bases. Notably, functional knowledge is not as strongly established as other resources for omics studies, like gene or protein databases, and is hence more subject to changes and updates [5].

Several knowledge bases exist, along with tools to query them, allowing the functional interpretation of large scale biological studies [7]. For example, the commercial resource QIAGEN's Ingenuity® Pathway Analysis (IPA®, QIAGEN Redwood City,

Jörg Reinders (ed.), *Proteomics in Systems Biology: Methods and Protocols*, Methods in Molecular Biology, vol. 1394,
DOI 10.1007/978-1-4939-3341-9_21, © Springer Science+Business Media New York 2016

www.qiagen.com/ingenuity) allows the query of large omics datasets and pathway analyses in a user friendly environment. However, the dynamic nature of the functional knowledge resources, always subject to evolution, makes it crucial to document the changes which can affect the analysis. Since this task is impossible in the case of private databases, open resources appear as an alternative of choice [5].

Among freely available academic resources, the most encountered are the Kyoto Encyclopedia of Genes and Genomes [8, 9] (KEGG, http://www.genome.jp/kegg), WikiPathways [10] (http://www.wikipathways.org), and Reactome [11] (http://www.reactome.org). These resources allow the browsing of pathways and analysis of data from their respective websites. KEGG can also be queried via the metabolic and physiological potential evaluator (MAPLE) [12] and Reactome using a plugin in Cytoscape [13]. Here, we present the use of Reactome for functional analyses. First, browsing pathways on the website are illustrated taking the JAK/STAT pathway [14] as an example. Second, the analysis of quantitative data is demonstrated with the processing of a quantitative proteomic dataset of different acute myeloid leukemia (AML) derived cell lines [15].

2   Material

1. The dataset here used for illustrative purposes is freely available through the ProteomeXchange [16] consortium via the PRIDE [17] partner repository under the accession number PXD000441. Here, the relative quantification of proteins from the cell line Molm-13 to an internal standard of five AML cell lines is used. Details can be found in Supplementary Table S2 of [15].

2. Reactome can be used directly from the Reactome website (http://www.reactome.org) or via the dedicated Cytoscape [13] plugin. Here, the use of the online version is demonstrated.

3   Methods

This section describes how to browse for a specific pathway and then navigate it, using the JAK/STAT pathway as an example.

1. Go to the homepage of Reactome (http://www.reactome.org) (Fig. 1).

2. In the *search* field type in the keywords related to the pathway you are looking for, here: "JAK STAT". The results of the search are then displayed (Fig. 2).

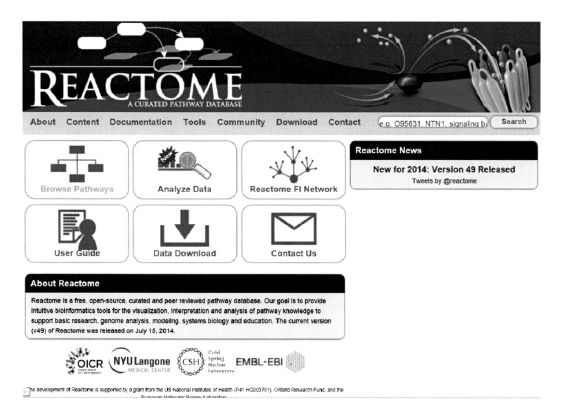

Fig. 1 Reactome homepage

3. In the results screen select the option corresponding to the pathway of interest for more details. Select: *Phosphorylation of STAT1 by JAK kinases* (*Homo sapiens*) (Fig. 3).

4. Click the "+" symbol to show the hierarchy of pathways corresponding to this function (Fig. 4). Note that you can select the pathway or reaction of interest in the Pathway Browser.

5. Go back to the Reactome homepage. Then go to the Pathway Browser by clicking the *Browse Pathways* button (Fig. 5). As an alternative, you can select the *Pathway Browser* option in the *Tools* menu (Fig. 6). The Pathway Browser will appear as displayed (Fig. 7).

6. The Hierarchy Panel on the left shows the available Reactome pathway topics sorted alphabetically. Find the topic associated with this pathway, i.e., *Immune System* (*Homo sapiens*), and click it to show the pathway diagram.

7. You can go deeper in the events hierarchy to find the sub-pathways related to JAK/STAT by clicking the "+" symbol

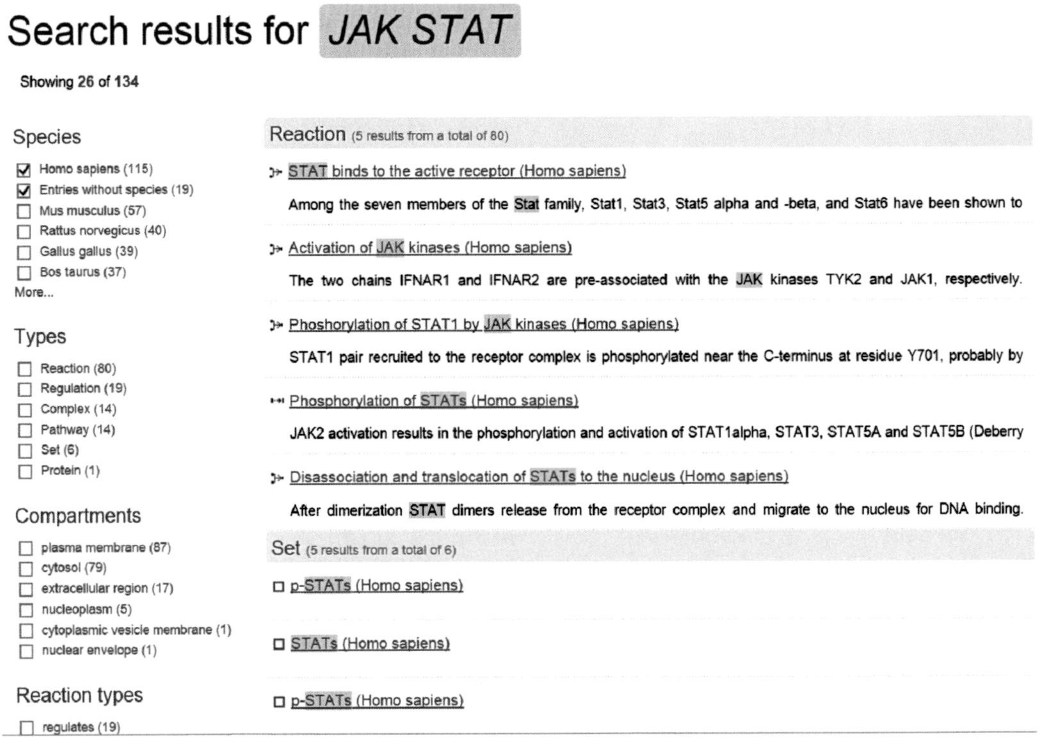

Search results for *JAK STAT*

Showing 26 of 134

Species

☑ Homo sapiens (115)
☑ Entries without species (19)
☐ Mus musculus (57)
☐ Rattus norvegicus (40)
☐ Gallus gallus (39)
☐ Bos taurus (37)
More...

Types

☐ Reaction (80)
☐ Regulation (19)
☐ Complex (14)
☐ Pathway (14)
☐ Set (6)
☐ Protein (1)

Compartments

☐ plasma membrane (87)
☐ cytosol (79)
☐ extracellular region (17)
☐ nucleoplasm (5)
☐ cytoplasmic vesicle membrane (1)
☐ nuclear envelope (1)

Reaction types

☐ regulates (19)

Reaction (5 results from a total of 80)

⊱ STAT binds to the active receptor (Homo sapiens)

   Among the seven members of the Stat family, Stat1, Stat3, Stat5 alpha and -beta, and Stat6 have been shown to

⊱ Activation of JAK kinases (Homo sapiens)

   The two chains IFNAR1 and IFNAR2 are pre-associated with the JAK kinases TYK2 and JAK1, respectively.

⊱ Phoshorylation of STAT1 by JAK kinases (Homo sapiens)

   STAT1 pair recruited to the receptor complex is phosphorylated near the C-terminus at residue Y701, probably by

⊷ Phosphorylation of STATs (Homo sapiens)

   JAK2 activation results in the phosphorylation and activation of STAT1alpha, STAT3, STAT5A and STAT5B (Deberry

⊱ Disassociation and translocation of STATs to the nucleus (Homo sapiens)

   After dimerization STAT dimers release from the receptor complex and migrate to the nucleus for DNA binding.

Set (5 results from a total of 6)

☐ p-STATs (Homo sapiens)

☐ STATs (Homo sapiens)

☐ p-STATs (Homo sapiens)

Fig. 2 Reactome search results of *JAK STAT*

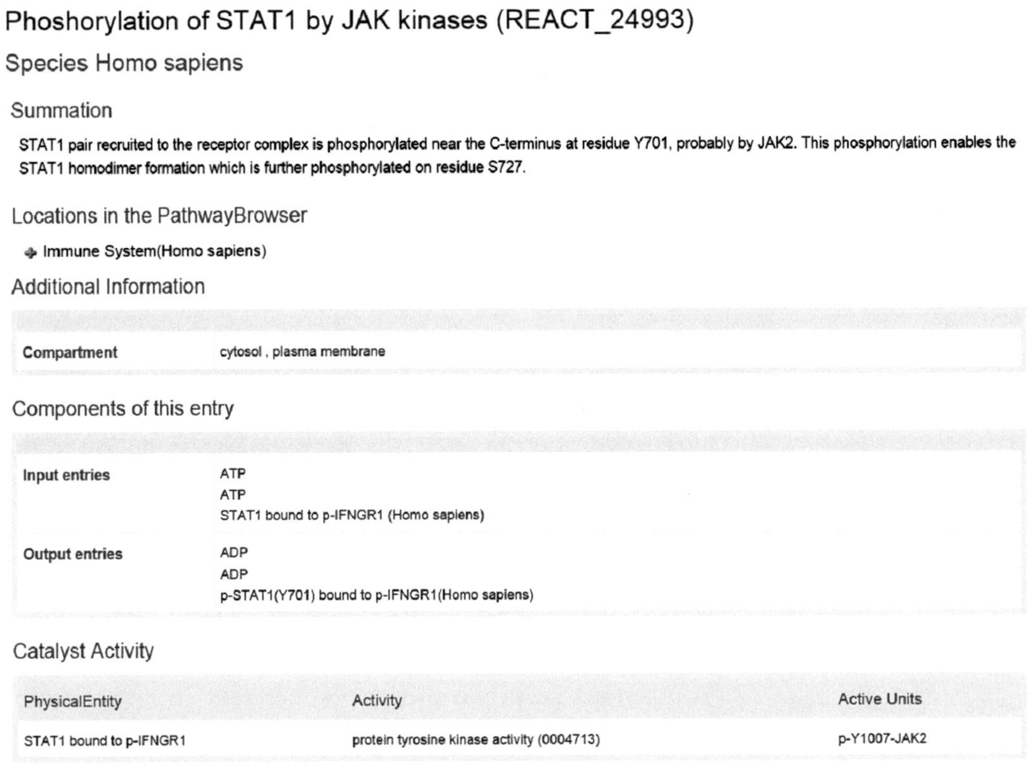

Phoshorylation of STAT1 by JAK kinases (REACT_24993)

Species Homo sapiens

Summation

STAT1 pair recruited to the receptor complex is phosphorylated near the C-terminus at residue Y701, probably by JAK2. This phosphorylation enables the STAT1 homodimer formation which is further phosphorylated on residue S727.

Locations in the PathwayBrowser

⊕ Immune System(Homo sapiens)

Additional Information

Compartment	cytosol , plasma membrane

Components of this entry

Input entries	ATP
	ATP
	STAT1 bound to p-IFNGR1 (Homo sapiens)
Output entries	ADP
	ADP
	p-STAT1(Y701) bound to p-IFNGR1(Homo sapiens)

Catalyst Activity

PhysicalEntity	Activity	Active Units
STAT1 bound to p-IFNGR1	protein tyrosine kinase activity (0004713)	p-Y1007-JAK2

Fig. 3 Details of pathway *Phosphorylation of STAT1 by JAK kinases* (*Homo sapiens*)

Phoshorylation of STAT1 by JAK kinases (REACT_24993)

Species Homo sapiens

Summation

STAT1 pair recruited to the receptor complex is phosphorylated near the C-terminus at residue Y701, probably by JAK2. This phosphorylation enables the STAT1 homodimer formation which is further phosphorylated on residue S727.

Locations in the PathwayBrowser

- Immune System(Homo sapiens)
  - Immune System(Homo sapiens)
    - Cytokine Signaling in Immune system(Homo sapiens)
      - Interferon Signaling(Homo sapiens)
        - Interferon gamma signaling(Homo sapiens)
          - Phoshorylation of STAT1 by JAK kinases(Homo sapiens)

Additional Information

Compartment	cytosol , plasma membrane

Components of this entry

Input entries	ATP
	ATP
	STAT1 bound to p-IFNGR1 (Homo sapiens)
Output entries	ADP
	ADP
	p-STAT1(Y701) bound to p-IFNGR1(Homo sapiens)

Catalyst Activity

Fig. 4 Events hierarchy for the pathway *Phosphorylation of STAT1 by JAK kinases* (*Homo sapiens*)

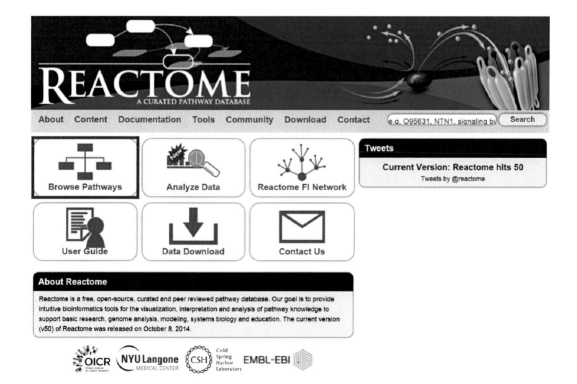

Fig. 5 Button on the homepage to go to the Pathway Browser of Reactome

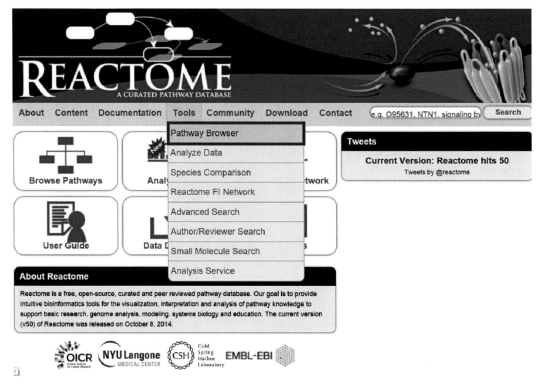

Fig. 6 The option on the *Tools* menu at the homepage to go to the Pathway Browser of Reactome

to the left of each pathway or sub-pathway. Here select *Cytokine Signaling in Immune system* to show the sub-pathway diagram (Fig. 8).

8. It is also possible to explore a sub-pathway by double clicking the appropriate diagram box, double click the *Interferon Signaling* box and then *Interferon gamma signaling*. The pathway diagram is now updated to show the selected sub-pathway and the hierarchy panel highlights the name of the pathway with its currently selected sub-pathways.

9. Select the desired pathway, *Phosphorylation of STAT1 by JAK kinases*, the pathway diagram focuses on that reaction highlighting in blue the components associated with that function (Fig. 9). Also note that the details panel at the bottom of the screen shows information about the selected pathway or reaction.

10. Click the diagram objects to display more information about specific elements of the reaction in the details panel.

11. Zoom in or out using the "+" and "−" symbols in the upper left corner of the diagram panel.

12. At any time, you can click and drag the pathway diagram to move to other areas, or use the small arrow buttons at the top left corner.

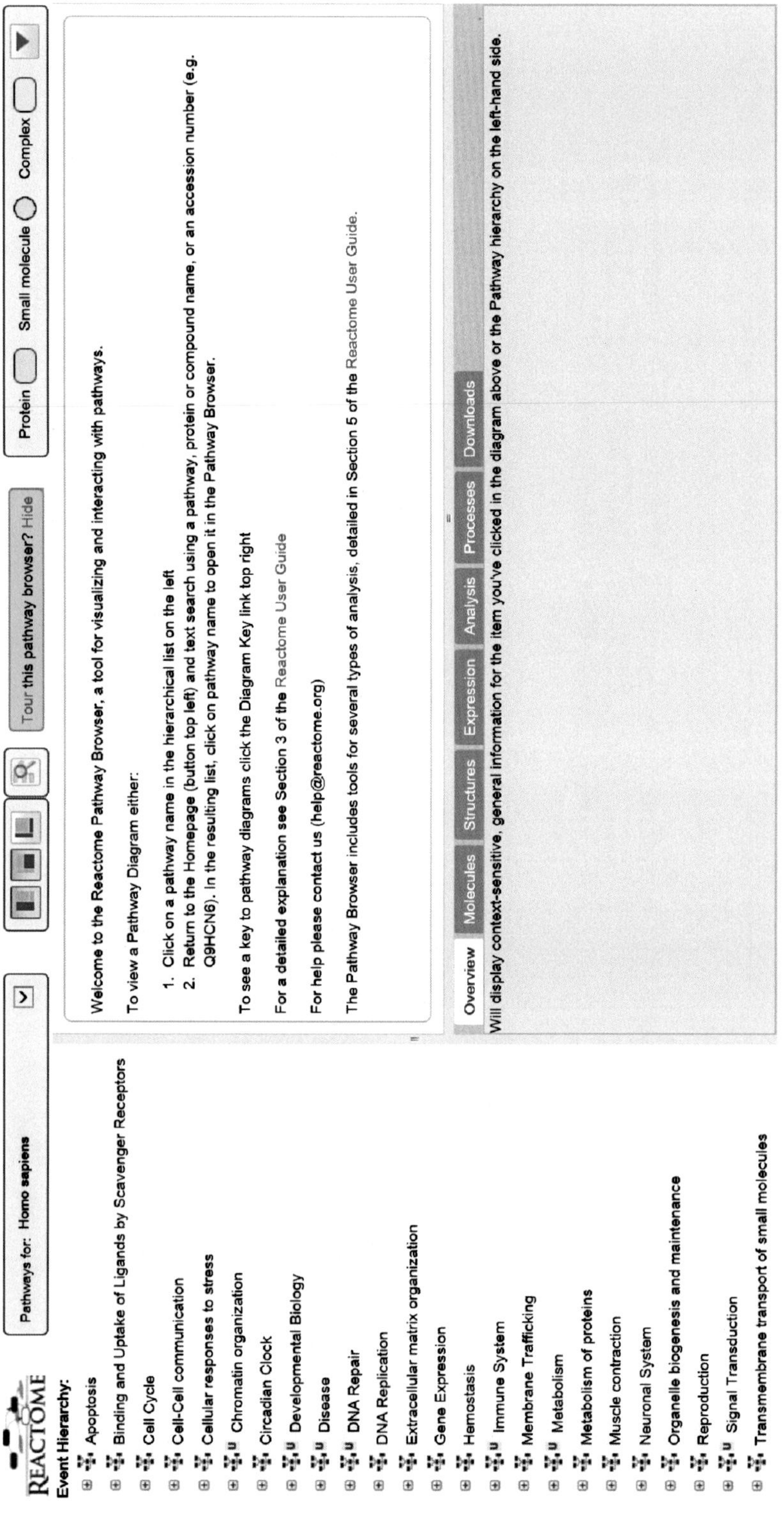

Fig. 7 The Pathway Browser of Reactome

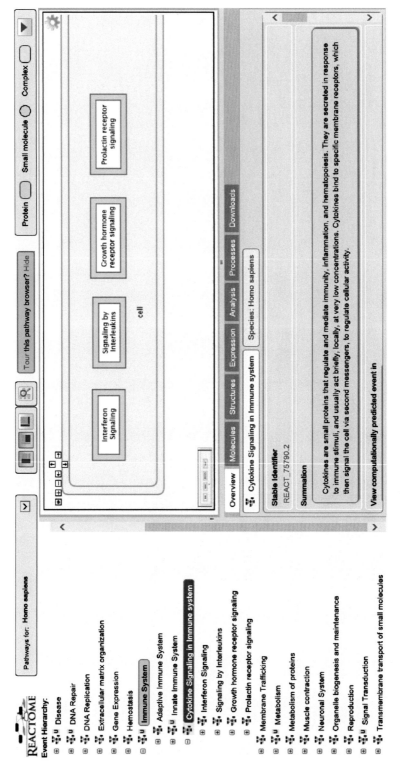

Fig. 8 The Pathway diagram for *Cytokine Signaling in Immune system*

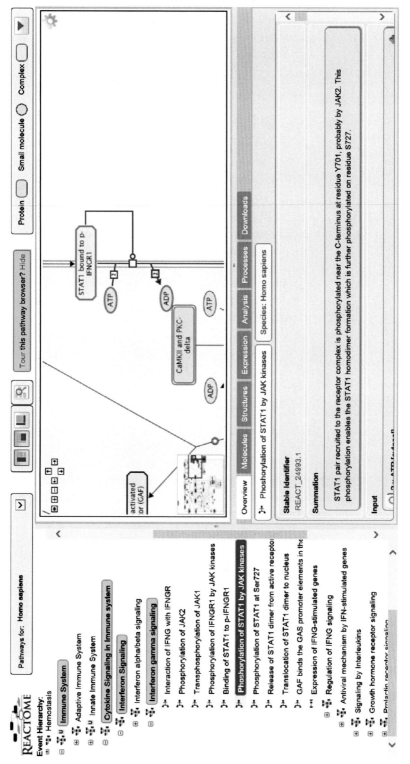

Fig. 9 The reaction diagram for *Phosphorylation of STAT1 by JAK kinases*

13. In the upper right corner, click the down arrow button to show the Diagram Key, describing the shapes corresponding to every type of entity in the diagram. To hide the key, click the arrow button again. Note that proteins are shown as rectangles with rounded corners.

14. In the *Hierarchy Panel* to the left, select the reaction called *Transphosphorylation of JAK1* inside the same sub-pathway *Interferon gamma signaling*.

15. Right click on the protein *PTPN6*, and select the option *Other pathways* to show the diagram of other pathways related to that protein. For the protein uniquely involved in this pathway, the label *No other pathways* is shown.

16. On the *Hierarchy Panel* on the left, select *Binding of STAT1 to p-IFNGR1*. Right click on the protein *STAT1-1* and choose *Display interactors* to show the compounds interacting with this protein in this pathway. Here 10 out of 52 interactors associated to this protein are highlighted with new entity boxes with blue thick borders. Note the small white square at the top right corner of the entity box of the protein indicating the number of interactors (Fig. 10).

17. You can switch to other pathway diagrams at any time by selecting another pathway, sub-pathway or reaction name in the hierarchy panel. The details panel at the bottom will be updated according to what is currently selected in the hierarchy panel or the pathway diagram.

This section describes how to analyze quantitative data using Reactome with the dataset material indicated in Subheading 2, **item 1**. First, the analysis of the list of protein accessions will be used to demonstrate how to find pathways of interest, a use case relevant to both qualitative and quantitative datasets. Second, the quantitative information will be provided to Reactome along with the protein identifiers.

1. From the Reactome homepage (Fig. 1), click *Analyze Data* to show the analysis tools (Fig. 11).

2. Prepare a data file containing the list of proteins according to the required format:

   (a) Create a new text file and name it *protein_list.txt*.

   (b) In Supplementary Table S2 of [15], go to the worksheet called *Results_Merged_Five_and_four_IS*. There you will find the table entitled *Supplementary Table 2: Merged data from the five and four cell lines experiments*.

   (c) Select the protein accessions, here the column called *Protein IDs*, along with the header, copy the data into the text file and save the file.

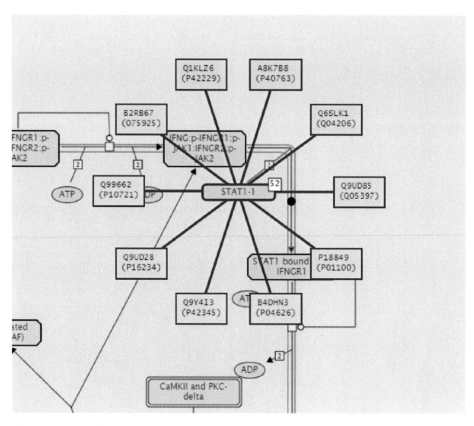

Fig. 10 The interactors of protein *STAT1-1* in the *Phosphorylation of STAT1 by JAK kinases* reaction

(d) Change the first row of the file from *Protein IDs* to *#GBM Uniprot* (*see* **Note 1 and 2**).

(e) Delete any empty lines after the protein list.

3. Upload the data file with the set of proteins to analyze:

(a) First click the *Browse* button to select your *protein_list.txt* file.

(b) Make sure that the checkbox *Project to human* is selected, as the identifiers to analyze are related to human pathways.

(c) Click *Analyze* to upload the file and start the analysis of the inserted data.

4. As an alternative, you can simply paste the information in the textbox within the *Analysis Tools* section.

(a) In the *Analyze Data* screen, click *Click here to paste you data or try example data sets…* to display the input field. At the right you can also select example datasets.

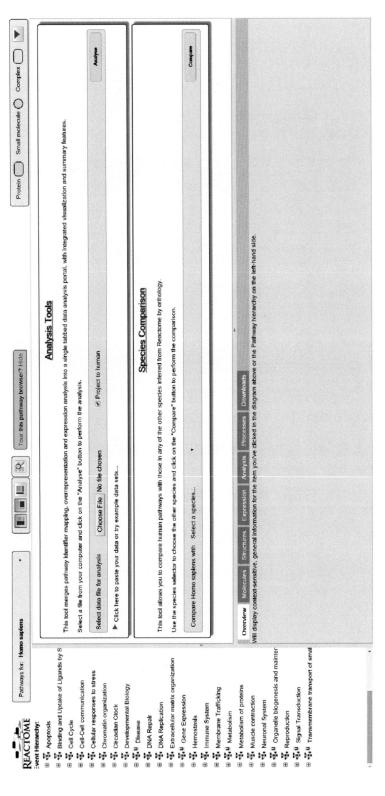

Fig. 11 The Data Analysis screen with the Analysis tools of Reactome

(b) Paste the set of proteins into the text field.

(c) Click on the *Analyze* button in the lower right corner.

5. Review the results in the details panel at the bottom of the screen, showing the pathways containing at least one protein from the dataset. There are columns with different types of information, for example you can see pathway names and how many entities were found in those pathways, as well as the confidence in the pathway identification (Fig. 12).

6. Choose the pathway *Translation*. In the diagram, the box entities are colored depending on the number of entities matched. Encapsulated pathways are represented by black border rectangles (Fig. 13).

7. Double click on *Eukaryotic Translation Elongation* to go deeper in the diagram and show the sub-pathway of interest (Fig. 14).

8. Sets of proteins are indicated by double border rectangle with rounded corners. Right click on *EEF1A1-like proteins* and choose the option *Display Participating Molecules*. This will show a small table with the components of the set colored in yellow (Fig. 15).

In the following, the procedure above will be repeated, but now including the quantitative information (Figs. 16, 17, 18, 19).

1. Prepare the data file to be uploaded in the Analysis screen of Reactome.

   (a) In Supplementary Table S2 of [15], go to the worksheet called *Results_Merged_Five_and_four_IS*. There you will find the table entitled *Supplementary Table 2*: *Merged data from the five and four cell lines experiments*.

   (b) Copy the columns *Protein Ids* and *Molm-13* under *Five cell lines as IS*, and paste them in a new worksheet. Rename the first column from *Protein IDs* to *#Probeset* (*see* **Note 3 and 4**).

   (c) Create a new text file called *expression_data.txt* and copy paste the two columns into the new text file.

2. Reopen the *Analysis tools* panel by clicking on the button representing a loupe over a pathway at the top of the screen (Fig. 20).

3. Click the *Browse* button and select the *expression_data.txt* file. Next, click the *Analyze* button. A table presenting the results of the analysis appears in the *Details Panel* upon completion. Note that the first eight columns are the same as in the previous example, but the table now also includes the quantitative results.

4. You can explore any pathway using the *Hierarchy Panel* or the *Details Panel*. Entities in the diagram are colored according to

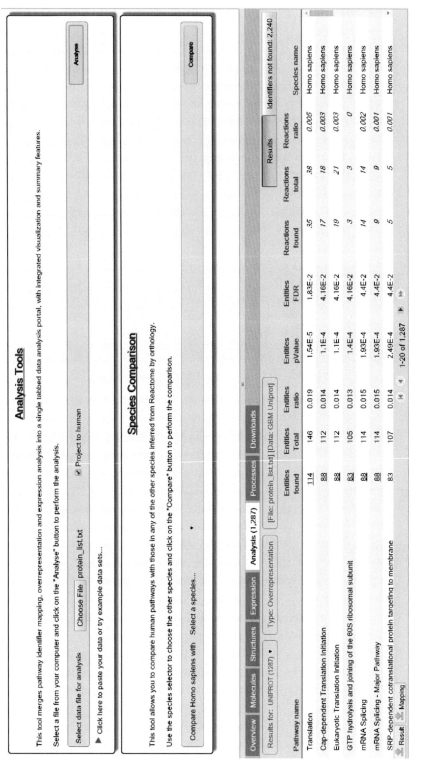

Fig. 12 The Analysis results of the protein sample data set

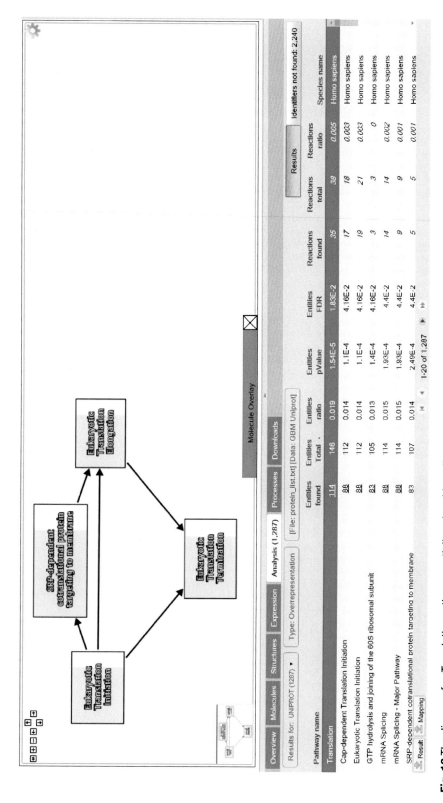

Fig. 13 The diagram for *Translation* pathway partially colored in *yellow* to indicate the percentage of structures in the sample data set found in this pathway

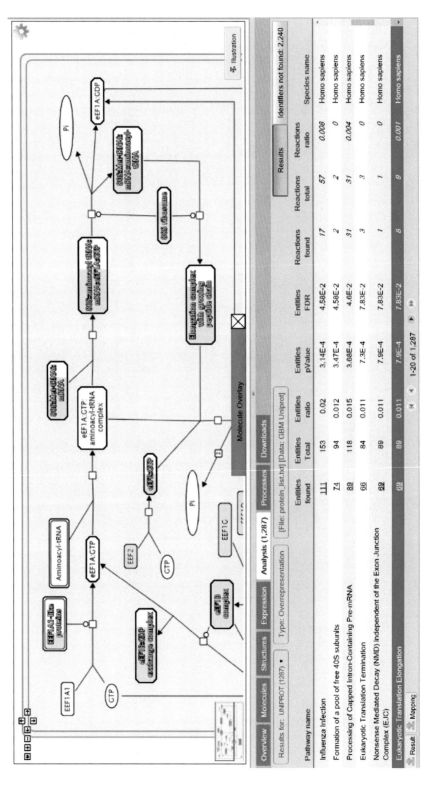

Fig. 14 The diagram for *Eukaryotic Translation Elongation* sub-pathway with *partially colored entity boxes*

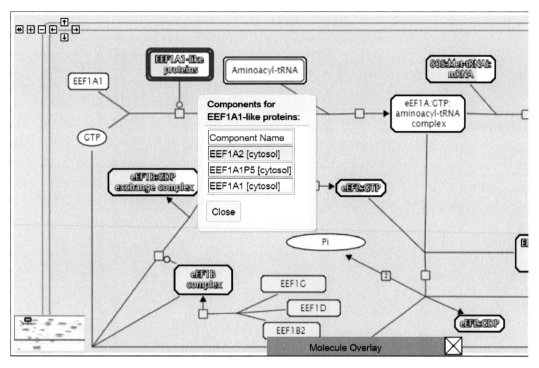

Fig. 15 The table of participating components of the protein set *EEF1A1-like proteins* highlighting in *yellow* the table cells of the components found in the sample protein data set

the quantitative results, ranging from blue to yellow for the lower and upper ratios, respectively. Only the entities with numerical data associated will be colored.

5. In the *Hierarchy Panel* select the *Metabolism* pathway. Note that next to its name, there is a number saying how many molecules were identified in the submitted data, in this case 343 out of 1588. Then select the sub-pathway called *Metabolism of lipids and lipoproteins*, then select *Lipid digestion, mobilization, and transport* and finally *Lipoprotein metabolism* (Fig. 21).

6. Note that some complex entities are partially colored. This means that only some participating molecules of that complex where present in the dataset.

7. Right click on the complex called *ApoB-48:TG:PL*. Then select *Display Participating Molecules*. Note, that only one out of three molecules where present in the dataset, the rectangle representing the complex therefore has one third of its area colored in gray. The gray is due to the numerical value of the protein of accession *P04114* (Fig. 22).

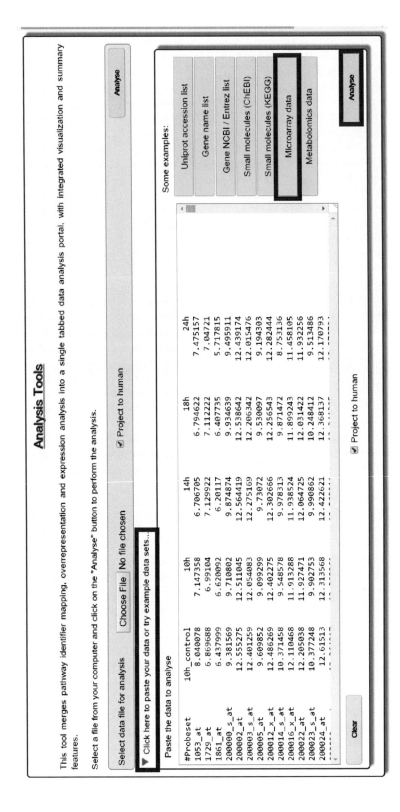

Fig. 16 Analysis Tools screen with expression data

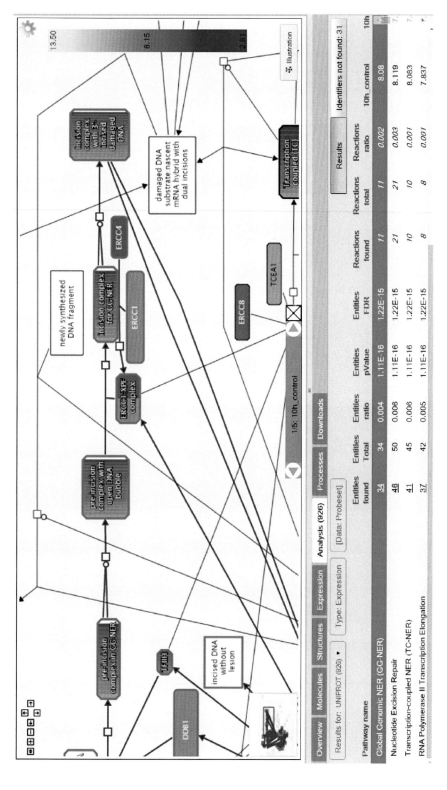

Fig. 17 Diagram panel with analysis results coloring entities according to the submitted numerical data

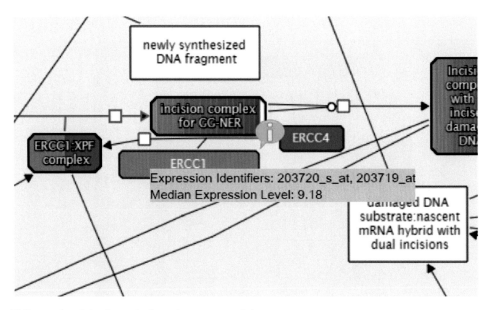

Fig. 18 Expression data shown by hovering over a protein

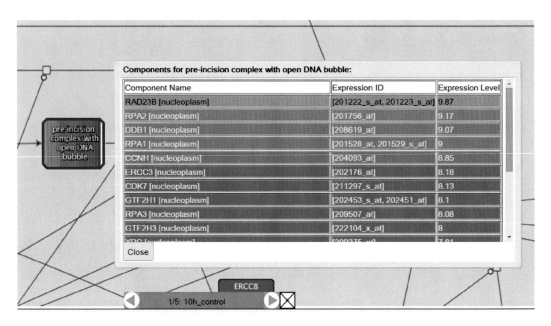

Fig. 19 Colored table of components of a complex entity

4   Notes

1. Proteomics dataset result files compatible with Reactome can be easily exported from most proteomic software like MaxQuant [18], OpenMS [19], the Trans-Proteomic Pipeline (TPP) [20], or PeptideShaker [21]. It is however recommended

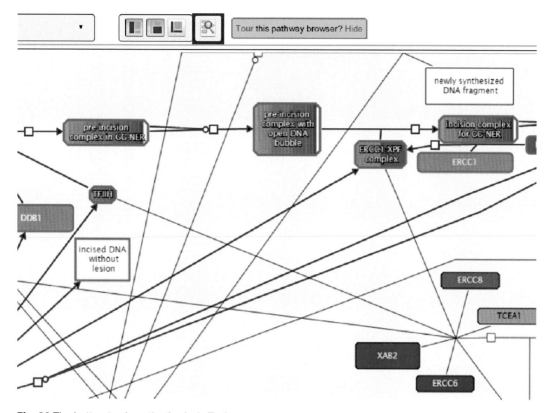

Fig. 20 The button to show the Analysis Tools

to post-process the results, for example conducting normalization and imputation of missing values. This can be easily conducted using the Perseus interface, available from the MaxQuant website (www.maxquant.org).

2. Reactome can be operated on a wide variety of omics datasets: metabolomics, genomics, transcriptomics, and proteomics. Data can be loaded separately or together in a unified dataset.

3. Make sure that column headers are recognized by Reactome. When the dada set is composed of only one column of data, Reactome can recognize different types of identifiers such as UniProt accessions for proteins and ChEBI IDs for small molecules. Quantitative information will be ignored if not indicated by the correct header (*see* Subheading 3).

4. Reactome also recognizes HGNC gene symbols, ENSEMBL IDs for DNA/RNA molecules, HUGO gene symbols, GenBank/EMBL/DDBJ, RefPep, RefSeq, EntrezGene, MIM, InterPro, EnsEMBL protein, EnsEMBL gene, EnsEMBL transcript, and some Affymetrix and Agilent probe IDs.

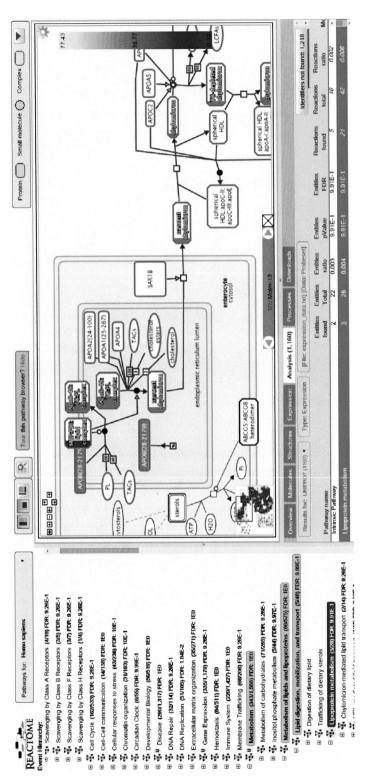

Fig. 21 Diagram of the *Lipoprotein metabolism* pathway after the analysis of the submitted data

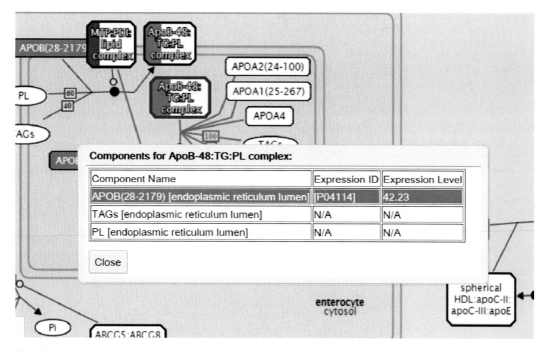

Fig. 22 Component molecules of the *ApoB-48:TG:PL* complex

Acknowledgements

F.S.B. and F.S. acknowledge support from the Norwegian Cancer Society. H.B. is supported by the Research Council of Norway.

References

1. Mouse Genome Sequencing Consortium, Waterston RH, Lindblad-Toh K et al (2002) Initial sequencing and comparative analysis of the mouse genome. Nature 420:520–562
2. Rosen R (1968) A means toward a new holism. Science 161:34–35
3. Barsnes H, Martens L (2013) Crowdsourcing in proteomics: public resources lead to better experiments. Amino Acids 44:1129–1137
4. Vaudel M, Sickmann A, Martens L (2014) Introduction to opportunities and pitfalls in functional mass spectrometry based proteomics. Biochim Biophys Acta 1844:12–20
5. Muller T, Schrotter A, Loosse C et al (2011) Sense and nonsense of pathway analysis software in proteomics. J Proteome Res 10:5398–5408
6. Khatri P, Sirota M, Butte AJ (2012) Ten years of pathway analysis: current approaches and outstanding challenges. PLoS Comput Biol 8, e1002375
7. Vizcaino JA, Mueller M, Hermjakob H et al (2009) Charting online OMICS resources: a navigational chart for clinical researchers. Proteomics Clin Appl 3:18–29
8. Kanehisa M (1997) A database for post-genome analysis. Trends Genet 13:375–376
9. Ogata H, Goto S, Sato K et al (1999) KEGG: Kyoto Encyclopedia of Genes and Genomes. Nucleic Acids Res 27:29–34
10. Pico AR, Kelder T, van Iersel MP et al (2008) WikiPathways: pathway editing for the people. PLoS Biol 6, e184
11. Croft D, O'Kelly G, Wu G et al (2011) Reactome: a database of reactions, pathways and biological processes. Nucleic Acids Res 39:D691–D697

12. Takami H, Taniguchi T, Moriya Y et al (2012) Evaluation method for the potential function-ome harbored in the genome and metage-nome. BMC Genomics 13:699

13. Shannon P, Markiel A, Ozier O et al (2003) Cytoscape: a software environment for inte-grated models of biomolecular interaction net-works. Genome Res 13:2498–2504

14. Kisseleva T, Bhattacharya S, Braunstein J et al (2002) Signaling through the JAK/STAT pathway, recent advances and future chal-lenges. Gene 285:1–24

15. Aasebo E, Vaudel M, Mjaavatten O et al (2014) Performance of super-SILAC based quantitative proteomics for comparison of dif-ferent acute myeloid leukemia (AML) cell lines. Proteomics 14(17-18):1971–6

16. Vizcaino JA, Deutsch EW, Wang R et al (2014) ProteomeXchange provides globally coordi-nated proteomics data submission and dissemi-nation. Nat Biotechnol 32:223–226

17. Martens L, Hermjakob H, Jones P et al (2005) PRIDE: the proteomics identifications data-base. Proteomics 5:3537–3545

18. Cox J, Mann M (2008) MaxQuant enables high peptide identification rates, individualized p.p.b.-range mass accuracies and proteome-wide protein quantification. Nat Biotechnol 26:1367–1372

19. Bertsch A, Gropl C, Reinert K et al (2011) OpenMS and TOPP: open source software for LC-MS data analysis. Methods Mol Biol 696:353–367

20. Deutsch EW, Mendoza L, Shteynberg D et al (2010) A guided tour of the Trans-Proteomic Pipeline. Proteomics 10:1150–1159

21. Vaudel M, Burkhart JM, Zahedi RP et al (2015) PeptideShaker enables reanalysis of MS-derived proteomics data sets. Nat Biotechnol 33:22–24

INDEX

A

Adipocytes...57, 58
Affinity purification-mass spectrometry (APMS), ..181–183
Antigen presentation 189–208
Assay design 47–49
Assay validation...50–51, 54

B

Biomarkers 44, 87, 88, 151, 275

C

Cell signaling pathways 181–186
Chemical cross-linking.....................109, 112, 114–115, 124
Chemical proteomics...211–218
Collision-induced dissociation (CID) 8, 60, 97, 98,
         120, 122, 171, 172, 176, 178, 240
Cotton 152, 160
Crop science 233–241
Cryoconserved tissue 131, 132, 135
Cultivar..234, 235, 238–240

D

Data interpretation..276, 279
Data post-processing 261
De Novo Sequencing..233–241
Drug discovery 212

E

Enrichment 2, 3, 30, 32, 33, 39, 152, 158,
         164–165, 167–169, 172, 177, 183, 212, 247, 261,
         267, 272, 276–279
Ethyl esterification (EE) 151–161

F

Formalin fixed paraffin embedded (FFPE) 130–132,
         135–137
Functional Networks 287–307
Functional proteomics 287

G

Gene expression..229
Gene/protein expression 221

G (continued)

Glycolysis 57–73
Glycopeptides.. 163–178, 275,
         277–279, 282
Glycoproteomics..275–284

H

High-performance liquid chromatography
        (HPLC).......................12, 16–18, 20–23, 27, 28,
         30, 31, 34, 35, 39, 45, 63, 78, 81, 89, 90, 93, 94,
         96, 97, 104–106, 183, 185, 194, 198, 199, 202,
         206, 215, 216
Human leukocyte antigen (HLA) 189–192, 194,
         199, 201, 204, 205
Hydrophilic interaction liquid chromatography
        (HILIC) 26, 28, 33–35, 40, 152,
         155–156, 207

I

ImagePrep 130, 131, 137–142, 145
Immobilized metal ion affinity chromatography
        (IMAC) 88, 90, 94
Immune response 189–208
In-gel digestion 113, 116–119, 165, 168–169,
         171–172, 174–175
In-solution digestions.......................113–114, 116, 119, 165,
         171, 174, 175
Isobaric labels 3, 248
Isoelectric focusing 16–18, 22, 220
iTRAQ.................................. 3, 15–40, 87, 97, 98, 101, 217,
         219, 248, 270

K

Kinases 59, 212, 245, 246, 289–292,
         295, 297

L

Label-free proteomics...102
Linkage-specific 152, 157
Liquid chromatography-mass spectrometry
        (LC-MS) 25–28, 30, 35, 48, 50, 55,
         79–80, 89–91, 94, 95, 97, 98, 114, 119–120, 185,
         192–194, 199–201, 205–207, 220, 222, 226, 236,
         251, 255, 276, 279

Jörg Reinders (ed.), *Proteomics in Systems Biology: Methods and Protocols*, Methods in Molecular Biology, vol. 1394,
DOI 10.1007/978-1-4939-3341-9, © Springer Science+Business Media New York 2016

M

Major histocompatibility complex (MHC) 189–192, 194–205, 207
MALDI imaging 129–149
Mass spectrometry (MS) 2–4, 8, 15, 18, 22–23, 37, 43, 44, 58, 78, 87, 102, 104, 110, 129–131, 143, 152, 164, 165, 183, 185, 186, 198, 212, 213, 215, 218, 221, 223, 224, 226, 234, 255, 276
Mass spectroscopy imaging (MSI) 129
Mathematical models 245, 246, 256
Matrix application 130, 134, 137–142
Matrix assisted laser desorption/ionization (MALDI), ... 151–161
Melanoma cell 106
Metabolic control 57
Metal oxide affinity chromatography (MOAC) 27, 248
Metaproteomics 221
Mixed-mode solid phase extraction, ... 247
Mode of action 211
Multiple reaction monitoring (MRM) 75–85, 87–99, 200, 202, 207, 239
Multiplexing 2, 16, 59, 102, 106, 223

N

NanoESI MS 164, 165, 171, 172, 176
N-glycan release 152–155
N-Glycosylation 163–178

P

Pathway Analysis 287
Pathway modelling 245–258
Peptide ligands 190, 191
Peptide-N-glycosidase F (PNGase F) 152–155, 157, 159, 164, 276–278
Peptides 2, 3, 7–11, 16–18, 20–22, 24, 30–33, 35–40, 43–55, 58–61, 63, 66–73, 75–79, 81–82, 88, 90, 106, 107, 110, 111, 116–122, 129–149, 161, 165, 167, 168, 172, 175, 177, 178, 186, 190–195, 198–207, 212, 219–222, 224–227, 229, 230, 234–236, 238, 239, 241, 246–252, 256–258, 262, 264, 270, 276–279, 282–284
Perseus 226, 227, 261–264, 270, 272, 307
Phosphatases 245, 246
Phosphopeptide-enrichment 26, 27, 32–34, 38–40, 88, 93, 247–249, 251, 254, 256–258
Phosphoproteomics 88
Photo-affinity labeling (PAL) 111
Photo-cross-linking 111, 112, 125
Pisum sativum 237, 239
Polymorphisms 189, 234, 239
Proteases 10, 16, 45, 46, 164, 171, 177, 220
Protein 3D-structure 109–127
Protein interaction 103, 111, 183
Protein quantitation 15, 67, 205
Proteogenomics 239
Proteomics 2, 10, 15–25, 44, 48, 88, 220–223, 227–229, 246, 272, 277, 279, 282, 306, 307
Proteomics informed by transcriptomics (PIT) 221

Q

Quadrupole mass spectrometer 59, 63, 76
Quantification 2, 3, 11, 16, 44, 47, 50, 53, 55, 58, 60, 61, 87, 88, 101–103, 106, 107, 117, 220, 223, 226, 230, 235, 239, 246–250, 252, 253, 255–257, 261, 288
Quantitation 12, 16, 52, 76, 83, 98, 99, 192, 194, 200, 202, 203, 249
Quantitative phosphoproteomics 26, 35
Quantitative proteomics 1–12, 75, 248

R

RNA sequencing 219–230

S

Selected reaction monitoring (SRM)
    assay design 47–49
    assay validation 50–51, 54
Sepharose 152–156, 160, 194, 197, 214, 216
Shotgun proteomics 43, 47, 48, 54, 220, 234, 238, 261–272, 276
Sialic acid (N-acetylneuraminic acid) 152, 157–159
SIS peptides 58, 60, 61, 66, 67, 72, 73
Small molecule profiling 211–218
Solid phase extraction (SPE) 152–156
Solid-phase extraction (SPE) 11, 27, 34, 39, 93, 94, 96, 164, 165, 247–249, 276
Selected/multiple reaction monitoring (SRM/MRM) 43–55, 57–73, 87–99, 102, 239
Stabilization 152
Stable isotope labeling in cell culture (SILAC) 2, 87, 97, 98, 101–107, 219, 248, 270
Subcellular fractionation 16–17
SunCollect 130, 131, 137–141, 148
SWATH-MS 102, 103
Systemic analysis 287–307

T

Tandem mass tags (TMT) 3, 7–8, 10, 12, 87, 217, 248, 270
Target deconvolution 212
Targeted proteomics 47, 192, 221

Tempo-spatial proteomics 75–85
Trypsin 3–6, 9, 10, 16, 18, 20, 26, 30, 37–39,
43, 45, 46, 49, 59, 62–64, 66, 81, 82, 84, 88, 92,
95, 96, 98, 104, 105, 113, 114, 118, 119, 121, 125,
130, 132, 138, 140, 144, 147, 164, 171, 177,
183–186, 195, 203, 206, 214, 220, 236, 250, 253,
257, 277, 278, 280

U

Unnatural amino acids..109